Advances in
Carbohydrate Chemistry and Biochemistry

Volume 69

Advances in Carbohydrate Chemistry and Biochemistry

Editor
DEREK HORTON
Ohio State University, Columbus, Ohio
American University, Washington, DC

Board of Advisors

DAVID C. BAKER
DAVID R. BUNDLE
STEPHEN HANESSIAN
JÉSUS JIMÉNEZ-BARBERO
YURIY A. KNIREL

TODD L. LOWARY
SERGE PÉREZ
PETER H. SEEBERGER
ARNOLD E. STÜTZ
J.F.G. VLIEGENTHART

Volume 69

AMSTERDAM • BOSTON • HEIDELBERG • LONDON
NEW YORK • OXFORD • PARIS • SAN DIEGO
SAN FRANCISCO • SINGAPORE • SYDNEY • TOKYO
Academic Press is an imprint of Elsevier

Academic Press is an imprint of Elsevier
The Boulevard, Langford Lane, Kidlington, Oxford, OX5 1GB, UK
32, Jamestown Road, London NW1 7BY, UK
Radarweg 29, PO Box 211, 1000 AE Amsterdam, The Netherlands
225 Wyman Street, Waltham, MA 02451, USA
525 B Street, Suite 1800, San Diego, CA 92101-4495, USA

First edition 2013

Copyright © 2013 Elsevier Inc. All rights reserved

No part of this publication may be reproduced, stored in a retrieval system or transmitted in any form
or by any means electronic, mechanical, photocopying, recording or otherwise without the prior
written permission of the publisher

Permissions may be sought directly from Elsevier's Science & Technology Rights
Department in Oxford, UK: phone (+44) (0) 1865 843830; fax (+44) (0) 1865 853333;
email: permissions @elsevier.com. Alternatively you can submit your request online by
visiting the Elsevier web site at http://elsevier.com/locate/permissions, and selecting
Obtaining permission to use Elsevier material

Notice
No responsibility is assumed by the publisher for any injury and/or damage to persons or property as
a matter of products liability, negligence or otherwise, or from any use or operation of any methods,
products, instructions or ideas contained in the material herein. Because of rapid advances in the
medical sciences, in particular, independent verification of diagnoses and drug dosages should be made

ISBN: 978-0-12-408093-5
ISSN: 0065-2318

British Library Cataloguing in Publication Data
A catalogue record for this book is available from the British Library

Library of Congress Cataloging-in-Publication Data
A catalog record for this book is available from the Library of Congress

For information on all Academic Press publications
visit our website at store.elsevier.com

Printed and bound in USA

13 14 15 16   11 10 9 8 7 6 5 4 3 2 1

CONTENTS

CONTRIBUTORS ix
PREFACE xi

STEPHEN JOHN CHARLES ANGYAL 1914–2012
JOHN D. STEVENS

JOHN GRANT BUCHANAN 1926–2012
RICHARD WIGHTMAN

SAUL ROSEMAN 1921–2011
SUBHASH C. BASU

De Novo Asymmetric Synthesis of the Pyranoses: From Monosaccharides to Oligosaccharides
ALHANOUF Z. ALJAHDALI, PEI SHI, YASHAN ZHONG, AND GEORGE A. O'DOHERTY

I. Introduction 57
    1. Background 57
II. Masamune–Sharpless Approach to the Hexoses 58
III. Danishefsky Hetereo-Diels–Alder Approach to Various Pyranoses 60
IV. MacMillan Iterative Aldol Approach to Various Pyranoses 67
V. Asymmetric Oxidative Biocatalytic Aldol Approach to Various Pyranoses 71
VI. Sharpless Dihydroxylation/Enzymatic Aldol Approach 2-Ketoses 73
VII. Non-*De Novo* Asymmetric Approaches to Pyranoses 74
    1. Dondoni Thiazole Approach 74
    2. Seeberger Approaches 76
    3. Reissig Approaches 80
VIII. O'Doherty *De Novo* Approach to Pyranoses 80
    1. *De Novo* Use of the Achmatowicz Approach to Pyranoses 83
    2. Iterative Dihydroxylation of Dienoates 84
    3. Palladium-Catalyzed Glycosylation 88
    4. Applications to Synthesis and Medicinal Chemistry 89
    Acknowledgments 115
    References 115

Recent Advances Toward the Development of Inhibitors to Attenuate Tumor Metastasis Via the Interruption of Lectin–Ligand Interactions
Rachel Hevey and Chang-Chun Ling

I.	Introduction	126
	1. Siglecs	127
	2. Galectins	130
	3. Selectins	134
II.	Roles of Carbohydrates in Tumor Development	137
	1. Abnormal Glycosylation in Tumors	137
	2. Abnormal Glycosylation and Cancer Metastasis	140
III.	Roles of Lectins in Tumor Development	141
	1. Siglecs	141
	2. Galectins	142
	3. Selectins	144
IV.	Developing Anticancer Approaches by Targeting Tumor-Associated Carbohydrate–Lectin Systems	145
	1. Efforts Toward Galectin Inhibitors	146
	2. Efforts Toward Selectin Inhibitors	171
V.	Conclusions	187
	References	187

Bacterial Cell-Envelope Glycoconjugates
Paul Messner, Christina Schäffer, and Paul Kosma

I.	Introduction	210
	1. Outline	210
	2. Background—Bacterial Protein Glycosylation	211
II.	Surface-Layer Glycoproteins	217
	1. Bacterial S-Layer Glycoproteins	218
	2. Archaeal S-Layer Glycoproteins	229
III.	Nonclassical Secondary Cell-Envelope Polysaccharides	231
	1. Background	231
	2. The Nonclassical Group of SCWPs	232
IV.	Structural Analysis	241
	1. Isolation of Cell-Envelope Polysaccharides and Glycopeptides	241
	2. Degradation Reactions	242
	3. Structure Elucidation by Nuclear Magnetic Resonance Spectroscopy	243
	4. Mass Spectrometry	245
V.	Cell-Envelope Glycan Biosynthesis	246
	1. Genetic Basis for S-Layer Glycoprotein Biosynthesis	246
	2. Nucleotide Sugar Biosynthesis	247
	3. Multispecific Glycosyltransferases	249
	4. Proposed Pathway for S-Layer Glycoprotein Biosynthesis	250
	5. SCWP Glycosylation Gene Clusters	252

VI.	Glycan Engineering and Applications.	254
	1. The S-Layer Glycobiology Toolbox	254
	2. Glycosylation Engineering.	255
VII.	Concluding Remarks.	257
	References.	257
	AUTHOR INDEX.	273
	SUBJECT INDEX	307

CONTRIBUTORS

Alhanouf Z. Aljahdali, Department of Chemistry and Chemical Biology, Northeastern University, Boston, Massachusetts, USA

Rachel Hevey, Alberta Glycomics Centre, Department of Chemistry, University of Calgary, Calgary, Alberta, Canada

Paul Kosma, Department of Chemistry, University of Natural Resources and Life Sciences, Vienna, Austria

Paul Messner, Department of NanoBiotechnology, NanoGlycobiology Unit, University of Natural Resources and Life Sciences, Vienna, Austria

Chang-Chun Ling, Alberta Glycomics Centre, Department of Chemistry, University of Calgary, Calgary, Alberta, Canada

George A. O'Doherty, Department of Chemistry and Chemical Biology, Northeastern University, Boston, Massachusetts, USA

Christina Schäffer, Department of NanoBiotechnology, NanoGlycobiology Unit, University of Natural Resources and Life Sciences, Vienna, Austria

Pei Shi, Department of Chemistry and Chemical Biology, Northeastern University, Boston, Massachusetts, USA

Yashan Zhong, Department of Chemistry and Chemical Biology, Northeastern University, Boston, Massachusetts, USA

PREFACE

This 69th volume of *Advances* features six contributions that cover a wide diversity of different aspects of new developments in the carbohydrate field, focusing variously on synthetic methodology, structural and functional aspects of carbohydrates in the bacterial cell envelope, and a glycobiology theme concerned with aberrant glycosylation on the surface of tumor cells and approaches to improved cancer therapies.

The Boston (Massachusetts)-based group led by O'Doherty and his coworkers Aljahdali, Shi, and Zhong provides a *tour de force* of organic synthetic virtuosity in the *de novo* synthesis of a wide range of monosaccharides and oligosaccharides, making use of newly introduced asymmetric catalysts to accomplish high enantiomeric purity in the products.

Early work by Lespieau in Volume 2 of this series and notably by Zamojski in Volume 34 provided pointers for the synthesis of alditols and aldoses "from scratch" employing simple organic precursors, but the methodology led mostly to racemic products. Important later work detailed here by the Boston group has focused on access to enantiopure products, as exemplified by the asymmetric epoxidation strategy of Sharpless and complementary work by Danishefsky, MacMillan, Dondoni, Seeberger, and others.

The O'Doherty group has synthesized an impressive series of targets by employing an extensive range of "name" reactions, specialized reagents, "fine-tuned" versions of traditional oxidative and reductive procedures, and in particular, the rearrangement reaction of enones introduced by Achmatowicz, utilizing asymmetric catalysts, along with numerous applications of palladium chemistry. These targets include the *C*-glycosyl framework of the papulacandins, the nojirimycin-type imino sugars, homoadenosine analogues, the pheromone daumone, the indolizidine alkaloids, analogues of trehalose and cyclitols, cardiac glycosides, the anthrax tetrasaccharide, and other carbohydrate structures of current biological interest.

The Vienna-based authors Messner, Schäffer, and Kosma detail the glycoconjugates of the cell envelope of bacterial and archaeal organisms, where glycosylation plays a critical role in preserving the integrity of the prokaryotic cell-wall and serves as a flexible adaption mechanism to evade environmental and host-induced pressure. Glycosylation is not limited to surface-layer proteins in the cell but is shown to occur also in pili, flagella, and in the form of secondary-cell-wall polysaccharides. In contrast to the glycan components in glycoproteins of eukaryotic organisms, these prokaryotic-associated glycans display a wide range of structural variations. With their incorporation of numerous unusual monosaccharide components, these glycans manifest a diversity in their constituent monosaccharides comparable to the complex range of sugar components found in the lipopolysaccharides and capsular polysaccharides of bacteria.

The discussion of these conjugates covers a range of aspects, from analysis and structure elucidation to function, biosynthesis, and genetic basis, and to biomedical and biotechnological applications. The S-layer glycoproteins form ordered structures through crystallization of the protein component, and a variety of techniques have helped to develop models for the structure of the cell envelope. Typically, the glycan component consists of complex linear or branched oligo- or polysaccharides, either O- or N-glycosylically linked, and the highly diversified structures have been elucidated by chemical and spectroscopic procedures. Their biosynthesis via the nucleotide-sugar pathway and elucidation of their genetic basis point the way to important applications in nanotechnology and biomedicine.

The article by Hevey and Ling (Calgary, Alberta) is focused on inhibitors designed to attenuate the metastasis of tumors through interruption of interaction between cell-surface lectins (carbohydrate-binding proteins) and those carbohydrate ligands (notably TF, sialyl Tn, and sialyl Le^x) that are typically overexpressed on tumor cells. These lectin–ligand interactions are correlated with such metastatic processes as cell adhesion, generation of new blood vessels (neoangiogenesis), and immune-cell evasion. The suppression of such interactions is a desired goal in seeking improved cancer therapies.

The authors discuss the structural features of the carbohydrate ligands in the tumor cell and the relation between abnormal glycosylation and tumor metastasis, and the role of lectins in tumor development. Focusing on the particular lectins involved, namely, the galectins, selectins, and the siglecs (sialic acid Ig-like lectins), their structures and binding sites are developed in detail. The key role of the carbohydrate–lectin interaction suggests a variety of approaches toward novel cancer therapies. They include the development of inhibitors of the carbohydrate–ligand interaction, the inhibition of glycosyltransferases and glycosidases involved in the aberrant glycosylation of tumor cells, and the targeting of the immune system against cancer cells. Although there have been promising results from clinical trials, especially from dual-mode therapies, as compared to traditional cancer chemotherapy agents, there remain significant challenges to be addressed, with a need for carbohydrate inhibitors that are more resistant to hydrolytic or metabolic breakdown. Promising results have been observed by the use of small-molecule inhibitors based on C-glycosyl structures, which are resistant to acid hydrolysis and glycosylase enzymes, and polyvalent ligands that have shown enhanced binding down to the nanomolar level.

The contributions to carbohydrate science of three towering figures in the field are recognized in this issue with obituary articles that detail their work in sufficient depth to provide a valuable source for citation. Stephen J. Angyal, who died at the age of 97, was a Hungarian-born Australian who literally "wrote the book" on the chemistry of the cyclitols, developed a quantitative understanding of the conformational behavior of pyranoses in solution, and provided a wealth of practical synthetic procedures based on the use of cationic complexes to modify the outcome of traditional reactions. His legacy is recorded here by his colleague John D. Stevens.

The scientific career of J. Grant Buchanan is presented by his colleague Richard Wightman. Grant's early studies in Alexander Todd's group in Cambridge provided a basis for his long-term interest in sugar phosphates and nucleotide chemistry, which developed further in his postdoctoral work with Calvin in Berkeley and later back in Britain at posts in London and in Newcastle. In conjunction with Baddiley, he studied the pneumococcal antigens and the ribitol teichoic acids. He also made major contributions to our understanding of epoxide migration and cleavage. In a subsequent move to his native Scotland, to Heriot-Watt University in Edinburgh, he extended his research to the chemistry and biochemistry of C-nucleosides, and the use of glycosylalkynes in synthesis. The list of his publications details his wide range of research accomplishments.

Saul Roseman was a pioneer American glycobiologist whose six decades of innovations in the field are documented by his student and coworker Subhash Basu. Very early in his career, Roseman established the correct structure of neuraminic acid (sialic acid), and subsequently, he made extensive contributions to our knowledge of glycoproteins and glycosphingolipids, and the glycosyltransferases involved in their biosynthesis. Critical to this work was the availability of ^{14}C-labeled nucleotide sugars produced in collaboration with Khorana in Vancouver. Roseman's groundbreaking work on the glycosyltransferases set the stage for development of this theme in many centers of research around the world.

The deaths are noted of several past contributors to this series. William George Overend (died on September 18, 2012, aged 90) of Birkbeck College, University of London, authored with Maurice Stacey an article on Deoxy Sugars in Volume 8, and with Finch wrote the obituary of Stacey in Volume 52. Robin Ferrier (after whom are named two "Ferrier Reactions"), of Victoria University, Wellington, New Zealand, contributed articles on Unsaturated Sugars in Volumes 24 and 58, and one on Boronates in Volume 35; he died on July 11, 2013, aged 81. Serge David of the University of Paris authored with Estramareix an article in Volume 52 on Sugars and Nucleotides and the Biosynthesis of Thiamine. He died on August 1, 2013, aged 92.

With this issue, Jésus Jiménez-Barbero is welcomed as a member of the Board of Advisors. Professor David C. Baker assisted with the Author Index.

DEREK HORTON

Washington, DC
August, 2013

STEPHEN JOHN CHARLES ANGYAL

1914–2012

Stephen John Charles Angyal was born in Budapest on November 21, 1914. His father, Professor Charles Engel, was born in a village in the north of Hungary in 1879. Charles studied in Budapest, received his medical degree in 1902, and became a highly respected specialist in syphilis. He published two books on the subject. As a keen reader of literature, Charles told young Stephen about his books while they were on Sunday morning walks, introducing him to the works of one of his favorite authors, Somerset Maugham. Charles was an accomplished pianist and regularly played chamber music with several string players (including some of Hungary's now famous 20th-century composers who came to the house). Stephen often turned the pages for his father. These sessions introduced Stephen to 19th-century chamber works and laid the foundation for his lifelong love of music. To speak two languages was important in Central Europe, and his parents engaged German nannies, usually young country girls. Stephen spoke German before Hungarian. Later, for his last three summer vacations at school, he was sent to a Swiss institution to learn French, and he also took lessons in English at home in the afternoons. This early facility with several European languages was to stand him in good stead throughout his long life in academia.

At 10 years of age, Stephen was enrolled in the prestigious high school of the Piarest order where he studied Latin and Greek, much literature, and essentially no chemistry. During this time, Stephen joined the school's scout group where he learned to swim and also the sport of skiing, two activities that he practiced with much enthusiasm until very late in life. He hung up his skis at the age of 92. After high school, he enrolled at the Royal Hungarian University of Science in Budapest, where he studied chemistry, physics, mathematics, geology, and mineralogy. For the Ph.D. degree, Stephen moved to the other university in Budapest, the University for Technology and Engineering, where he enrolled under Professor Géza Zemplén, although most of the supervision was done by an assistant, Zoltán Csűrös. The research work on glucose derivatives, which led to the award of the degree of Doctor *summa cum laude* in June 1937, did not spark the imagination of Stephen, and he vowed never to work with carbohydrates again.

Prior to enrolling with Zemplén, Stephen changed his name from the German spelling (Engel) to the Hungarian version, Angyal. His formal education was completed by one year of military service, graduating from officers' school as a lieutenant in the reserve.

Stephen's first job after graduation was as a research chemist with Chinoin, one of Hungary's leading pharmaceutical manufacturers. Here he worked on the production of synthetic estrogens and sulfathiazole. Knowledge of the latter would later lead to his first job when he arrived in Australia. He decided to go to Australia, where he had a cousin, to forward his research career and thus rise up the ladder faster when he returned to Chinoin. With great foresight, Stephen obtained a landing permit for Australia, which he received on July 3, 1939. September 1939 saw the start of the Second World War, and Stephen was called up for army duty. Again, luck was with him as Hungary kept out of the conflict for some time, and Stephen was demobilized. He immediately made preparations for his travels and left Hungary early in 1940, traveling to Milan and boarding in Genoa the passenger ship Viminale. This happened to be the last Italian ship to leave for Australia; soon afterwards, Italy entered the war.

After arriving in Sydney, Stephen met a fellow Hungarian, Dr. Andrew Ungar, with whom he started a small company, named the Andrews Laboratories. Although little came of this venture at the time, the association was of considerable significance in later years. To keep in touch with the local chemistry community, Stephen used to visit the Chemistry Department of the University of Sydney, where he met Dr. Frank Lions, who told him of their synthesis of sulfathiazole as part of the war effort. Stephen noted that he had also made that drug, and after this information had been reported in a local newspaper, Stephen received a phone call from a Melbourne pharmaceutical company, Nicholas Pty Ltd, offering him a job as a research chemist. During the five years that Stephen worked in Melbourne, he became aware of a compound, inositol, which was thought at the time to be a member of the vitamin B group. As very little was known about the chemistry of inositol, he decided that it might be worth studying some day. In connection with the synthesis of a new sulfa drug, marfanil, Stephen noted an unusual reaction that had been discovered by the French chemist, Sommelet. He decided that this reaction warranted further study, and the results led to a series of papers and a review.[202;*]

A particularly happy event that had lifelong ramifications for Stephen was his meeting in Melbourne with a lovely young girl, Helga Steininger. Originally from

---

*Superscript reference numbers in this memoir denote numbered citations in the appended grouped list of Angyal's publications.

Vienna, Helga had arrived from England six months earlier. She worked as a dress designer. They were married in February 1942 and remained an inseparable couple for the rest of Stephen's life. They had two children, Annette (1944, now Associate Professor Annette Gero, a biochemist) and Robert (1949, a barrister at law).

In 1946, Stephen secured a lectureship in chemistry at the University of Sydney, and here he started a study on the Sommelet reaction, as well as his early research on inositols. During that year, he was admitted as an Associate of the Royal Australian Chemical Institute (RACI) and later (1953) was elected as a Fellow of that institute. As a result of Australia's being geographically isolated from the countries of the northern hemisphere, where most chemical research was being carried out, it was usual for Australian academics to be given a year's paid leave every seventh year to allow them to study overseas. During his sixth year at the University of Sydney, Stephen was invited by Alexander Todd, who was a Visiting Professor at that university, to spend his study leave at Cambridge University. As Todd was one of the trustees of the prestigious Nuffield Dominion Travelling Fellowship, he suggested to Stephen that he should apply for a Fellowship, which in due course he was awarded. During his stay in Cambridge, Stephen met an Australian chemist, John Mills. From their mutual interest in conformational analysis, they wrote a review[196] for *Reviews of Pure and Applied Chemistry*. They later collaborated on several aspects of metal-ion complexes.

In 1952, Stephen obtained a grant-in-aid from the Carnegie Corporation of New York. This allowed him to visit a number of institutions in the USA on his way back to Australia in February, 1953. Among the chemists that Stephen met were Hermann O. L. Fischer (son of Emil Fischer), with whom he spent his next study leave, and Ernest Eliel, a fellow contributor to the well-known book, *Conformational Analysis*.[52]

While Stephen was on his study leave in Cambridge, an advertisement appeared from the New South Wales University of Technology seeking an associate professor in organic chemistry. He applied for that position and was successful, taking up the appointment in March 1953. Some years later, when a medical school was established, the name was changed to The University of New South Wales (UNSW hereafter). In 1960, Stephen was successful in applying for the newly created chair of organic chemistry, and he became the first professor of organic chemistry at UNSW.

Stephen was now in a position to start work on the inositols. Few aspects of cyclitol chemistry escaped the attention of the Angyal research group. Over 60 papers record their work on cyclic acetals, sulfonic esters, phosphoric esters, aminocyclitols, and their deamination products, 5-carbon cyclitols, acyl migration, tritium labeling, and selective protection. Of the nine possible isomers of inositol, only the common one, *myo*-inositol, which is widespread in nature, was available at a reasonable price from commercial sources. The preparation of acetal derivatives of inositols, acetone derivatives in particular, proved to be the key to the interconversion of the various stereoisomers. Satisfactory acetonation of *myo*-inositol was achieved only after a detailed study[13] of the reaction conditions. Two of the nine isomers had never been prepared at the beginning of this research. The first one, *neo*-inositol, was prepared[10] from the tosylate of a diacetal of *levo*-inositol. The remaining isomer, *cis*-inositol, proved to be the most difficult to prepare. Initially isolated[16] by chromatography over cellulose powder from the multitude of products formed by hydrogenation of hexahydroxybenzene, it was later prepared[84] in 25% yield in seven steps from *epi*-inositol and finally obtained[171] in 31% yield from hydrogenation of tetrahydroxyquinone, using an improved catalyst and separating the products on an ion-exchange column.

A most significant result that emerged from these early studies involved the interaction between the inositol hydroxyl groups and sodium borate in aqueous solution.[12,14] The products were shown to be tridentate boric esters. From the equilibrium constants determined for this reaction, the free-energy changes of complex formation were calculated. As the different nonbonded interactions determined the extent of complex formation, it was possible to calculate these various energies of interaction. Another reaction discovered at this time involved treatment of inositols with acetic acid containing sulfuric acid. For those inositols in which a *cis–trans* sequence of hydroxyl groups occurs, it was found[32] that epimerization occurs at the middle carbon atom. A reasonable mechanism for this reaction was described. Again, nonbonded interaction energies could be calculated[42] from the differences between the free energies of the epimers.

A detailed review on the cyclitols was written with Laurens Anderson[21] at this time, and the Royal Australian Chemical Institute recognized[23] the research by the award of the H. G. Smith Memorial Medal in 1958. Few chemists, even carbohydrate specialists, have the structures of the nine stereoisomeric inositols committed to memory, and the accompanying scheme depicts them in the convenient projection attributable to John Mills, a close friend and collaborator of Stephen (Scheme 1).

With the values for interaction energies for inositols in hand, Stephen applied these values to the pyranose forms of reducing sugars in order to explain the α:β ratios for aqueous solutions[64] and the ratios of reducing sugars and their 1,6-anhydrides in equilibrium,[65] and he later extended this work to acyclic compounds.[76] At this time, Stephen published an article on the composition and conformations of monosaccharide sugars in solution.[66] He regarded this as his most successful paper, as it was used in several universities as a teaching aid, with every student being given a copy. A much more detailed review on conformational analysis as applied to carbohydrates was published in the well-known textbook *Conformational Analysis*.[52] Stephen was

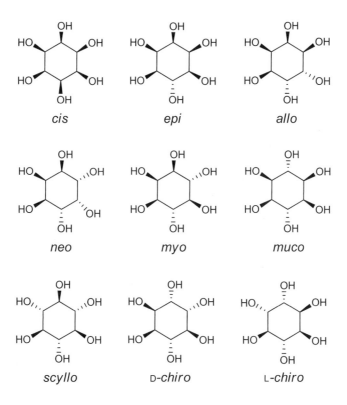

SCHEME 1. The nine inositols, in the Mills projection.

awarded the Archibold D. Ollé Prize by the Royal Australian Chemical Institute for this publication.

During a study leave in 1962, which was spent at the Imperial College in London, Stephen attended the first International Symposium on Carbohydrate Chemistry (that meeting, in Birmingham, is now regarded as the second in the series; an earlier, smaller meeting had been held in Gif-sur-Yvette, near Paris). Subsequently, he attended almost all these biannual meetings, and after his retirement from UNSW in 1979, he organized and hosted the 10th meeting, which was held at UNSW in Sydney. Participants remember this meeting with much fondness. In 2008, Stephen published an historical account[183] of the International Symposia on Carbohydrate Chemistry.

Stephen's interest in carbohydrates was rekindled when he was invited to review the topic for *Annual Reviews of Biochemistry*.[43] At this time, the award of a research grant enabled the purchase of a Varian A60 nuclear magnetic resonance (NMR) spectrometer. Now a method was available to him for determining the tautomeric composition of many aldoses in aqueous solution, using ^{1}H NMR spectroscopy. These studies were extended to ketoses[116] when a ^{13}C NMR spectrometer became available. This work led to a comprehensive review[146] of sugar compositions in solution, followed by a supplementary review[166] seven years later. As a prelude to this experimental work, Stephen published a detailed paper[64] on the conformational free energies of aldopyranoses in solution, in which the interaction energies already mentioned were applied to the reducing sugars.

Following up on electrophoresis studies by John Mills which showed that various polyols complexed with a variety of salts, Stephen commenced[82] a study, using ^{1}H NMR spectroscopy, of the complexing of sugars with metal cations. This led to a significant series of papers that covered a diverse range of topics. The classic Fischer glycosidation using acidified methanol was modified by incorporating various salts, which enabled him to prepare some less-readily obtained glycosides[97,110,133] and dimethyl acetals,[111] and the application of cation-exchange resins,[126,179] commonly in the Ca^{2+} form, was used to effect the separation of glycosides[133] and inositols.[171] In his last experimental paper, the procedure was applied to the preparation of L-ribose.[181] These studies were the subject of several reviews,[101,102,104,160] and formed his topic for the Haworth Memorial Lecture[137] delivered on March 31, 1980 and also for the C. S. Hudson Award Address.[154] Support for the proposed structures of metal cation complexes deduced from NMR spectra was obtained by single-crystal X-ray diffraction studies.[120,121,168]

In a detailed paper[129] on methyl aldofuranosides, Stephen extended his studies on conformations to five-membered ring compounds. In this paper, extensive ^{1}H NMR data were collected and the conformations deduced therefrom were correlated with optical rotation values.

In a short series of papers, the application of chromium trioxide in acetic acid was described in which methoxyl groups were converted[74] into formic esters. Acetylated methyl glycosides gave[75] acetylated methyl 5-hexulosonates, acetylated acetals of alditols gave[83] derivatives of 3-hexuloses, and cyclic methylene acetals gave cyclic carbonates.[89] Building upon the discovery by Koch and Stuart that the methine hydrogen atom of secondary alcohols can be exchanged for deuterium using deuterated Raney nickel in deuterium oxide, Stephen applied the procedure to a variety of compounds, including inositols and inositol methyl ethers,[142] methyl glycosides,[151] and 1,6-anhydrohexoses.[153]

Many research students and collaborators contributed over the years to the realization of the large body of experimental work recorded in Stephen's publications. In addition to the work of numerous researchers in his Sydney laboratory, Stephen's visits to other laboratories led to a number of collaborative ventures with various institutions in Europe and North America. His joint work with John Mills and Laurens Anderson documents important understanding of the inositols, and the work of Dennis McHugh, to select just one of the many research contributors in Stephen's group, on the determination of interaction energies, played a key role in predicting the conformations of polysubstituted six-membered ring systems.

The Andrews Lectures

During the course of the first International Symposium on the Chemistry of Natural Products, which was held in Australia in 1960, Stephen decided to invite one of the participants to give a series of lectures over a two-week period. Professor E. R. H. Jones accepted the invitation and became the first Andrews Lecturer. Funds for the lectures were provided by Dr. Andrew Ungar and the lectures were named after his company, Andrews Laboratories. Over the next 50 years, 26 eminent organic chemists have been appointed as Andrews Lecturer, each one delivering three lectures, generally over a period of one to two weeks. It was an essential part of the Lectureship that the lecturer spends some time between lectures at UNSW and be available for discussions with staff and advanced students. Until his retirement in 1979, Stephen made all the arrangements for these lectureships, with significant help from Helga for the social engagements. After he retired, management of the Andrews Lectures was

Andrews Lecturer (1970). John Cornforth and Stephen discuss a stereochemistry problem.

passed to the University of New South Wales Chemical Society and they continue to arrange biannual lectureships.

Administration

Stephen served as Head of the School of Chemistry at UNSW from 1968 to 1970. He was appointed as Dean of the Faculty of Science in 1970 and kept that position until he retired in 1979. Stephen enjoyed being involved in the organization of conferences. After the Natural Products meeting in 1966, the next major conference was the XXII International Congress of Pure and Applied Chemistry, held in 1969 at the University of Sydney. Stephen was the convener of the Social Program and Helga chaired the Ladies' Program Committee. Finally, after his retirement, Stephen organized the Xth International Symposium on Carbohydrate Chemistry, held in 1980.

Throughout his academic career, Stephen was a keen traveler. He was Visiting Professor at the University of California, Berkeley in 1957, Visiting Lecturer at London's Imperial College of Science and Technology in 1962, Visiting Professor at the Universities of Oxford and Grenoble in 1968, Visiting Lecturer at Eidgenössische Technische Hochschule (ETH), Zurich, in 1972, and Visiting Professor at the University of Grenoble in 1977. Besides attending the International Symposia on Carbohydrate Chemistry, Stephen attended a number of the Bürgenstock Conferences on Stereochemistry and several of the Gordon Conferences on Carbohydrates.

Although his early papers were published in the *Journal of the Chemical Society* (London), Stephen later favored the local publication, the *Australian Journal of Chemistry*. He served as a member and, in turn, chairman of its editorial board. Despite early reservations on the introduction of a journal devoted solely to carbohydrates, Stephen supported *Carbohydrate Research*, serving on the editorial board for many years and publishing more than 40 papers there. He headed the IUPAC panel that developed the official 1973 Recommendations for Nomenclature of Cyclitols.[99]

Stephen was elected as a Fellow of the Australian Academy of Science in 1962, awarded the first UNSW D.Sc. degree in 1964, appointed as the Haworth Memorial Lecturer of the Chemical Society, London, in 1980, and received the Claude S. Hudson Award of the American Chemical Society in 1987. In 1990, he was honored by the Hungarian Academy of Science, being elected as an External Member. His contributions to science in Australia were acknowledged in the award of Officer of the Order of the British Empire (OBE) in 1977. More recently, his daughter Annette and son Robert have endowed UNSW with the Angyal Medal, which is awarded each year to the top honors student of the year in chemistry, as a permanent memory of Stephen in the School of Chemistry at UNSW.

Stephen enjoyed a number of interests besides chemistry. He had learned to ski while at high school and this became a lifelong activity until his early nineties. Not surprisingly, most of his study leaves were taken at places close to snowfields, and he also made full use of the facilities in the Australian ski resorts. Another lifelong pleasure for Stephen was swimming, which he enjoyed year round in the mild climate of Sydney. Both Stephen and Helga were lovers of music. They regularly attended performances of the Sydney Symphony Orchestra and were staunch supporters of the Australia Ensemble (resident at UNSW), and it was rare for them to miss a performance.

Stephen, a slim and tall man, enjoyed robust health for most of his life. He died peacefully on May 14, 2012, aged 97. His beloved wife, Helga, after 70 years of marriage, died peacefully and unexpectedly six months later.

JOHN D. STEVENS

ACKNOWLEDGMENTS

In preparing this obituary, I have drawn upon extensive biographical notes prepared by Stephen in 2000 and on an interview by Mr. David Salt on behalf of the Australian Academy of Science in 2003. I am grateful to Professor Laurens Anderson for permission to use his introduction to the special issue of *Carbohydrate Research* in honor of Stephen on the occasion of his retirement.

LIST OF PUBLICATIONS BY S. J. ANGYAL

A. Inositols and Sugars

1. G. Zemplén, Z. Csűrös, and S. J. Angyal, Über benzylierte derivative des lävoglucosans und der glucose, *Ber.*, 70 (1937) 1848–1856.
2. G. Zemplén, Z. Csűrös, and S. J. Angyal, Über 2,4-dibenzyllävoglucosan, *Arbeiten Biol. Forschungsinst.*, 10 (1938) 429.
3. S. J. Angyal and J. V. Lawler, 1,3:2,4-Dibenzylidene-D-sorbitol, *J. Am. Chem. Soc.*, 66 (1944) 837–838.
4. S. J. Angyal and N. K. Matheson, Dimorphism of esters of *scyllo*-inosose, *J. Chem. Soc.*, (1950) 3349.
5. S. J. Angyal and C. G. Macdonald, Cyclitols. Part I. Isopropylidene derivatives of inositols and quercitols, *J. Chem. Soc.*, (1952) 686–695.
6. C. L. Angyal and S. J. Angyal, Cyclitols. Part II. Dipole moments of some cyclitol acetates, *J. Chem. Soc.*, (1952) 695–697.
7. N. K. Matheson and S. J. Angyal, The replacement of secondary tosyloxy-groups by iodine in polyhydroxy-compounds, *J. Chem. Soc.*, (1952) 1133–1138.
8. S. J. Angyal, C. G. Macdonald, and N. K. Matheson, The structure of the di-*O*-isopropylidene derivatives of (-)-inositol and pinitol, *J. Chem. Soc.*, (1953) 3321–3323.

9. S. J. Angyal, Ring-opening of anhydrosugars of the ethylene oxide type, *Chem. Ind. (Lond.)*, (1954) 1230–1231.
10. S. J. Angyal and N. K. Matheson, Cyclitols. Part III. Some tosyl esters of inositols. Synthesis of a new inositol, *J. Am. Chem. Soc.*, 77 (1955) 4343–4346.
11. S. J. Angyal and D. J. McHugh, cis-Inositol, a 1,3,5-triaxially substituted cyclohexane derivative, *Chem. Ind. (Lond.)*, (1955) 947–948.
12. S. J. Angyal and D. J. McHugh, Interaction energies of axial hydroxyl groups, *Chem. Ind. (Lond.)*, (1956) 1147–1148.
13. S. J. Angyal, P. T. Gilham, and C. G. Macdonald, Cyclitols. Part IV. Methyl ethers of *myo*-inositol, *J. Chem. Soc.*, (1957) 1417–1422.
14. S. J. Angyal and D. J. McHugh, Cyclitols. Part V. Paper ionophoresis, complex formation with borate, and the rate of periodic acid oxidation, *J. Chem. Soc.*, (1957) 1423–1431.
15. S. J. Angyal, D. J. McHugh, and P. T. Gilham, The paper chromatography of cyclitols, *J. Chem. Soc.*, (1957) 1432–1433.
16. S. J. Angyal and D. J. McHugh, Cyclitols. Part VI. The hydrogenation of hexahydroxybenzene, *J. Chem. Soc.*, (1957) 3682–3691.
17. S. J. Angyal and P. T. Gilham, Cyclitols. Part VII. Anhydroinositols and the "Epoxide Migration", *J. Chem. Soc.*, (1957) 3691–3699.
18. S. J. Angyal, The inositols, *Quart. Rev.*, 11 (1957) 212–226.
19. S. J. Angyal and P. T. Gilham, Cyclitols. Part VIII. Elimination of vicinal sulphonyloxy-groups by iodide ion, *J. Chem. Soc.*, (1958) 375–379.
20. L. Anderson, R. Takeda, S. J. Angyal, and D. J. McHugh, Cyclitol oxidation by *Acetobacter suboxydans*. II. Additional cyclitols and the "Third Specificity Rule", *Arch. Biochem. Biophys.*, 78 (1958) 518–531.
21. S. J. Angyal and L. Anderson, The cyclitols, *Adv. Carbohydr. Chem.*, 14 (1959) 135–212.
22. S. J. Angyal, J. S. Murdoch, and M. E. Tate, Myoinositol 4- and 5-phosphate, *Proc. Chem. Soc. (Lond.)*, (1960) 416.
23. S. J. Angyal, Inositols: Chemistry, stereochemistry and biochemistry, *Proc. R. Aust. Chem. Inst.*, 27 (1960) 32–38. (Text of the H. G. Smith Memorial Lecture, 16 July, 1959).
24. S. J. Angyal, M. E. Tate, and (in part) S. D. Gero, Cyclitols. Part IX. Cyclohexylidene derivatives of *myo*-Inositol, *J. Chem. Soc.*, (1961) 4116–4122.
25. S. J. Angyal and M. E. Tate, Cyclitols. Part X. *myo*-inositol phosphates, *J. Chem. Soc.*, (1961) 4122–4128.
26. S. J. Angyal and V. J. Bender, Cyclitols. Part XI. The constitution of liriodendritol, *J. Chem. Soc.*, (1961) 4718–4720.
27. S. J. Angyal and R. M. Hoskinson, Cyclitols. Part XII. The formation of isopropylidene derivatives by ketal interchange, *J. Chem. Soc.*, (1962) 2985–2991.
28. S. J. Angyal and R. J. Young, The elimination reaction of vicinal disulphonyloxy compounds with sodium iodide, *Aust. J. Chem.*, 14 (1961) 8–14.
29. S. J. Angyal and J. E. Klavins, Some aspects of the oxidation of sugars and inositols by periodate: The formation of intermediary esters, *Aust. J. Chem.*, 14 (1961) 577–585. Correction, *Aust. J. Chem.*, 14 (1961) 479–479.
30. Z. S. Krzeminski and S. J. Angyal, Separation of inositols and their monomethyl ethers by gas chromatography, *J. Chem. Soc.*, (1962) 3251–3252.
31. S. J. Angyal and R. M. Hoskinson, Cyclitols. Part XIII. The conformation of cyclic diketals: The skew conformation, *J. Chem. Soc.*, (1962) 2991–2995.
32. S. J. Angyal, P. A. J. Gorin, and M. E. Pitman, A stereospecific epimerisation of cyclitols, *Proc. Chem. Soc.*, (1962) 337–338.
33. S. J. Angyal, J. L. Garnett, and R. M. Hoskinson, Occurrence of configurational inversion during tritiation by the Wilzbach method, *Nature*, 197 (1963) 485–486.

34. S. J. Angyal, J. L. Garnett, and R. M. Hoskinson, The stereochemistry of tritium substitution by the Wilzbach method. I. (−)-Inositol, *Aust. J. Chem.*, 16 (1963) 252–257.
35. S. J. Angyal and B. Shelton, Structure and synthesis of "Manninositose", *Proc. Chem. Soc.*, (1963) 57.
36. S. J. Angyal and R. M. Hoskinson, Cyclitols. Part XIV. Formation of 1,4-anhydro-*epi*-inositol by dehydration of *myo*-inositol, *J. Chem. Soc.*, (1963) 2043–2047.
37. S. J. Angyal, Cyclitols, In: M. Florkin and E. H. Stotz, (Eds.), *Comprehensive Biochemistry*, Elsevier, Amsterdam, 1963, pp. 297–303.
38. S. J. Angyal and R. M. Hoskinson, L-Mannitol, *Methods Carbohydr. Chem.*, 2 (1963) 87–89.
39. S. J. Angyal and C. Fernandez, Specific tritium labelling by the Wilzbach method, *Nature*, 202 (1964) 176–177.
40. S. J. Angyal and T. S. Stewart, The reaction of tosyl derivatives of inositols with sodium benzoate in dimethylformamide, *Proc. Chem. Soc.*, (1964) 331.
41. S. J. Angyal, C. M. Fernandez, and J. L. Garnett, The stereochemistry of tritium substitution by the Wilzbach method. II, *Aust. J. Chem.*, 18 (1965) 39–45.
42. S. J. Angyal, P. A. J. Gorin, and M. E. Pitman, Cyclitols. Part XV. A stereospecific epimerization of cyclitols: Conformational free energies, *J. Chem. Soc.*, (1965) 1807–1816.
43. S. J. Angyal and D. Rutherford, Carbohydrates—mono- and oligo-saccharides, *Annu. Rev. Biochem.*, 34 (1965) 77–100.
44. S. J. Angyal, P. T. Gilham, and G. J. H. Melrose, Cyclitols. Part XVI. Toluene-*p*-sulphonyl derivatives of *myo*-inositol. Acetyl migration in anhydrous pyridine solution, *J. Chem. Soc.*, (1965) 5252–5255.
45. S. J. Angyal and S. D. Gero, Cyclitols. Part XVII. The selectivity of protecting groups: Tetrahydropyranyl ethers, *J. Chem. Soc.*, (1965) 5255–5258.
46. S. J. Angyal and G. J. H. Melrose, Cyclitols. Part XVIII. Acetyl migration: Equilibrium between axial and equatorial acetates, *J. Chem. Soc.*, (1965) 6494–6500.
47. S. J. Angyal and G. J. H. Melrose, Cyclitols. Part XIX. Control of acetyl migration during methylation of partially acetylated cyclitols, *J. Chem. Soc.*, (1965) 6501–6509.
48. S. J. Angyal, G. C. Irving, D. Rutherford, and M. E. Tate, Cyclitols. Part XX. Cyclohexylidene ketals of inositols, *J. Chem. Soc.*, (1965) 6662–6664.
49. S. J. Angyal and M. E. Tate, Cyclitols. Part XXI. Benzyl ethers of *myo*-inositol. Aromatisation of a tosyl derivative of *myo*-inositol, *J. Chem. Soc.*, (1965) 6949–6955.
50. S. J. Angyal, Liversidge Lecture—Conformations of molecules in solution, *Aust. J. Sci.*, 28 (1965) 173–178.
51. S. J. Angyal and S. D. Gero, Convenient preparation of a crystalline derivative of *meso*-tartraldehyde, *Aust. J. Chem.*, 18 (1965) 1973–1976.
52. S. J. Angyal, Conformational analysis in carbohydrate chemistry, in *Conformational Analysis*, Interscience Publishers, New York, 1965, pp. 351–432.
53. S. J. Angyal and B. Shelton, Cyclitols. Part XXII. Synthesis of some mannosyl- and mannosylmannosyl-*myo*-inositols and of galactinol, *J. Chem. Soc., (C)*, (1966) 433–438.
54. S. J. Angyal, V. A. Pickles, and R. Ahluwahlia, The interaction energy between an axial methyl and an axial hydroxyl group in pyranoses, *Carbohydr. Res.*, 1 (1966) 365–370.
55. S. J. Angyal, V. J. Bender, and J. H. Curtin, Cyclitols. Part XXIII. Suppression of epoxide migration. Synthesis of *muco*-inositol, *J. Chem. Soc., (C)*, (1966) 798–800.
56. S. J. Angyal and T. S. Stewart, Cyclitols. Part XXV. Benzyl ethers of (-)-inositol. Lack of selectivity in *O*-substitution, *Aust. J. Chem.*, 19 (1966) 1683–1691.
57. S. J. Angyal, M. H. Randall, and M. E. Tate, Cyclitols. Part XXIV. Selective acetolysis of *myo*-inositol benzyl ethers, *J. Chem. Soc., (C)*, (1967) 919–922.
58. S. J. Angyal, V. A. Pickles, and R. Ahluwalia, Equilibrium between pyranoid and furanoid forms. Part 1. 2,3-*O*-isopropylidene-L-rhamnose, *Carbohydr. Res.*, 3 (1967) 300–307.

59. S. J. Angyal and V. A. Pickles, Equilibria between furanoses and pyranoses, *Carbohydr. Res.*, 4 (1967) 269–270.
60. S. J. Angyal, V. J. Bender, P. T. Gilham, R. M. Hoskinson, and M. E. Pitman, Cyclitols. Part XXVI. The solvolysis of tosylinositols in acetic acid, *Aust. J. Chem.*, 20 (1967) 2109–2116. Corrections: *Aust. J. Chem.*, 21 (1968) 273–273.
61. S. J. Angyal and T. S. Stewart, Cyclitols. Part XXVII. The reaction of tosylinositols with sodium benzoate in dimethylformamide. Anomalous opening of epoxides, *Aust. J. Chem.*, 20 (1967) 2117–2136.
62. R. Ahluwalia, S. J. Angyal, and M. H. Randall, Synthesis of the methyl D-allopyranosides and of D-allose from 1,2:5,6-di-*O*-isopropylidene-3-*O*-*p*-tolylsulfonyl-α-D-glucofuranose, *Carbohydr. Res.*, 4 (1967) 478–485.
63. S. J. Angyal, C. M. Fernandez, and J. L. Garnett, Catalytic deuterium exchange reactions with organics. XXXVI. The labelling of (-)-inositol on self-activated platinum oxide with and without radiation, *Aust. J. Chem.*, 20 (1967) 2647–2653.
64. S. J. Angyal, Conformational analysis in carbohydrate chemistry. I. Conformational free energies. The conformations and α:β ratios of aldopyranoses in aqueous solution, *Aust. J. Chem.*, 21 (1968) 2737–2746.
65. S. J. Angyal and K. Dawes, Conformational analysis in carbohydrate chemistry. II. Equilibria between reducing sugars and their glycosidic anhydrides, *Aust. J. Chem.*, 21 (1968) 2747–2760.
66. S. J. Angyal, The composition and conformation of sugars in solution, *Angew. Chem.*, 8 (Int. ed.), (1969) 157–166.
67. S. J. Angyal and A. F. Russell, Cyclitols. XXVIII. Methyl esters of inositol phosphates. The structure of phytic acid, *Aust. J. Chem.*, 22 (1969) 383–390.
68. S. J. Angyal and A. F. Russell, Cyclitols. XXIX. Polyphosphorylation of polyols. The synthesis of *myo*-inositol pentaphosphates, *Aust. J. Chem.*, 22 (1969) 391–404.
69. S. J. Angyal and K. James, Oxidative ring opening of pyranosides and furanosides, *J. Chem. Soc. Chem. Commun.*, (1969) 617–618.
70. S. J. Angyal and J. S. Murdoch, Cyclitols. XXX. The deamination of inosamines by nitrous acid, *Aust. J. Chem.*, 22 (1969) 2417–2428.
71. S. J. Angyal, B. M. Luttrell, A. F. Russell, and D. Rutherford, Inositol phosphates, glycosides and some 5-carbon cyclitols, *Ann. N. Y. Acad. Sci.*, 165 (1969) 533–540.
72. D. E. Dorman, S. J. Angyal, and J. D. Roberts, Nuclear magnetic resonance spectroscopy: ^{13}C spectra of unsubstituted inositols, *Proc. Natl. Acad. Sci. U.S.A.*, 63 (1969) 612–614.
73. S. J. Angyal and K. James, A general method for the synthesis of hex-3-uloses, *J. Chem. Soc. Chem. Commun.*, (1970) 320–321.
74. S. J. Angyal and K. James, Oxidative demethylation with chromium trioxide in acetic acid, *Carbohydr. Res.*, 12 (1970) 147–149.
75. S. J. Angyal and K. James, Oxidation of carbohydrates with chromium trioxide in acetic acid. I. Glycosides, *Aust. J. Chem.*, 23 (1970) 1209–1221.
76. S. J. Angyal and K. James, Conformations of acyclic sugar derivatives. Acetylated methyl 5-hexulosonates and keto-hexuloses, *Aust. J. Chem.*, 23 (1970) 1223–1228. (Part I of the series).
77. S. J. Angyal and B. M. Luttrell, Stereochemistry of the addition of protons to nitronate ions, *Aust. J. Chem.*, 23 (1970) 1485–1489.
78. R. Ahluwalia, S. J. Angyal, and B. M. Luttrell, Cyclitols. XXXI. Synthesis of amino- and nitrocyclopentanetetrols, *Aust. J. Chem.*, 23 (1970) 1819–1829.
79. S. J. Angyal and B. M. Luttrell, Cyclitols. XXXII. Cyclopentanepentols, *Aust. J. Chem.*, 23 (1970) 1831–1838.
80. D. E. Dorman, S. J. Angyal, and J. D. Roberts, Carbon-13 spectra of some inositols and their *O*-methylated derivatives, *J. Am. Chem. Soc.*, 92 (1970) 1351–1354.

81. S. J. Angyal and K. James, 2,4-*O*-Benzylidene-L-xylose and 2,4-*O*-methylene-L-xylose: Sugars which do not readily form pyranose rings, *Carbohydr. Res.*, 15 (1970) 91–100.
82. S. J. Angyal and K. P. Davies, Complexing of sugars with metal ions, *J. Chem. Soc. Chem. Commun.*, (1971) 500–501.
83. S. J. Angyal and K. James, Oxidation of carbohydrates with chromium trioxide in acetic acid. II. Acetals. A new synthesis of ketoses, *Aust. J. Chem.*, 24 (1971) 1219–1227.
84. S. J. Angyal and R. J. Hickman, Cyclitols. XXXIII. A practical synthesis of *cis*-inositol, *Carbohydr. Res.*, 20 (1971) 97–104.
85. S. J. Angyal, Inositols—Chemistry, In: W. H. Sebrell and R. S. Harris, (Eds.), 2nd ed. *The Vitamins*, Vol. 3, Academic Press, New York, 1971, pp. 345–353.
86. S. J. Angyal, Conformations of sugars, In: W. Pigman and D. Horton, (Eds.), *The Carbohydrates—Chemistry and Biochemistry*, Vol. 1A, Academic Press, New York, 1972, pp. 195–215.
87. S. J. Angyal and M. E. Evans, DL-*ribo*-3-Hexulose and DL-*arabino*-3-hexulose, *Aust. J. Chem.*, 25 (1972) 1347–1350.
88. S. J. Angyal and M. E. Evans, Oxidation of carbohydrates with chromium trioxide in acetic acid. III. Synthesis of the 3-hexuloses, *Aust. J. Chem.*, 25 (1972) 1495–1512.
89. S. J. Angyal and M. E. Evans, Oxidation of carbohydrates with chromium trioxide in acetic acid. IV. Conversion of cyclic methylene acetals into cyclic carbonates, *Aust. J. Chem.*, 25 (1972) 1513–1520.
90. S. J. Angyal, R. LeFur, and D. Gagnaire, Conformations of acyclic sugar derivatives. II. Determination of the conformations of alditol acetates in solution by the use of 250-MHz N.M.R. spectra, *Carbohydr. Res.*, 23 (1972) 121–134.
91. S. J. Angyal, R. LeFur, and D. Gagnaire, Conformations of acyclic sugar derivatives. III. 3,4,5,6,7-Pentaacetoxy-*trans*-1-nitro-1-heptenes, *Carbohydr. Res.*, 23 (1972) 135–138.
92. S. J. Angyal and V. A. Pickles, Equilibria between pyranoses and furanoses. II. Aldoses, *Aust. J. Chem.*, 25 (1972) 1695–1710.
93. S. J. Angyal and V. A. Pickles, Equilibria between pyranoses and furanoses. III. Deoxyaldoses. The stability of furanoses, *Aust. J. Chem.*, 25 (1972) 1711–1718.
94. S. J. Angyal, Complexes of carbohydrates with metal cations. I. Determination of the extent of complexing by N.M.R. spectroscopy, *Aust. J. Chem.*, 25 (1972) 1957–1966.
95. K. James and S. J. Angyal, New syntheses of 1-deoxyhexuloses, *Aust. J. Chem.*, 25 (1972) 1967–1977.
96. M. H. Randall and S. J. Angyal, Formation of 3,6-anhydro-4,5-*O*-isopropylidene-D-allose dimethyl acetal in the methanolysis of 1,2:5,6-di-*O*-isopropylidene-3-*O*-tolylsulphonyl-α-D-glucofuranose. Synthesis of 3,6-anhydro-D-allose, *J. Chem. Soc. Perkin Trans. 1*, (1972) 346–351.
97. M. E. Evans and S. J. Angyal, Complexes of carbohydrates with metal cations. II. Glycosidations of D-allose in the presence of strontium and calcium ions, *Carbohydr. Res.*, 25 (1972) 43–48.
98. J. F. McConnell, S. J. Angyal, and J. D. Stevens, Cyclitols. XXXIV. X-Ray crystal and molecular structure of 1,2:5,6-di-*O*-isopropylidene-3,4-di-*O*-tosyl-L-*chiro*-inositol and its conformation in solution by nuclear magnetic resonance, *J. Chem. Soc. Perkin Trans. 2*, (1972) 2039–2044.
99. S. J. Angyal (Chairman), L. Anderson, R. S. Cahn, R. M. C. Dawson, O. Hoffmann-Ostenhof, W. Klyne, and T. Posternak IUPAC Joint Cyclitol Nomenclature Sub-committee, Nomenclature of Cyclitols, IUPAC Recommendations, 1973, *Biochem. J.*, 153 (1976) 23–31. *Eur. J. Biochem.*, 57 (1975) 1–7; *Pure Appl. Chem.*, 37 (1974) 283–297; http:/www.chem.qmul.ac.uk/iupac/cyclitol/.
100. S. J. Angyal, Shifts induced by lanthanide ions in the N.M.R. spectra of carbohydrates in aqueous solution, *Carbohydr. Res.*, 26 (1973) 271–273.
101. S. J. Angyal, Complex formation between sugars and metal ions, *Pure Appl. Chem.*, 35 (1973) 131–146.
102. S. J. Angyal, Complexes of sugars with cations, *Adv. Chem. Ser.*, 117 (1973) 106–120.
103. S. J. Angyal, V. J. Bender, and B. J. Ralph, Structure of polysaccharides from the *Polyporus tumulosus* cell wall, *Biochim. Biophys. Acta*, 362 (1974) 175–187.
104. S. J. Angyal, Complexing of polyols with cations, *Tetrahedron*, 30 (1974) 1695–1702.

105. S. J. Angyal, D. Greeves, and J. A. Mills, Complexes of carbohydrates with metal cations. III. Conformations of alditols in aqueous solution, *Aust. J. Chem.*, 27 (1974) 1447–1456. Conformations of acyclic sugar derivatives. IV.
106. S. J. Angyal, D. Greeves, and V. A. Pickles, The stereochemistry of complex formation of polyols with borate and periodate anions and with metal cations, *Carbohydr. Res.*, 35 (1974) 165–173.
107. S. J. Angyal, J. E. Klavins, and J. A. Mills, Cyclitols. XXXV. Tridentate complexing of some C-methyl and C-hydroxymethyl-cyclitols with borate ions, *Aust. J. Chem.*, 27 (1974) 1075–1086.
108. S. J. Angyal, D. Greeves, and V. A. Pickles, Stereospecific contact interactions in the nuclear magnetic resonance spectra of polyol–lanthanide complexes, *J. Chem. Soc. Chem. Commun.*, (1974) 589–590.
109. S. J. Angyal and R. J. Hickman, Complexes of carbohydrates with metal cations. IV. Cyclitols, *Aust. J. Chem.*, 28 (1975) 1279–1287. (Cyclitols. XXXVI.). Correction: *Aust. J. Chem.*, 28 (1975) 1864–1864.
110. S. J. Angyal, C. L. Bodkin, and F. W. Parrish, Complexes of carbohydrates with metal cations. V. Synthesis of methyl glycosides in the presence of metal ions, *Aust. J. Chem.*, 28 (1975) 1541–1549.
111. F. W. Parrish, S. J. Angyal, M. E. Evans, and J. A. Mills, Complexes of carbohydrates with metal cations. VI. Formation of dimethylacetals of D-glucose, D-xylose, and L-idose in the presence of strontium ions, *Carbohydr. Res.*, 45 (1975) 73–83.
112. S. J. Angyal, R. T. Gallagher, and P. M. Pojer, The synthesis of pinpollitol (1,4-di-*O*-methyl-D-*chiro*-inositol), *Aust. J. Chem.*, 29 (1976) 219–222.
113. S. J. Angyal and D. Greeves, Complexes of carbohydrates with metal cations. VII. Lanthanide-induced shifts in the P.M.R. spectra of cyclitols, *Aust. J. Chem.*, 29 (1976) 1223–1230.
114. S. J. Angyal, D. Greeves, L. Littlemore, and V. A. Pickles, Complexes of carbohydrates with metal cations. VIII. Lanthanide induced shifts in the P.M.R. spectra of some methyl glycosides and 1,6-anhydrohexoses, *Aust. J. Chem.*, 29 (1976) 1231–1237.
115. S. J. Angyal, G. S. Bethell, D. E. Cowley, and V. A. Pickles, Equilibria between pyranoses and furanoses. IV. 1-Deoxyhexuloses and 3-hexuloses, *Aust. J. Chem.*, 29 (1976) 1239–1247.
116. S. J. Angyal and G. S. Bethell, Conformational analysis in carbohydrate chemistry. III. The ^{13}C N.M.R. spectra of the hexuloses, *Aust. J. Chem.*, 29 (1976) 1249–1265.
117. P. M. Pojer and S. J. Angyal, Methylthiomethyl ethers: general synthesis and mild cleavage. Protection of hydroxyl groups, *Tetrahedron Lett.*, 35 (1976) 3067–3068.
118. S. J. Angyal, C. L. Bodkin, J. A. Mills, and P. M. Pojer, Complexes of carbohydrates with metal cations. IX. Synthesis of the methyl D-tagatosides, D-psicosides, D-apiosides, and D-erythrosides, *Aust. J. Chem.*, 30 (1977) 1259–1268.
119. L. D. Hayward and S. J. Angyal, A symmetry rule for the circular dichroism of reducing sugars, and the proportion of carbonyl forms in aqueous solutions thereof, *Carbohydr. Res.*, 53 (1977) 13–20.
120. R. A. Wood, V. J. James, and S. J. Angyal, The crystal structure of the *epi*-inositol–strontium chloride complex, *Acta Cryst. B*, 33 (1977) 2248–2251.
121. J. Ollis, V. J. James, S. J. Angyal, and P. M. Pojer, An X-ray crystallographic study of α-D-allopyranosyl α-D-allopyranoside.CaCl$_2$.5H$_2$O (a pentadentate complex), *Carbohydr. Res.*, 60 (1978) 219–228.
122. S. J. Angyal and R. J. Beveridge, Intramolecular acetal formation by primary versus secondary hydroxyl groups, *Carbohydr. Res.*, 65 (1978) 229–234. (Conformational analysis in carbohydrate chemistry. IV.)
123. S. J. Angyal and R. J. Beveridge, The synthesis of 1,6-anhydro-hexofuranoses, *Aust. J. Chem.*, 31 (1978) 1151–1155.
124. P. M. Pojer and S. J. Angyal, Methylthiomethyl ethers: their use in the protection and methylation of hydroxyl groups, *Aust. J. Chem.*, 31 (1978) 1031–1040.
125. S. J. Angyal, John Archer Mills—Obituary, *Adv. Carbohydr. Chem. Biochem.*, 36 (1979) 1–8.

126. S. J. Angyal, G. S. Bethell, and R. J. Beveridge, The separation of sugars and polyols on cation-exchange resin in the calcium form, *Carbohydr. Res.*, 73 (1979) 9–18. (Complexes of carbohydrates with metal cations. X).
127. S. J. Angyal, Composition of reducing sugars in solution, *J. Carbohydr. Nucleosides, Nucleotides*, 6 (1979) 15–30.
128. S. J. Angyal, D. Range, J. Defaye, and A. Gadelle, The behaviour of inososes in neutral and basic aqueous solution, *Carbohydr. Res.*, 76 (1979) 121–130. (Cyclitols. XXXVII.).
129. S. J. Angyal, Hudson's rules of isorotation as applied to furanosides and the conformations of methyl aldofuranosides, *Carbohydr. Res.*, 77 (1979) 37–50.
130. S. J. Angyal and J. A. Mills, Complexes of carbohydrates with metal cations. XI. Paper electrophoresis of polyols in solutions of calcium ions, *Aust. J. Chem.*, 32 (1979) 1993–2001.
131. S. J. Angyal and L. Odier, An easy synthesis of *muco*-inositol, *Carbohydr. Res.*, 80 (1980) 203–206.
132. S. J. Angyal and R. G. Wheen, The composition of reducing sugars in aqueous solution: glyceraldehyde, erythrose, threose, *Aust. J. Chem.*, 33 (1980) 1001–1011.
133. S. J. Angyal, M. E. Evans, and R. J. Beveridge, Complexes of carbohydrates with metal cations—methyl β-D-mannofuranoside, *Methods Carbohydr. Chem.*, 8 (1980) 233–235.
134. S. J. Angyal and Y. Kondo, Complexes of carbohydrates with metal cations. XII. 2,5-Anhydroallitol and 2,5-anhydrogalactitol, *Aust. J. Chem.*, 33 (1980) 1013–1019.
135. S. J. Angyal and Y. Kondo, The 4,6-benzylidene acetals and the conformation of methyl α-D-idopyranoside, *Carbohydr. Res.*, 81 (1980) 35–48.
136. S. J. Angyal and R. Le Fur, The ^{13}C N.M.R. spectra of alditols, *Carbohydr. Res.*, 84 (1980) 201–209. (Conformations of acyclic sugar derivatives. V.). Correction: *Carbohydr. Res.*, 88 (1981) C28–C28.
137. S. J. Angyal, Haworth Memorial Lecture—Sugar–cation complexes: Structure and applications, *Chem. Soc. Rev.*, 9 (1980) 415–428.
138. S. J. Angyal and Q. T. Tran, Conformational analysis in carbohydrate chemistry. V. Formation of glycosidic anhydrides from heptoses, *Can. J. Chem.*, 59 (1981) 379–383.
139. S. J. Angyal and L. Odier, The ^{13}C N.M.R. spectra of inositols and cyclohexanepentols. The validity of rules correlating chemical shifts with configuration, *Carbohydr. Res.*, 100 (1982) 43–54. (Cyclitols. XXXIX.).
140. S. J. Angyal and L. Odier, The preparation of cyclohexanepentols from inositols by deoxygenation, *Carbohydr. Res.*, 101 (1982) 209–219. (Cyclitols. XXXVIII.).
141. S. J. Angyal and T. Q. Tran, Equilibria between pyranoses and furanoses. V. The composition in solution and the ^{13}C N.M.R. spectra of the heptoses and heptuloses, *Aust. J. Chem.*, 36 (1983) 937–946.
142. S. J. Angyal and L. Odier, Selective deuteration. The rate of protium–deuterium exchange in inositols with Raney-nickel catalyst and the effect thereon of *O*-methylation, *Carbohydr. Res.*, 123 (1983) 13–22.
143. S. J. Angyal and L. Odier, The effect of *O*-methylation on chemical shifts in the ^{1}H and ^{13}C N.M.R. spectra of cyclic polyols, *Carbohydr. Res.*, 123 (1983) 23–29. (Cyclitols. XL).
144. S. J. Angyal and R. Le Fur, The ^{13}C N.M.R. spectra and the conformations of heptitols in solution, *Carbohydr. Res.*, 126 (1984) 15–26. (Conformations of acyclic sugar derivatives. VI).
145. J. Defaye, A. Gadelle, and S. J. Angyal, An efficient synthesis of L-fucose and L-(4-^{2}H)fucose, *Carbohydr. Res.*, 126 (1984) 165–169.
146. S. J. Angyal, The composition of reducing sugars in solution, *Adv. Carbohydr. Chem. Biochem.*, 42 (1984) 15–68.
147. S. J. Angyal, L. Littlemore, and P. A. J. Gorin, Lanthanide-induced shifts in the ^{13}C N.M.R. spectra of *epi*-inositol and some anhydrohexoses. The anomalous behaviour of the heavy lanthanides, *Aust. J. Chem.*, 38 (1985) 411–418. (Complexes of carbohydrates with metal cations. XIII).
148. S. J. Angyal and J. A. Mills, Complexes of carbohydrates with metal cations. XIV. Separation of sugars and alditols by means of their lanthanum complexes, *Aust. J. Chem.*, 38 (1985) 1279–1285.

149. S. J. Angyal, D. Greeves, and L. Littlemore, Complexes of muellitol [1,3,5-tri(3-methylbut-2-enyl)-*scyllo*-inositol] with metal cations, *Aust. J. Chem.*, 38 (1985) 1561–1566.
150. S. J. Angyal, J. K. Saunders, C. T. Grainger, R. Le Fur, and P. G. Williams, Heptitol conformations revisited: C,O *versus* O,O parallel 1,3-interactions, *Carbohydr. Res.*, 150 (1986) 7–21. (Conformations of acyclic sugar derivatives. VII).
151. S. J. Angyal, J. D. Stevens, and L. Odier, Selective deuteration over Raney nickel in deuterium oxide: methyl glycosides, *Carbohydr. Res.*, 157 (1986) 83–94.
152. S. J. Angyal, J. D. Stevens, and L. Odier, Selective deuteration over Raney nickel in deuterium oxide: 1,6-anhydrohexoses, *Carbohydr. Res.*, 169 (1987) 151–157.
153. S. J. Angyal, Topics in carbohydrate stereochemistry, *Pure Appl. Chem.*, 59 (1987) 1521–1528.
154. S. J. Angyal, The effect of cations on the properties of sugars and polyols (C. S. Hudson Award Address), *Abs. Papers Am. Chem. Soc.*, 193 (1987) 28.
155. P. Dittrich and S. J. Angyal, 2-*C*-Methyl-D-erythritol in leaves of *Liriodendron tulipifera*, *Phytochemistry*, 27 (1988) 935.
156. S. J. Angyal, D. Greeves, and L. Littlemore, The structure of the complex formed from D-galacturonate ions and cations in solution, *Carbohydr. Res.*, 174 (1988) 121–131.
157. P. Köll and S. J. Angyal, ^{1}H und ^{13}C N.M.R. spectren der 1,6-anhydrohexofuranosen, *Carbohydr. Res.*, 179 (1988) 1–5.
158. S. J. Angyal, D. C. Craig, and J. Kuszmann, The composition and conformation of D-*threo*-3,4-hexodiulose in solution and the X-ray crystal structure of its ββ-anomer, *Carbohydr. Res.*, 194 (1989) 21–29.
159. S. J. Angyal and R. Le Fur, Conformation of acyclic sugar derivatives. VIII. Partially acetylated alditols, *J. Org. Chem.*, 54 (1989) 1927–1931.
160. S. J. Angyal, Complexes of metal cations with carbohydrates in solution, *Adv. Carbohydr. Chem. Biochem.*, 47 (1989) 1–43.
161. D. Lewis and S. J. Angyal, ^{1}H N.M.R. spectra and conformations of three heptitols in deuterium oxide, *J. Chem. Soc. Perkin Trans. 2*, (1989) 1763–1765.
162. S. J. Angyal, D. C. Craig, J. Defaye, and A. Gadelle, Complexes of carbohydrates with metal cations. XVI. Di-D-fructose and di-L-sorbose dianhydrides, *Can. J. Chem.*, 68 (1990) 1140–1144.
163. S. J. Angyal, Complexes of carbohydrates with copper ions: A reappraisal, *Carbohydr. Res.*, 200 (1990) 181–188. (Complexes of carbohydrates with metal cations. XV).
164. P. Köll, H. Komander, S. J. Angyal, M. Morf, B. Zimmer, and J. Kopf, Crystal and molecular structure of D-*glycero*-D-*manno*-heptitol (α-sedoheptitol, volemitol) and D-*glycero*-D-*gluco*-heptitol (β-sedoheptitol), *Carbohydr. Res.*, 218 (1991) 55–62.
165. J. Kopf, P. Köll, and S. J. Angyal, Structure of D-*glycero*-L-*galacto*-heptitol, *Acta Cryst. C*, 47 (1991) 1503–1506.
166. S. J. Angyal, The composition of reducing sugars in solution: current aspects, *Adv. Carbohydr. Chem. Biochem.*, 49 (1991) 19–35.
167. S. J. Angyal, A new type of anhydro sugar: 1,3'-anhydro-3-*C*-hydroxymethyl aldoses, *Carbohydr. Res.*, 216 (1991) 171–178. (Conformational analysis in carbohydrate chemistry. VI).
168. S. J. Angyal and D. C. Craig, Complex formation between polyols and rare earth cations. The crystal structure of galactitol·2PrCl$_3$·14H$_2$O, *Carbohydr. Res.*, 241 (1993) 1–8.
169. S. J. Angyal, The composition of reducing sugars in dimethyl sulfoxide solution, *Carbohydr. Res.*, 263 (1994) 1–11.
170. S. J. Angyal and D. C. Craig, The unusually stable crystal structure of *neo*-inositol, *Carbohydr. Res.*, 263 (1994) 149–154. (Cyclitols. XLI.).
171. S. J. Angyal, L. Odier, and M. E. Tate, A simple synthesis of *cis*-inositol, *Carbohydr. Res.*, 266 (1995) 143–146.

172. J. E. Anderson, S. J. Angyal, and D. C. Craig, Eclipsed exocyclic carbon–oxygen bonds in the hexamethyl ether of *scyllo*-inositol. X-Ray crystallographic and N.M.R. studies, *Carbohydr. Res.*, 272 (1995) 141–148.
173. S. J. Angyal and J. C. Christofides, Intramolecular hydrogen bonds in monosaccharides in dimethyl sulfoxide solution, *J. Chem. Soc. Perkin Trans. 2*, (1996) 1485–1491.
174. S. J. Angyal, A short note on the epimerization of aldoses, *Carbohydr. Res.*, 300 (1997) 279–281. (Complexes of carbohydrates with metal cations. XVIII).
175. J. E. Anderson, S. J. Angyal, and D. C. Craig, Eclipsed conformations for carbon–oxygen bonds in methoxy derivatives of inositols, sugars, and similar compounds, *J. Chem. Soc. Perkin Trans. 2*, (1997) 729–734.
176. S. J. Angyal, Complexes of carbohydrates with metal cations. XIX. The effect of the size of the cation on the inter-oxygen distance, *Aust. J. Chem.*, 53 (2000) 567–570.
177. S. J. Angyal, *myo*-Inositol 4,6-carbonate: An easily prepared small molecule with three *syn*-axial hydroxyl groups, *Carbohydr. Res.*, 325 (2000) 313–320.
178. S. J. Angyal, The Lobry de Bruyn—Alberda van Ekenstein transformation and related reactions, *Top. Curr. Chem.*, 215 (2001) 1–14.
179. S. J. Angyal, Chromatography on cation columns: a much-neglected method of separation, *Aust. J. Chem.*, 55 (2002) 79–81.
180. S. J. Angyal, Carbohydrates Special Issue: An introduction, *Aust. J. Chem.*, 55 (2002) 1–2.
181. S. J. Angyal, L-Ribose, an easily prepared rare sugar, *Aust. J. Chem.*, 58 (2005) 58–59.
182. S. J. Angyal, J. E. Anderson, and D. C. Craig, Inositols: The effect of bulky substituents on conformations, *Aust. J. Chem.*, 60 (2007) 572–577.
183. S. J. Angyal, International Carbohydrate Symposia—A history, *Adv. Carbohydr. Chem. Biochem.*, 61 (2008) 29–58.
184. S. J. Angyal, Carbohydrate research in the 20th century, *Aust. J. Chem.*, 62 (2009) 501–502.

B. Other Papers

185. S. J. Angyal, Analysis of mixtures of β-picoline, γ-picoline, and 2,6-lutidine, *Aust. Chem. Inst. J. Proc.*, 14 (1947) 12–16.
186. S. J. Angyal and R. C. Rassack, The Sommelet reaction, *Nature*, 161 (1948) 723–724.
187. S. J. Angyal and R. C. Rassack, The Sommelet reaction. Part I. The course of the reaction, *J. Chem. Soc.*, (1949) 2700–2704.
188. S. J. Angyal, P. J. Morris, R. C. Rassack, and J. A. Waterer, The Sommelet reaction. Part II. The ortho-effect, *J. Chem. Soc.*, (1949) 2704–2706.
189. S. J. Angyal, P. J. Morris, R. C. Rassack, J. A. Waterer, and J. G. Wilson, New toluene-*p*-sulphonamides, *J. Chem. Soc.*, (1949) 2722.
190. S. J. Angyal and S. R. Jenkin, Sulphonamides. I. Marfanil and its *o*- and *m*-isomer, *Aust. J. Sci. Res. A*, 3 (1950) 463–465.
191. S. J. Angyal, J. R. Tetaz, and J. G. Wilson, 1-Naphthaldehyde, *Org. Synth.*, 30 (1950) 67–69.
192. S. J. Angyal, P. J. Morris, J. R. Tetaz, and J. G. Wilson, The Sommelet reaction. Part III. The choice of solvent and the effect of substituents, *J. Chem. Soc.*, (1950) 2141–2145.
193. S. J. Angyal and W. K. Warburton, Sulphonamides. II. The structure and tautomerism of sulphapyridine, sulphathiazole and sulphanilylbenzamidine, *Aust. J. Sci. Res. A*, 4 (1951) 93–106.
194. S. J. Angyal, G. B. Barlin, and P. C. Wailes, Fluorene-2-aldehyde, *J. Chem. Soc.*, (1951) 3512–3513.
195. S. J. Angyal and W. K. Warburton, The basic strengths of methylated guanidines, *J. Chem. Soc.*, (1951) 2492–2494.
196. S. J. Angyal and J. A. Mills, The shape and reactivity of the cyclo-hexane ring, *Rev. Pure Appl. Chem.*, 2 (1952) 185–202.

197. S. J. Angyal and C. L. Angyal, The tautomerism of N-heteroaromatic amines. Part I, *J. Chem. Soc.*, (1952) 1461–1466.
198. S. J. Angyal, P. J. Morris, R. C. Rassack, J. A. Waterer, and J. G. Wilson, Sulphonamides. IV. The reaction of N-heterocyclic amines with sulphonyl halides, *Aust. J. Sci. Res. A*, 5 (1952) 374–378.
199. S. J. Angyal, D. R. Penman, and G. P. Warwick, The Sommelet reaction. Part IV. Preparation of aliphatic aldehydes, *J. Chem. Soc.*, (1953) 1737–1739.
200. S. J. Angyal, G. B. Barlin, and P. C. Wailes, The Sommelet reaction. Part V. N-Heteroaromatic aldehydes, *J. Chem. Soc.*, (1953) 1740–1741.
201. S. J. Angyal, D. R. Penman, and G. P. Warwick, The Sommelet reaction. Part VI. Methyleneamines. A proposed mechanism for the reaction, *J. Chem. Soc.*, (1953) 1742–1747.
202. S. J. Angyal, The Sommelet reaction, *Org. React.*, 8 (1954) 197–217.
203. S. J. Angyal, E. Bullock, W. G. Hanger, and A. W. Johnson, The chromophore of actinomycin, *Chem. Ind. (Lond.)*, (1955) 1295–1296.
204. S. J. Angyal, E. Bullock, W. G. Hanger, W. C. Howell, and A. W. Johnson, Actinomycin. III. The reaction of actinomycin with alkali, *J. Chem. Soc.*, (1957) 1592–1602.
205. S. J. Angyal, The hydrolysis of phosphoric esters, *Proc. R. Aust. Chem. Inst.*, 25 (1958) 427–429.
206. S. J. Angyal and R. J. Young, Glycol fission in rigid systems. I. The camphane-2,3-diols, *J. Am. Chem. Soc.*, 81 (1959) 5467–5472.
207. S. J. Angyal and R. J. Young, Glycol fission in rigid systems. II. The cholestane-3β,6,7-triols. Existence of a cyclic intermediate, *J. Am. Chem. Soc.*, 81 (1959) 5251–5255.
208. S. J. Angyal, Organic chemistry in the University of New South Wales, *Nature*, 186 (1960) 357.
209. S. J. Angyal, R. G. Nicholls, and J. T. Pinhey, Steric effects on the iodide-induced elimination of 2- and 3-methanesulfonyloxy groups from 5α-cholestane derivatives, *Aust. J. Chem.*, 32 (1979) 2433–2440.
210. S. J. Angyal, D. C. Craig, and T. Q. Tran, The configurations of the *trans*-camphane-2,3-diols, *Aust. J. Chem.*, 37 (1984) 661–666.

JOHN GRANT BUCHANAN

1926–2012

Grant Buchanan—always known to his family, friends, and colleagues by his second given name—was born as the elder of two sons of Robert and Mary Buchanan, on September 26, 1926, in Dumbarton, Scotland, a town on the northern shore of the Firth of Clyde some 20 km west of Glasgow. This is a part of Scotland historically closely associated with the name Buchanan, the traditional lands of the Clan Buchanan lying in the parts of Stirlingshire on the eastern shores of Loch Lomond, a few kilometers to the south of which lies Dumbarton. Grant's initial schooling was locally, but for his secondary education this "lad o' pairts" studied at Glasgow Academy, where he gained a typically wide range of Scottish "Highers," but with an emphasis on scientific subjects.

The Buchanan family business was concerned with running a glue works, but although this can be perceived as an exercise in applied organic and biological chemistry, it did not seem to appeal to Grant. Thus in 1944, Grant, on the strength of an impressive school record, was accepted to read Natural Sciences at the University of Cambridge, and he took the long road south, assisted by a Scholarship from Glasgow Academy.

The beginning of Grant Buchanan's undergraduate studies coincided fortuitously with the arrival in Cambridge of his fellow Scotsman Professor Alexander Todd (later Lord Todd, and Nobel Laureate of 1957) as Professor of Organic Chemistry. Indeed, Grant became a junior member of the same College, Christ's, at which Todd was elected to a Professorial Fellowship, and of which he later served as a distinguished Master for some 15 years. During his undergraduate years, Grant became particularly interested in the interaction of chemistry with the life sciences, and it was therefore hardly surprising that after graduating with his BA in 1947 he was attracted by the prospect of doing research in the exciting intellectual atmosphere of Todd's group.

Grant was asked to work on the structure of Vitamin B_{12}, which at the time (the late 1940s) had just become available in strictly limited quantities of somewhat dubious purity. One can imagine this as being a topic that would be highly challenging, and not a little frightening, to a beginning research student with only the techniques which were available at that time! Nonetheless, under Todd's direction, and with the assistance of Alan Johnson (later Professor of Organic Chemistry at the University of Sussex) and Cedric Hassall (subsequently Professor at University College Swansea and Director of Research at Roche Products Ltd.), Grant's studies established the correct location of the phosphate unit on either the 2'- or 3'-hydroxyl group of the α-D-ribofuranosyl-dimethylbenzimidazole ligand of the vitamin. For this work, Grant was awarded his PhD in 1951, and this initial grounding in the chemistry of nucleotides and sugar phosphates was to prove highly relevant in his subsequent career. Todd clearly recognized Grant's scientific abilities and potential, and maintained a lively interest in his later activities; indeed it was Grant Buchanan who after Todd's death contributed a substantial memoir of Todd that was published in *Advances* (Vol. 55, 2000).

From Cambridge, Grant moved to a postdoctoral research fellowship in the laboratory of another future Nobel Laureate, Melvin Calvin, at the University of California, Berkeley. Grant's work there was concerned with establishing the involvement of uridine 5'-(α-D-glucopyranosyl diphosphate) (UDP-Glc), one of the nucleoside sugar diphosphates then only recently identified by Leloir, in the biosynthesis of sucrose, and with the intermediacy of a sucrose phosphate in this process. It was shown that, in this sucrose phosphate, the phosphate group was attached to the D-fructose unit, although the precise location could not be determined. This material is now well established as sucrose 6'-phosphate, and many years later Grant was responsible for the first unequivocal synthesis of this compound.

Grant may well have wished to stay longer in Berkeley, but this was the time of the Korean War, and staying in the US any longer may have resulted in Grant's call-up in the service of Uncle Sam. Thus after just 1 year, in 1952, he returned to Britain to take up a research post at the Lister Institute for Preventive Medicine in London, working with Dr. James Baddiley (later Professor Sir James Baddiley). In 1954, Baddiley was appointed to the Chair of Organic Chemistry at King's College Newcastle, an institution which was at the time a College of the University of Durham, but later became the independent University of Newcastle-upon-Tyne. Grant moved to Newcastle with Baddiley, and in 1955 was appointed to a Lectureship at King's College Newcastle.

However, during his time in London, Grant had met his future wife Sheila, and they were married in July 1956, shortly after Grant's move north. Sheila was a tremendous rock of support to Grant throughout his later career, and this very happy marriage led to Grant and Sheila having three sons, Andrew, John, and Neil.

Grant Buchanan's time at the Lister Institute and in Newcastle was one of great scientific productivity. Although his original brief with Baddiley had been to work on the synthesis of coenzyme A, Grant soon became involved in another project in Baddiley's laboratory concerned with the characterization and synthesis of two nucleotide derivatives, cytidine 5′-(glycerol diphosphate) and cytidine 5′-(ribitol diphosphate), from *Lactobacillus arabinosus*. Although the biological role of these compounds was unclear when they were first identified, it soon became apparent that they were biosynthetic precursors of a novel type of cell-wall polymer. In a series of definitive papers in the late 1950s and early 1960s, Baddiley and Buchanan established the major structural features of these macromolecules, for which in their first publication they coined the name "teichoic acids," in a range of Gram-positive bacteria. Alongside this work on teichoic acids and their components, Baddiley and Buchanan also collaborated on studies of the biosynthetic pathway to purine nucleotides, including the chemical synthesis of some of the intermediates involved. Another interest at this time was in the chemistry of "active sulfate" (adenosine 3′-phosphate-5′-sulfatophosphate, 3′-phospho-5′-adenylyl sulfate, PAPS), and the synthesis of this extremely labile compound was reported in 1957.

In the 1960s, Baddiley and Buchanan increasingly turned their attention to the structures of pneumococcal antigens, particularly to those type-specific capsular polymers that contain ribitol phosphate residues within the repeating unit, and thus have some features in common with the ribitol teichoic acids. Their work in this area led to the successful determination of the primary structures of the polymers from a number of different serotypes.

Alongside this large amount of collaborative work with Baddiley, Grant Buchanan also carried out extensive studies on the chemistry of carbohydrate epoxides, an area that had first attracted his attention in the early 1950s. Grant's first foray into this area was a study which he carried out himself on derivatives of 3,4-anhydrogalactopyranose, demonstrating the ease with which such compounds can isomerize to 2,3-anhydrogulopyranose systems. Later, in association with a series of able coworkers, a number of other significant advances were reported to help clarify the whole area of epoxide migration in carbohydrate systems, and also to identify and study cases in which neighboring-group participation occurred in the ring-opening of acetoxyepoxides. In addition to the original papers, written with Grant's typical care and thoroughness, it is perhaps appropriate to note an excellent and much-cited review, co-authored with H.Z. Sable and published in 1972, which surveyed with great clarity the factors involved in the stereoselectivity of epoxide cleavages in both carbohydrate examples and other systems.

It is perhaps a little unfortunate that all this work was being carried out at a time when carbohydrate chemistry was perceived by many in the wider organic chemistry community as something of an esoteric specialist area. Thus, although Grant's work was much read and praised by the aficionados, it did not perhaps gain the general appreciation that it warranted. Grant was maybe just a little peeved when epoxide rearrangements began to be associated among the larger organic chemical community with the name of G.B. Payne, who studied such processes in acyclic systems, although some of Grant's work well preceded this.

The quality of Grant's work was recognized by his promotion to Senior Lecturer in 1962, and to a Readership in 1965. A year later, he was awarded the degree of Sc.D. by his alma mater, the University of Cambridge, and in 1969 Grant was invited to return to Scotland to take up the newly created Chair of Organic Chemistry at Heriot-Watt University in Edinburgh.

Heriot-Watt was an institution that, although it had a long tradition in the teaching of chemistry, had only gained University status 3 years earlier. One of Grant's main responsibilities, together with the Head of Department, Professor Brian Gowenlock, also recently appointed from outside, was to build up the research activities of the Chemistry Department. A considerable impetus in this was given in 1973 by the move of the Department from its rather antiquated premises in the center of Edinburgh (which maybe reminded Grant of the old Pembroke Street labs in Cambridge in which he had worked) to brand-new laboratories on the Riccarton Campus just to the west of the city. Although Grant continued with some work on sugar epoxides at Heriot-Watt, a new major line of research was instigated directed to the synthesis of C-nucleoside antibiotics, the structures of which were becoming known in the late 1960s. The approach used by Grant's group involved the intermediacy of glycosylalkynes, and over the years this work led to more than 20 full papers in the 1970s and 1980s describing the synthesis of most of the naturally-occurring C-nucleosides and a range of analogues. In addition to the synthetic work, studies on the biosynthesis of C-nucleoside antibiotics were also carried out, which led to the delineation of the principal biosynthetic building blocks, and gave some insight into the mechanism of their linkage.

A research area that was particularly in vogue in the 1980s was the use of sugars as chiral synthons ("chirons") for the preparation of compounds of other classes, and Grant's group became involved in work in this field, directed particularly towards the synthesis of hydroxylated pyrrolidine and pyrrolizidine alkaloids. Work was also done on the structure and stereochemistry of carbohydrate acetals, where the application of ^{13}C NMR techniques proved particularly helpful in structural elucidation.

Grant was elected as a Fellow of the Royal Society of Edinburgh (FRSE) in 1972, and was also active in professional matters; he was a member of the Council of the Royal Society of Chemistry between 1982 and 1985, and of the Council of the Royal Society of Edinburgh, 1980–1982. He was President of the European Carbohydrate Organization 1989–1993, and served as UK Representative to the International Carbohydrate Organization. Grant also served Heriot-Watt University in many capacities, particularly as Head of the Chemistry Department from 1987 to 1991.

Grant retired from Heriot-Watt University at the end of September 1991, shortly after his 65th birthday. Also in September 1991, the 6th European Symposium on Carbohydrate Chemistry (Eurocarb VI) was held at Heriot-Watt University, and at the Conference Dinner Grant was presented, in honor of his retirement, with a Special Issue of *Carbohydrate Research* dedicated to him, and to which many of his professional friends and colleagues throughout the world contributed over 40 papers.

After retirement from Heriot-Watt, Grant and Sheila moved to Bath, in the southwest of England, a city with which they had family connections. In Bath, Grant took up an honorary position of Visiting Professorial Fellow in the Department of Chemistry at the University of Bath. Grant's appointment was greatly facilitated by the presence at Bath of Malcolm Campbell as Professor of Organic Chemistry, who was the Head of Department at the time; Malcolm had previously been at Heriot-Watt, where he had been appointed as a Lecturer at the same time as Grant's arrival. This appointment meant that Grant had office space and access to library facilities, which were very necessary, as after moving to Bath he took over from Allan Foster as one of the UK Editors of *Carbohydrate Research*. He undertook this role until 1996, having been a member of the Editorial Board since 1966.

Sadly, this first period of Grant's time in Bath, which should have been a happy period of more relaxed life, was marred by the serious illness of Sheila, who tragically died in September 1996.

Following his term of editorship and Sheila's death, Grant began to turn his energies once again to his research ideas. He completed some work on the equilibration of a pair of anhydro-tagatose and -fructose derivatives, with the help of X-ray and NMR support from colleagues in Bath, and his renewed attention to epoxide migration led him somewhat tangentially to suspect that earlier classic work on the mechanisms of halogen addition to 2,3-dimethylmaleate and 2,3-dimethylfumarate may be incorrect. This hunch, which proved accurate, led to a fruitful collaboration with Professor Ian Williams, a computational organic chemist, and between them they sorted out misinterpretations that had been in the literature since 1937. Grant also interacted with various other staff in the Departments of Chemistry, Biochemistry, and Pharmacy and Pharmacognosy, which led to his name appearing on ten papers since the year 2000. Grant also returned to teaching, giving a course of lectures on carbohydrate chemistry to final-year undergraduates until 2008, and he assisted with tutorials and with the supervision of undergraduate research projects.

Grant Buchanan's major interest and enthusiasm beyond the lab and his family was almost certainly his particularly Scottish addiction to the game of golf. When in Edinburgh, Grant was a keen member of a local club near the family home in Barnton, and on moving to Bath he joined Bath Golf Club, where he was a similarly active participant until a degree of infirmity in his last years prevented him from playing any longer. There are clubs for golfers at both Newcastle and Heriot-Watt Universities, and one of the first things that Grant did on moving from one to the other was to instigate an annual Heriot-Watt—Newcastle golf match, which celebrated its 21st anniversary in the year Grant retired from Heriot-Watt, and which continues to this day.

Sometimes (at least in the UK) a person is described as "a scholar and a gentleman." Often this phrase is used flippantly and casually, but surely this is a term that can be used very accurately and quite genuinely to describe Grant Buchanan. Throughout his career Grant's work was always characterized not only by a high degree of scientific quality, but also it was imbued with real scholarship. When Grant wrote a paper, not only was there no hyperbole about the findings reported, but the work was set in the context of previous work, where relevant earlier results from others were accurately and appropriately acknowledged. Perhaps these are old-fashioned virtues that are sometimes not always fully observed in our present times, with the frenetic rush to publish and perhaps to make inflated claims about the significance of one's work, but they are surely highly commendable and the mark of a true scholar. Grant was also someone who commanded respect and great affection from his colleagues because of the attitude he took and showed towards them. He was extremely collegial and would be prepared to see all sides of an issue. Grant may have had reservations about the significance or relevance or quality or whatever of a piece of scientific work, but this would never be manifest in any sort of malicious or derogatory manner, but rather by a considered appraisal expressed in as positive a way as possible. His enthusiasm for his subject was infectious toward the research workers who studied under his direction, and motivation was achieved by Grant's combination of encouragement and respect, coupled with his extensive knowledge of his subject which could be brought to bear on any problem.

Grant Buchanan passed away on April 17, 2012, after a period of increasingly serious illness. He was nonetheless in contact over chemical matters with colleagues in Bath almost to the end. Since his passing, Grant's contributions to carbohydrate chemistry, particularly in the UK, have been honored by the Carbohydrate Group of the Royal Society of Chemistry by the establishment of an award in his name, which will be given annually for a particularly meritorious presentation by a postgraduate student at a meeting of the Group.

There is an (as yet) unofficial "National Anthem" in Scotland, sung for example at National sporting occasions, the first line of which reads: "O Flower o' Scotland, when will we see your like again?" In the case of Grant Buchanan, indeed, when?

The assistance of Professor Ian H. Williams (University of Bath) in providing information for this memoir is gratefully acknowledged.

RICHARD WIGHTMAN

LIST OF PUBLICATIONS BY J. GRANT BUCHANAN

1. J. G. Buchanan, A. W. Johnson, J. A. Mills, and A. R. Todd, Isolation of a phosphorus-containing degradation product from vitamins B_{12} and B_{12c}, *Chem. Ind. (London)* (1950) 426.
2. J. G. Buchanan, C. A. Dekker, and A. G. Long, Detection of glycosides and non-reducing carbohydrate derivatives in paper-partition chromatography, *J. Chem. Soc.* (1950) 3162–3167.
3. J. G. Buchanan, A. W. Johnson, J. A. Mills, and A. R. Todd, Chemistry of the vitamin B_{12} group. 1. Acid-hydrolysis studies. Isolation of a phosphorus-containing degradation product, *J. Chem. Soc.* (1950) 2845–2855.
4. J. G. Buchanan, Detection of deoxyribonucleosides on paper chromatograms, *Nature*, 168 (1951) 1091.
5. J. G. Buchanan, V. H. Lynch, A. A. Benson, D. F. Bradley, and M. Calvin, The path of carbon in photosynthesis. XVIII. The identification of nucleotide coenzymes, *J. Biol. Chem.*, 203 (1953) 935–945.
6. J. G. Buchanan, The path of carbon in photosynthesis. XIX. The identification of sucrose phosphate in sugar beet leaves, *Arch. Biochem. Biophys.*, 44 (1953) 140–149.
7. J. G. Buchanan, V. H. Lynch, A. A. Benson, and M. Calvin, The path of carbon in photosynthesis. XVIII. The identification of nucleotide coenzymes, *J. Biol. Chem.*, 203 (1953) 935–945.
8. J. Baddiley, J. G. Buchanan, and E. M. Thain, The polysaccharide of *Penicillium islandicum*, *J. Chem. Soc.* (1953) 1944–1946.
9. J. Baddiley, J. G. Buchanan, and L. Szabó, Sugar phosphates. Part 1. Derivatives of glucose 4:6-(hydrogen phosphate), *J. Chem. Soc.* (1954) 3826–3832.
10. J. G. Buchanan, The reactions of derivatives of 3,4-anhydrogalactose, *Chem. Ind. (London)* (1954) 1484–1485.
11. J. Baddiley, J. G. Buchanan, B. Carss, and A. P. Matthias, Cytidine diphosphate ribitol, *Biochim. Biophys. Acta*, 21 (1956) 191–192.
12. J. Baddiley, J. G. Buchanan, and R. Letters, Phosphorylation through glyoxalines (imidazoles) and its significance in enzymic transphosphorylation, *J. Chem. Soc.* (1956) 2812–2817.
13. J. Baddiley, J. G. Buchanan, R. E. Handschumacher, and J. F. Prescott, Biosynthesis of purine nucleotides. I. Preparation of *N*-glycylglycosylamines, *J. Chem. Soc.* (1956) 2818–2823.
14. J. Baddiley, J. G. Buchanan, A. P. Mathias, and A. R. Sanderson, Cytidine diphosphate glycerol, *J. Chem. Soc.* (1956) 4186–4190.
15. J. Baddiley, J. G. Buchanan, B. Carss, and A. P. Mathias, Cytidine diphosphate ribitol from *Lactobacillus arabinosus*, *J. Chem. Soc.* (1956) 4583–4588.
16. J. Baddiley, J. G. Buchanan, F. J. Hawker, and J. E. Stephenson, Synthesis of '6-succinoaminopurine' [6-(1,2-dicarboxyethylamino)purine], *J. Chem. Soc.* (1956) 4659–4661.
17. J. Arris, J. Baddiley, J. G. Buchanan, and E. M. Thain, Coenzyme A. X. Model experiments on synthesis of pyrophosphates of *N*-pantoylamines. New methods for debenzylation of esters of pyrophosphoric acid, *J. Chem. Soc.* (1956) 4968–4973.
18. J. Baddiley, J. G. Buchanan, B. Carss, A. P. Matthias, and A. R. Sanderson, The isolation of cytidine diphosphate glycerol, cytidine diphosphate ribitol and mannitol 1-phosphate from *Lactobacillus arabinosus*, *Biochem. J.*, 64 (1956) 599–603.

19. J. Baddiley, J. G. Buchanan, and R. Letters, Synthesis of 5'-sulfatophosphate, a degradation product of an intermediate in the enzymic synthesis of sulfuric esters, *J. Chem. Soc.* (1957) 1067–1071.
20. R. Bonnett, J. G. Buchanan, A. W. Johnson, and A. Todd, Chemistry of the vitamin B_{12} group. VI. Isomeric 5,6-dimethylbenzimidazole nucleotides produced by hydrolysis of vitamin B_{12}, *J. Chem. Soc.* (1957) 1162–1172.
21. J. Baddiley, J. G. Buchanan, and R. Letters, Synthesis of 'active sulfate', *Proc. Chem. Soc.* (1957) 147–148.
22. J. Baddiley, J. G. Buchanan, R. Hodges, and J. F. Prescott, Synthesis of the *N*-glycyl-D-ribofuranosylamines, *Proc. Chem. Soc.* (1957) 148–149.
23. J. Baddiley, J. G. Buchanan, and J. Stewart, Synthesis of 5-amino-1-(β-D-ribofuranosyl) glyoxaline-4-carboxamide, *Proc. Chem. Soc.* (1957) 149.
24. J. Baddiley, J. G. Buchanan, and B. Carss, Configuration of the ribitol phosphate residue in cytidine diphosphate ribitol, *J. Chem. Soc.* (1957) 1869–1876.
25. J. Baddiley, J. G. Buchanan, and B. Carss, Hydrolysis of ribitol 1(5)-phosphate, riboflavine-5'-phosphate, and related compounds, *J. Chem. Soc.* (1957) 4058–4063.
26. J. Baddiley, J. G. Buchanan, R. Hodges, and J. F. Prescott, Chemical studies in the biosynthesis of purine nucleotides. II. Synthesis of *N*-glycyl-D-ribofuranosylamines, *J. Chem. Soc.* (1957) 4769–4774.
27. J. Baddiley, J. G. Buchanan, and B. Carss, Method for the identification of pentitols and hexitols, *J. Chem. Soc.* (1957) 4138–4139.
28. R. T. Williams, K. D. Gibson, W. J. Whelan, J. Baddiley, J. G. Buchanan, and H. R. V. Arnstein, Biological chemistry, *Annu. Rep. Prog. Chem.*, 54 (1957) 306–352.
29. J. Baddiley, J. G. Buchanan, and B. Carss, The presence of ribitol phosphate in bacterial cell walls, *Biochim. Biophys. Acta*, 27 (1958) 220.
30. J. G. Buchanan, The behaviour of derivatives of 3,4-anhydrogalactose towards acidic reagents. Part I, *J. Chem. Soc.* (1958) 995–1000.
31. J. Baddiley, J. G. Buchanan, and R. Letters, Separation and characterization of 2'-5'- and 3',5'-diphosphate and the 2',3',5'-triphosphate of adenosine, *J. Chem. Soc.* (1958) 1000–1007.
32. J. Baddiley, J. G. Buchanan, and G. O. Osborne, The preparation of 7- and 9-glucopyranosyl and -xylopyranosyl derivatives of 8-azaxanthine (5,7-dihydroxy-v-triazolo[*d*]pyrimidine), *J. Chem. Soc.* (1958) 1651–1657.
33. J. G. Buchanan, The behaviour of derivatives of 3,4-anhydrogalactose towards acidic reagents. Part II, *J. Chem. Soc.* (1958) 2511–2516.
34. J. Baddiley, J. G. Buchanan, and A. R. Sanderson, Synthesis of cytidine diphophate glycerol, *J. Chem. Soc.* (1958) 3107–3110.
35. J. Baddiley, J. G. Buchanan, and G. O. Osborne, The preparation of 7- and 9-ribofuranosyl derivatives of 8-azaxanthine. A note on the preparation of 9-glucopyranosylxanthine, *J. Chem. Soc.* (1958) 3606–3610.
36. J. J. Armstrong, J. Baddiley, J. G. Buchanan, B. Carss, and G. R. Greenberg, Isolation and structure of ribitol phosphate derivatives (teichoic acids) from bacterial cell walls, *J. Chem. Soc.* (1958) 4344–4354.
37. J. J. Armstrong, J. Baddiley, J. G. Buchanan, and B. Carss, Nucleotides and the bacterial cell wall, *Nature*, 181 (1958) 1692–1693.
38. J. Baddiley and J. G. Buchanan, Recent developments in the biochemistry of nucleotide coenzymes, *Quart. Rev. Chem. Soc.*, 12 (1958) 152–172.
39. J. Baddiley, J. G. Buchanan, and J. E. Stevenson, The decomposition of 6-succinoaminopurine and its derivatives in the presence of metal ions, *Arch. Biochem. Biophys.*, 83 (1959) 54–59.
40. J. J. Armstrong, J. Baddiley, J. G. Buchanan, A. L. Davison, M. V. Kelemen, and F. C. Neuhaus, Composition of teichoic acids from a number of bacterial walls, *Nature*, 184 (1959) 247–248.

41. J. J. Armstrong, J. Baddiley, and J. G. Buchanan, Structure of teichoic acid from the walls of *Bacillus subtilis*, *Nature*, 184 (1959) 248–249.
42. J. Baddiley, J. G. Buchanan, R. Letters, and A. R. Sanderson, Synthesis of 'active sulphate' (adenosine 3'-phosphate-5'-sulphatophosphate), *J. Chem. Soc.* (1959) 1731–1734.
43. J. Baddiley, J. G. Buchanan, and C. P. Fawcett, Synthesis of cytidine diphosphate ribitol, *J. Chem. Soc.* (1959) 2192–2196.
44. J. Baddiley, J. G. Buchanan, F. E. Hardy, and J. Stewart, Chemical studies in the biosynthesis of purine nucleotides. Part III. The synthesis of 5-amino-1-(β-D-ribofuranosyl)glyoxaline-4-carboxyamide and 5-amino-1-(β-D-ribofuranosyl)glyoxaline-5-carboxyamide, *J. Chem. Soc.* (1959) 2893–2901.
45. J. J. Armstrong, J. Baddiley, and J. G. Buchanan, Structure of the ribitol teichoic acid from the walls of *Bacillus subtilis*, *Biochem. J.*, 76 (1960) 610–621.
46. J. G. Buchanan and K. J. Miller, The action of ammonia on methyl 2,3-anhydro-4,6-*O*-benzylidene -α-D-glucoside and -taloside, *J. Chem. Soc.* (1960) 3392–3394.
47. D. A. Applegarth and J. G. Buchanan, Paper chromatography of triphenylmethyl ethers of carbohydrate derivatives, *J. Chem. Soc.* (1960) 4706–4707.
48. J. Baddiley, J. G. Buchanan, and F. E. Hardy, Synthesis of 4-*O*-(β-D-glucopyranosyl)-D-ribitol, a degradation product of the ribitol teichoic acid from the walls of *Bacillus subtilis*, *J. Chem. Soc.* (1961) 2180–2186.
49. J. J. Armstrong, J. Baddiley, and J. G. Buchanan, Further studies on the teichoic acid from *Bacillus subtilis* walls, *Biochem. J.*, 80 (1961) 254–261.
50. A. R. Archibald, J. Baddiley, and J. G. Buchanan, The ribitol teichoic acid from *Lactobacillus arabinosus* walls: Isolation and structure of ribitol glucosides, *Biochem. J.*, 81 (1961) 124–134.
51. J. Baddiley, J. G. Buchanan, F. E. Hardy, R. O. Martin, U. L. Rajbhandary, and A. R. Sanderson, The structure of the ribitol teichoic acid of *Staphylococcus aureus* H, *Biochim. Biophys. Acta*, 52 (1961) 406–407.
52. J. Baddiley, J. G. Buchanan, R. O. Martin, and U. L. Rajbhandary, Teichoic acid from the walls of *Staphylococcus aureus* H. II. Location of the phosphate and alanine residues, *Biochem. J.*, 82 (1962) 49–56.
53. J. Baddiley, J. G. Buchanan, U. L. Rajbhandary, and A. R. Sanderson, Teichoic acid from the walls of *Staphylococcus aureus* H. Structure of the *N*-acetylglucosaminylribitol residues, *Biochem. J.*, 82 (1962) 439–448.
54. W. K. Roberts, J. G. Buchanan, and J. Baddiley, Galactofuranose units in the specific substance from type 34 Pneumococcus, *Biochem. J.*, 82 (1962) 42P.
55. L. J. Sargent, J. G. Buchanan, and J. Baddiley, Synthesis of 4-*O*-α-D-glucopyranosyl-D-ribitol, a degradation product of the ribitol teichoic acid from the walls of *Lactobacillus arabinosus*, *J. Chem. Soc.* (1962) 2184–2187.
56. J. G. Buchanan and J. C. P. Schwarz, Methyl 2,3-anhydro-α-D-mannoside and 3,4-anhydro-α-D-altroside and their derivatives. Part 1, *J. Chem. Soc.* (1962) 4770–4777.
57. Z. A. Shabarova, J. G. Buchanan, and J. Baddiley, The composition of pneumococcus type-specific substances containing phosphorus, *Biochim. Biophys. Acta*, 57 (1962) 146–148.
58. F. E. Hardy, J. G. Buchanan, and J. Baddiley, Synthesis of 4-O-(2-amino-2-deoxy-β-D-glucosyl)-D-ribitol and 4-O-(2-amino-2-deoxy-α-D-glucosyl)-D-ribitol, degradation products of the ribitol teichoic acid from *Staphylococcus aureus*, *J. Chem. Soc.* (1963) 3360–3366.
59. P. W. Austin, F. E. Hardy, J. G. Buchanan, and J. Baddiley, The separation of isomeric glycosides on basic ion-exchange resins, *J. Chem. Soc.* (1963) 5350–5353.
60. F. E. Hardy and J. G. Buchanan, The identification of carbohydrates giving derivatives of malondialdehyde on oxidation with sodium periodate, *J. Chem. Soc.* (1963) 5881–5885.
61. W. K. Roberts, J. G. Buchanan, and J. Baddiley, The specific substance from Pneumococcus type 34(41). The structure of a phosphorus-free repeating unit, *Biochem. J.*, 88 (1963) 1–7.

62. J. G. Buchanan and R. M. Saunders, Methyl 2,3-anhydro-α-D-mannoside and 3-4-anhydro-α-D-altroside and their derivatives. II, *J. Chem. Soc.* (1964) 1791–1795.
63. J. G. Buchanan and R. M. Saunders, Methyl 2,3-anhydro-α-D-mannoside and 3-4-anhydro-α-D-altroside and their derivatives. III, *J. Chem. Soc.* (1964) 1796–1803.
64. P. W. Austin, F. E. Hardy, J. G. Buchanan, and J. Baddiley, 2,3,4,6-Tetra-*O*-benzyl-D-glucosyl chloride and its use in the synthesis of the α- and β-anomers of 2-*O*-D-glucosylglycerol and 4-*O*-D-glucosyl-D-ribitol, *J. Chem. Soc.* (1964) 2128–2137.
65. J. G. Buchanan and E. M. Oakes, 3,5-Anhydro-1,2-*O*-isopropylidene-α-DD-glucofuranose and -β-L-idofuranose, two new carbohydrate oxetanes, *Tetrahedron Lett.* (1964) 2013–2017.
66. J. G. Buchanan, N. A. Hughes, and G. A. Swan, Monohydric alcohols, their ethers and esters, In: S. Coffey, (Ed.), 2nd ed. *Rodd's Chemistry of Carbon Compounds*, Vol. 1, Elsevier, Amsterdam, 1964, pp. 1–72. Chapter 4; Part B.
67. J. G. Buchanan, N. A. Hughes, F. J. McQuillan, and G. A. Swan, Aldehydes and ketones, In: S. Coffey, (Ed.), 2nd ed. *Rodd's Chemistry of Carbon Compounds*, Vol. 1, Elsevier, Amsterdam, 1964, pp. 1–91. Chapter 8; Part C.
68. J. G. Buchanan and J. Conn, The acid hydrolysis of methyl 2,3-anhydro-D-hexopyranosides, *J. Chem. Soc.* (1965) 201–208.
69. D. A. Applegarth, J. G. Buchanan, and J. Baddiley, Synthesis of ribitol 1,5-diphosphate and a polymeric ribitol phosphodiester, *J. Chem. Soc.* (1965) 1213–1219.
70. P. W. Austin, F. E. Hardy, J. G. Buchanan, and J. Baddiley, 2,3,4,6-Tetra-*O*-benzyl-D-galactosyl chloride and its use in the synthesis of α- and β-D-galactopyranosides, *J. Chem. Soc.* (1965) 1419–1424.
71. P. W. Austin, J. G. Buchanan, and R. M. Saunders, Solvolysis of some carbohydrate *p*-nitrobenzenesulfonates, *Chem. Commun.* (1965) 146–147.
72. P. W. Austin, J. G. Buchanan, and E. M. Oakes, Reactions of methyl 2,3-anhydro-D-ribofuranosides with nucleophiles, *Chem. Commun.* (1965) 374–375.
73. J. G. Buchanan and E. M. Oates, Oxetanes. I. 3,5-Anhydro-1,2-*O*-isopropylidene-α-D-glucofuranose and -β-L-idofuranose, *Carbohydr. Res.*, 1 (1965) 242–253.
74. R. Fletcher and J. G. Buchanan, The behaviour of derivatives of 3,4-anhydro-D-galactose towards acidic reagents. III, *J. Chem. Soc.* (1965) 6316–6323.
75. R. Dixon, J. G. Buchanan, and J. Baddiley, The specific substance from *Pneumococcus* type 34: The configuration of the glycosidic linkages, *Biochem. J.*, 100 (1966) 507–511.
76. E. V. Rao, J. G. Buchanan, and J. Baddiley, The type-specific substance from *Pneumococcus* type 10A (34). Structure of the dephosphorylated repeating unit, *Biochem. J.*, 100 (1966) 801–810.
77. E. V. Rao, J. G. Buchanan, and J. Baddiley, The type-specific substance from *Pneumococcus* type 10A (34). The phosphodiester linkages, *Biochem. J.*, 100 (1966) 811–814.
78. J. G. Buchanan and R. Fletcher, The methyl 2,3-anhydrolyxopyranosides and 3,4-anhydroarabinopyranosides, *J. Chem. Soc. C* (1966) 1926–1931.
79. J. G. Buchanan and A. R. Edgar, A new synthesis of methyl 3,4-*O*-ethylidene-β-D-arabinopyranoside by reduction of an acetoxonium ion salt, *Chem. Commun.* (1967) 29–30.
80. P. W. Austin, J. G. Buchanan, and R. M. Saunders, Rearrangement in the solvolysis of some carbohydrate nitrobenzene-*p*-sulphonates, *J. Chem. Soc. C* (1967) 372–377.
81. P. W. Austin, J. G. Buchanan, and D. G. Large, Ring contraction in the hydrolysis of methyl 4-*O*-nitrobenzene-*p*-sulphonyl-α-D-glucopyranoside, *Chem. Commun.* (1967) 418–419.
82. G. J. F. Chittenden, W. K. Roberts, J. G. Buchanan, and J. Baddiley, The specific substance from *Pneumococcus* type 34 (41). The phosphodiester linkages, *Biochem. J.*, 109 (1968) 597–602.
83. J. R. Dixon, W. K. Roberts, G. T. Mills, J. G. Buchanan, and J. Baddiley, The *O*-acetyl groups of the specific substance from *Pneumococcus* type 34 (U.S. type 41), *Carbohydr. Res.*, 8 (1968) 262–265.
84. E. V. Rao, M. J. Watson, J. G. Buchanan, and J. Baddiley, The type-specific substance from *Pneumococcus* type 29, *Biochem. J.*, 111 (1969) 547–556.

85. D. A. Kennedy, J. G. Buchanan, and J. Baddiley, The type-specific substance from Pneumococcus type 11A(43), *Biochem. J.*, 115 (1969) 37–45.
86. J. G. Buchanan and A. R. Edgar, Neighbouring-group effects in the chemistry of 3,4-anhydro-D-altritol, *Carbohydr. Res.*, 10 (1969) 295–305.
87. G. J. F. Chittenden and J. G. Buchanan, Conversion of benzyl 3-*O*-benzoyl-4,6-*O*-benzylidene-β-D-galactopyranoside into the 2-benzoate by acyl migration, *Carbohydr. Res.*, 11 (1969) 379–385.
88. A. R. Archibald and J. G. Buchanan, Use of the periodate Schiff spray reagents in the linkage analysis of oligosaccharides, *Carbohydr. Res.*, 11 (1969) 558–560.
89. J. G. Buchanan, R. Fletcher, K. Parry, and W. A. Thomas, Conformational analysis of some methyl 2,3- and 3,4-anhydroglucopyranosides, *J. Chem. Soc. B* (1969) 377–385.
90. J. G. Buchanan, A. R. Edgar, and D. G. Large, Methoxy-group migtation in the hydrolysis of the 4-nitro-*p*-sulphonates of methyl β-D-xylopyranoside and methyl β-D-glucopyranoside, *J. Chem. Soc. D Chem. Commun.* (1969) 558–559.
91. A. J. Trejo, G. J. Chittenden, J. G. Buchanan, and J. Baddiley, Uridine diphosphate α-D-galactofuranose, an intermediate in the biosynthesis of galactofuranosyl residues, *Biochem. J.*, 117 (1970) 637–639.
92. J. G. Buchanan, J. Conn, A. R. Edgar, and R. Fletcher, The acetolysis and benzolysis of some carbohydrate oxirans, *J. Chem. Soc. C* (1971) 1515–1521.
93. J. G. Buchanan, Chemical synthesis of sucrose phosphates, Sugar: Chemical, Biological and Nutritional Aspects of Sucrose [Paper Symposium], 1971, pp. 80–84.
94. M. J. Watson, J. M. Tyler, J. G. Buchanan, and J. Baddiley, The type specific substance of *Pneumococcus* type 13, *Biochem. J.*, 130 (1972) 45–54.
95. J. G. Buchanan, D. A. Cummerson, and D. M. Turner, The synthesis of sucrose-6'-phosphate, *Carbohydr. Res.*, 21 (1972) 283–292.
96. J. G. Buchanan and D. A. Cummerson, 1',4':3',6'-dianhydrosucrose, *Carbohydr. Res.*, 21 (1972) 293–296.
97. J. G. Buchanan, A. R. Edgar, and M. J. Power, Synthesis of 2,3,5-tri-*O*-benzyl-α- (and β-)-D-ribofuranosylethyne, potential intermediates for the synthesis of *C*-nucleosides, *J. Chem. Soc., Chem. Commun.* (1972) 346–347.
98. J. G. Buchanan and H. Z. Sable, Stereoselective epoxide cleavages, *Select. Org. Transf.*, 2 (1972) 1–95.
99. J. G. Buchanan, Migration of epoxide rings and stereoselective ring opening of acetoxyepoxides. Methyl 2,3-anhydro-6-*O*-triphenylmethyl-α-D-gulopyranoside, methyl 3-*O*-acetyl-α-D-gulopyranoside, methyl 4-*O*-acetyl-α-D-arabinopyranoside, 3,4-anhydro-α-D-arabinopyranoside, and others, *Methods Carbohydr. Chem.*, 6 (1972) 135–141.
100. J. G. Buchanan, Displacement, elimination, and rearrangement reactions, In: G. O. Aspinall, (Ed.), *MTP International Review of Science: Organic Chemistry, Series One*, Vol. 7, Butterworths, London, 1973, pp. 31–70.
101. J. G. Buchanan and D. M. Clode, Synthesis and properties of 2,3-anhydro-D-mannose and 3,4-anhydro-D-altrose, *J. Chem. Soc., Perkin Trans. 1* (1974) 388–394.
102. J. G. Buchanan, A. R. Edgar, and M. J. Power, *C*-Nucleoside studies I. Synthesis of (2,3,5-tri-*O*-benzyl-α (and β-)-D-ribofuranosyl)ethyne, *J. Chem. Soc., Perkin Trans. 1* (1974) 1943–1949.
103. S. A. S. Al-Janabi, J. G. Buchanan, and A. R. Edgar, Base-catalyzed equilibration and conformational analysis of some methyl 2,3- and 3,4-anhydro-6-deoxy-β-D-hexopyranosides, *Carbohydr. Res.*, 35 (1974) 151–164.
104. J. G. Buchanan, A. D. Dunn, and A. R. Edgar, Reaction of ethynylmagnesium bromide with 2,3-O-isopropylidene-D-ribose and 2,3:5,6-di-O-isopropylidene-D-mannofuranose. Syntheses of glycofuranosylethynes, *Carbohydr. Res.*, 36 (1974) C5–C7.
105. J. G. Buchanan, A. R. Edgar, M. J. Power, and P. D. Theaker, Synthesis of D-ribofuranosyl derivatives of dimethyl maleate and of ethyl acetate as *C*-nucleoside precursors, *Carbohydr. Res.*, 38 (1974) C22–C24.

106. J. G. Buchanan, A. D. Dunn, and A. R. Edgar, C-Nucleoside studies. Part II. Pentofuranosylethynes from 2,3-O-isopropylidene-D-ribose, *J. Chem. Soc., Perkin Trans. 1* (1975) 1191–1200.
107. J. G. Buchanan, A. R. Edgar, M. J. Power, and G. C. Williams, Synthesis of D-ribofuranosyl derivatives of methyl propiolate as C-nucleoside precursors, *J. Chem. Soc., Chem. Commun.* (1975) 501–502.
108. J. G. Buchanan, A. R. Edgar, M. J. Power, and G. C. Williams, C-Nucleoside studies: Part IV. Reduction of 2,3,5-tri-O-benzyl-α- (and β-)-D-ribofuranosylethyne, *Carbohydr. Res.*, 45 (1975) 312–316.
109. J. G. Buchanan, A. R. Edgar, M. J. Power, and G. C. Williams, Studies in C-nucleoside synthesis, *Nucleic Acids Res., Spec. Pub. 1, Symp.Chem. Nucleic Acids Components, 3rd*, 1975, pp. S69–S71.
110. J. G. Buchanan, A. D. Dunn, and A. R. Edgar, C-Nucleoside studies. Part III. Glycofuranosylethynes from 2,3:5,6-di-O-isopropylidene-D-mannose, *J. Chem. Soc., Perkin Trans. 1* (1976) 68–75.
111. J. G. Buchanan, D. M. Clode, and N. Vethaviyasar, Potential hexokinase inhibitors. Synthesis and properties of 2,3-anhydro-D-allose, 2,3-anhydro-D-ribose, and 2-O-methylsulfonyl-D-mannose, *J. Chem. Soc., Perkin Trans. 1* (1976) 1449–1453.
112. J. G. Buchanan and A. R. Edgar, Acetoxonium ions from acetoxyoxiranes and orthoesters: their conversion into ethylidene acetals, rearrangement, and solvolysis, *Carbohydr. Res.*, 49 (1976) 289–304.
113. J. G. Buchanan, A. R. Edgar, M. J. Power, and G. C. Williams, C-Nucleoside studies. Part 5. The synthesis of D-ribofuranosyl derivatives of methyl propiolate and a study of the activating influence of the ester group in cycloaddition reactions, *Carbohydr. Res.*, 55 (1977) 225–238.
114. J. G. Buchanan, A. D. Dunn, A. R. Edgar, R. J. Hutchison, M. J. Power, and G. C. Williams, C-Nucleoside studies. Part 6. Synthesis of 3-[2,3,5-tri-O-benzyl-β-(and α-)-D-ribofuranosyl]-prop-2-yn-1-ol and related compounds: A new synthesis of 3(5)-(2,3,5-tri-O-benzyl-β-D-ribofuranosyl)pyrazole, *J. Chem. Soc., Perkin Trans. 1* (1977) 1768–1791.
115. J. G. Buchanan and D. R. Clark, Action of ammonia on the methyl 2,3-anhydro-D-ribofuranosides, and treatment of the products with nitrous acid, *Carbohydr. Res.*, 57 (1977) 85–93.
116. J. G. Buchanan and D. R. Clark, Studies on the interconversion of 2,3'-anhydro-β-D-xylofuranosyluracil and 2,2'-anhydro-1-β-D-arabinofuranosyluracil, *Carbohydr. Res.*, 68 (1979) 331–341.
117. J. G. Buchanan, A. R. Edgar, M. J. Power, and C. T. Shanks, C-Nucleoside studies. Part 7. A new synthesis of showdomycin, 2-β-D-ribofuranosylmaleimide, *J. Chem. Soc., Perkin Trans. 1* (1979) 225–227.
118. J. G. Buchanan, M. E. Chacón-Fuertes, and R. H. Wightman, C-Nucleoside studies. Part 8. Synthesis of 3-β-D-arabinofuranosylpyrazole from D-mannose, *J. Chem. Soc., Perkin Trans. 1* (1979) 244–248.
119. J. G. Buchanan, M. E. Chacón-Fuertes, A. Stobie, and R. H. Wightman, C-Nucleoside studies. Part 9. Synthesis of 3(5)-α-D-ribofuranosylpyrazole and related compounds, *J. Chem. Soc., Perkin Trans. 1* (1980) 2561–2566.
120. J. G. Buchanan, A. R. Edgar, R. J. Hutchison, A. Stobie, and R. H. Wightman, C-Nucleoside studies. Part 10. A new synthesis of 3-(2,3,5-tri-O-benzyl-β-D-ribofuranosyl)pyrazole and its conversion into 4-nitro-3(5)-β-D-ribofuranosylpyrazole, *J. Chem. Soc., Perkin Trans. 1* (1980) 2567–2571.
121. J. G. Buchanan, A. R. Edgar, R. J. Hutchison, A. Stobie, and R. H. Wightman, A new synthesis of formycin *via* nitropyrazole derivatives, *J. Chem. Soc., Chem. Commun.* (1980) 237–238.
122. J. G. Buchanan, A. Stobie, and R. H. Wightman, A new synthesis of pyrazofurin, *J. Chem. Soc., Chem. Commun.* (1980) 916–917.
123. J. G. Buchanan, M. R. Hamblin, G. R. Sood, and R. H. Wightman, *J. Chem. Soc., Chem. Commun.* (1980) 917–918.

124. J. G. Buchanan, A. Stobie, and R. H. Wightman, C-Nucleoside studies. Part XI. Cine-substitution in 1.4-dinitropyrazoles: Application to he synthesis of formycin via nitropyrazole derivatives, Can. J. Chem., 58 (1980) 2624–2627.
125. G. Aslani-Shotorbani, J. G. Buchanan, A. R. Edgar, D. Henderson, and P. Shahidi, Application of ^{13}C-N.M.R. in a re-examination of the isopropylidenation of D-ribose diethyldithioacetal and erythritol, Tetrahedron Lett., 21 (1980) 1791–1792.
126. J. G. Buchanan, M. E. Chacón-Fuertes, A. R. Edgar, S. J. Moorhouse, D. I. Rawson, and R. H. Wightman, Assignment of ring size in isopropylidene acetals by ^{13}C N.M.R, Tetrahedron Lett., 21 (1980) 1793–1796.
127. J. G. Buchanan, S. J. Moorhouse, and R. H. Wightman, C-Nucleoside studies. Part 12. Synthesis of 3-α- and 3-β-(D-xylofuranosyl)pyrazoles, J. Chem. Soc., Perkin Trans. 1 (1981) 2258–2266.
128. G. Aslani-Shotorbani, J. G. Buchanan, A. R. Edgar, C. T. Shanks, and G. C. Williams, C-Nucleoside studies. Part 13. A new synthesis of 2,3,5-tri-O-benzyl-α (and β-)-D-ribofuranosylethyne involving benzyloxy participation, and a synthesis of α-showdomycin, J. Chem. Soc., Perkin Trans. 1 (1981) 2267–2272.
129. J. G. Buchanan, A. Stobie, and R. H. Wightman, C-Nucleoside studies. Part 14. A new synthesis of pyrazofurin, J. Chem. Soc., Perkin Trans. 1 (1981) 2374–2378.
130. L. N. Chamberlain, I. A. S. Edwards, H. P. Stadler, J. G. Buchanan, and A. Thomas, The conformation of methyl 4,6-O-(S)-benzylidene-2-chloro-2-deoxy-α-D-idopyranoside in the crystalline state and in chloroform solution, Carbohydr. Res., 90 (1981) 131–137.
131. J. G. Buchanan, A. R. Edgar, D. I. Rawson, P. Shahidi, and R. H. Wightman, Assignment of ring size in isopropylidene acetals by carbon-13 N. M. R. spectroscopy, Carbohydr. Res., 100 (1982) 75–86.
132. J. G. Buchanan and R. H. Wightman, The chemistry of nucleoside antibiotics, In: P. G. Sammes, (Ed.), Topics in Antibiotic Chemistry, Vol. 6, Ellis Horwood, Chichester, 1982, pp. 229–339.
133. J. G. Buchanan, K. A. MacLean, H. Paulsen, and R. H. Wightman, A new chiral synthesis of (-)-anisomycin and its demethoxy analogue, J. Chem. Soc., Chem. Commun. (1983) 486–488.
134. J. G. Buchanan, G. Singh, and R. H. Wightman, An enantiospecific synthesis of (+)-retronecine and related alkaloids, J. Chem. Soc., Chem. Commun. (1984) 1299–1300.
135. J. G. Buchanan, M. R. Hamblin, A. Kumar, and R. H. Wightman, The biosynthesis of showdomycin: Studies with stable isotopes and the determination of principal precursors, J. Chem. Soc., Chem. Commun. (1984) 1515–1517.
136. J. G. Buchanan, A. Kumar, R. H. Wightman, S. J. Field, and D. W. Young, The biosynthesis of showdomycin: Stereochemical aspects of maleimide ring formation, J. Chem. Soc., Chem. Commun. (1984) 1517–1518.
137. J. G. Buchanan, D. Smith, and R. H. Wightman, C-Nucleoside studies—15. Synthesis of 3-β-D-arabinofuranosylpyrazoles and the D-arabinofuranosyl analogue of formycin, Tetrahedron, 40 (1984) 119–123.
138. J. G. Buchanan, N. K. Saxena, and R. H. Wightman, C-Nucleoside studies. Part 17. The synthesis of 3 (5)-carbamoyl-5(3)-β-D-ribofuranosylpyrazole (4-deoxypyrazofurin) and 4-amino 3(5)-carbamoyl-5 (3) β-D-ribofuranosylpyrazole, J. Chem. Soc., Perkin Trans. 1 (1984) 2367–2370.
139. G. Aslani-Shotorbani, J. G. Buchanan, and A. R. Edgar, C-Nucleoside studies, part 16. The isopropylidenenation of D-ribose diethyl thioacetal and ribitol. A new synthesis of α- and β-D-ribofuranosylethyne via 2,3:4,5-di-O-isopropylidene-aldehydo-D-ribose, Carbohydr. Res., 136 (1985) 37–52.
140. J. G. Buchanan, A. Millar, R. H. Wightman, and M. R. Harnden, C-Nucleoside studies. Part 18. The synthesis of C-nucleoside analogues of the antiviral agent (S)-9-(2,3-dihydroxypropyl)adenine, J. Chem. Soc., Perkin Trans. 1 (1985) 1425–1430.

141. J. G. Buchanan, K. A. MacLean, R. H. Wightman, and H. Paulsen, A new synthesis of (-)-anisomycin and its demethoxy analogue from D-ribose, *J. Chem. Soc., Perkin Trans. 1* (1985) 1463–1470.
142. J. G. Buchanan, Studies in the synthesis and biosynthesis of C-nucleosides, *Nucleosides Nucleotides*, 4 (1985) 13–19.
143. N. Baggett, J. G. Buchanan, M. Y. Fatah, C. H. Lachut, K. J. McCullough, and J. M. Webber, Benzylidene acetals of the D-ribonolactones: A structural reassessment, *J. Chem. Soc., Chem. Commun.* (1985) 1826–1827.
144. J. G. Buchanan, D. Smith, and R. H. Wightman, C-Nucleoside studies. Part 19. The synthesis of the β-D-xylofuranosyl analogue of formycin, *J. Chem. Soc., Perkin Trans. 1* (1986) 1267–1271.
145. J. G. Buchanan, A. Flinn, P. H. C. Mundill, and R. H. Wightman, Approaches to the synthesis of sinefungin via nitroaldol reactions, *Nucleosides Nucleotides*, 5 (1986) 313–323.
146. J. G. Buchanan, A. R. Edgar, and B. D. Hewitt, A new route to chiral hydroxypyrrolidines from D-erythrose via intramolecular 1,3-cycloaddition, *J. Chem. Soc., Perkin Trans. 1* (1987) 2371–2376.
147. J. G. Buchanan, V. B. Jigajinni, G. Singh, and R. H. Wightman, Enantiospecific synthesis of (+)-retronecine, (+)-crotonecine, and related alkaloids, *J. Chem. Soc., Perkin Trans. 1* (1987) 2377–2384.
148. K. V. Sastry, E. V. Rao, J. G. Buchanan, and R. J. Sturgeon, Cleistanthoside B, a diphyllin glycoside from *Cleitanthus patulus* heartwood, *Phytochemistry*, 26 (1987) 1153–1154.
149. J. G. Buchanan, A. R. Edgar, and R. J. Hutchison, Rearrangement of derivatives of bis(ethylsulfonyl)-α-D-lyxopyranosylmethane, *Carbohydr. Res.*, 164 (1987) 403–414.
150. J. G. Buchanan, A. R. Edgar, B. D. Hewitt, V. B. Jigajinni, G. Singh, and R. H. Wightman, Synthesis of chiral pyrrolidines from carbohydrates, *ACS Symp. Ser.*, 386 (1989) 107–108.
151. J. G. Buchanan, M. Harrison, R. H. Wightman, and M. R. Harnden, C-Nucleoside studies. Part 20. Synthesis of some pyrazolo[4,3-*d*]pyrimidine acyclonucleosides related to (*S*)-9-(2,3-dihydroxypropyl)adenine; a direct method for double functionalisation of the pyrazole ring, *J. Chem. Soc., Perkin Trans. 1* (1989) 925–930.
152. J. G. Buchanan, J. Stoddart, and R. H. Wightman, Synthesis of the indole nucleoside antibiotics neosidomycin and SF-2140: Structural revision of neosidomycin, *J. Chem. Soc., Chem. Commun.* (1989) 823–824.
153. J. G. Buchanan, K. W. Lumbard, R. J. Sturgeon, D. K. Thompson, and R. H. Wightman, Potential glycosidase inhibitors: Synthesis of 1,4-dideoxy-1,4-imino derivatives of D-glucitol, D- and L-xylitol, D- and L-alltol, D- and L-talitol, and D-gulitol, *J. Chem. Soc., Perkin Trans. 1* (1990) 699–706.
154. J. G. Buchanan, A. E. McCaig, and R. H. Wightman, The synthesis of 4-alkylsulfonyl-5-amino- and 5-amino-4-phosphonoimidazole nucleosides as potential inhibitors of purine biosynthesis, *J. Chem. Soc., Perkin Trans. 1* (1990) 955–963.
155. I. Robina, R. P. Gearing, J. G. Buchanan, and R. H. Wightman, Diastereoselective conjugate addition of ammonia in the synthesis of chiral pyrrolidines, *J. Chem. Soc., Perkin Trans. 1* (1990) 2622–2624.
156. A. Awal, A. S. F. Boyd, J. G. Buchanan, and A. R. Edgar, The formation of isopropylidene acetals of erythritol and ribitol under conditions of kinetic control, *Carbohydr. Res.*, 205 (1990) 173–179.
157. J. G. Buchanan, 2nd Clemo Memorial Lecture: C-nucleosides; synthesis and biosynthesis, In: C. Bleasdale and B. T. Golding, (Eds.), *Molecular Mechanisms of Bioorganic Processes*, Royal Society of Chemistry, London, 1990, pp. 225–243.
158. J. G. Buchanan, D. A. Craven, R. H. Wightman, and M. R. Harnden, C-Nucleoside studies. Part 21. Synthesis of some hydroxyalkylated pyrrolo and thieno[3,2-*d*]pyrimidines related to known acyclonucleosides, *J. Chem. Soc., Perkin Trans. 1* (1991) 195–202.
159. J. G. Buchanan and R. H. Wightman, Synthesis of nucleosides as potential inhibitors of purine biosynthesis, *Nucleic Acids Symp. Ser.*, 25 (1991) 53–54.
160. J. G. Buchanan, A. O. Jumaah, G. Kerr, R. R. Talekar, and R. H. Wightman, C-Nucleoside studies. Part 22. *cine*-Substitution in 1,4-dinitropyrazoles: Further model studies, an improved synthesis of

formycin and pyrazofurin, and the synthesis of some 3(5)-alkylsulfonyl-4-amino-5(3)-β-D-ribofuranosylpyrazoles, *J. Chem. Soc., Perkin Trans. 1* (1991) 1077–1083.
161. J. G. Buchanan, M. L. Quijano, and R. H. Wightman, *C*-Nucleoside studies. Part 23. New and more direct synthesis of 3-(β-D-xylofuranosyl)pyrazole, *J. Chem. Soc., Perkin Trans. 1* (1992) 1573–1576.
162. J. G. Buchanan, A. P. W. Clelland, T. Johnson, R. A. C. Rennie, and R. H. Wightman, Synthesis of *C*-glycosyltetrazoles related to 3-deoxy-D-arabino-heptulosonic acid 7-phosphate (DAHP); potential inhibitors of early steps in the shikimate pathway, *J. Chem. Soc., Perkin Trans. 1* (1992) 2593–2601.
163. J. G. Buchanan, A. P. W. Clelland, R. H. Wightman, T. Johnson, and R. A. C. Rennie, Synthesis of glycosidic and 2-doexyglycosidic ortholactones from 1-bromoglycosyl cyanides, *Carbohydr. Res.*, 237 (1992) 295–301.
164. J. G. Buchanan, J. Stoddart, and R. H. Wightman, Synthesis of the indole nucleoside antibiotics neosidomycin and SF-2140, *J. Chem. Soc., Perkin Trans. 1* (1994) 1417–1426.
165. J. G. Buchanan, D. G. Hill, R. H. Wightman, I. K. Boddy, and B. D. Hewitt, A stereoselective route to the sugar-cinnamate unit of hygromycin A, *Tetrahedron*, 51 (1995) 6033–6050.
166. M. Ataie, J. G. Buchanan, A. R. Edgar, R. G. Kinsman, M. Lyssikatou, M. F. Mahon, and P. M. Welch, 3,4-Anhydro-1,2-*O*-isopropylidene-β-D-tagatopyranose and 4,5-anhydro-1,2-*O*-isopropylidene-β-D-fructopyranose, *Carbohydr. Res.*, 323 (2000) 36–43.
167. S. W. Johnson, D. Angus, C. Taillefumier, J. H. Jones, D. J. Watkin, E. Floyd, J. G. Buchanan, and G. W. J. Fleet, Two epimerisations in the formation of oxetanes from L-rhamnose: Towards oxetane-containing peptidomimetics, *Tetrahedron: Asymmetry*, 11 (2000) 4113–4125.
168. J. G. Buchanan, Lord Todd 1907-1997, *Adv. Carbohydr. Chem. Biochem.*, 55 (2000) 1–13.
169. J. J. Robinson, J. G. Buchanan, M. H. Charlton, R. G. Kinsman, M. F. Mahon, and I. H. Williams, Evidence for α-lactone intermediates in addition of aqueous bromine to disodium dimethyl-maleate and -fumarate, *Chem. Commun.* (2001) 485–486.
170. C. Jones, A. Begona, J. H. van Boom, and J. G. Buchanan, Confirmation of the D-configuration of the 2-substituted arabinitol 1-phosphate residue in the capsular polysaccharide from *Streptococcus pneumoniae* Type 17F, *Carbohydr. Res.*, 337 (2002) 2353–2358.
171. J. G. Buchanan, M. H. Charlton, M. F. Mahon, J. J. Robinson, G. D. Ruggiero, and I. H. Williams, Experimental and computational studies of α-lactones: Structure and bonding in the three-membered ring, *J. Phys. Org. Chem.*, 15 (2002) 642–646.
172. J. G. Buchanan, R. A. Diggle, G. D. Ruggiero, and I. H. Williams, The Walden cycle revisited: A computational study of competitive ring closure to α- and β-lactones, *Chem. Commun.* (2006) 1106–1108.
173. N. Pirinccioglu, J. J. Robinson, M. F. Mahon, J. G. Buchanan, and I. H. Williams, Experimental and computational evidence for α-lactone intermediates in the addition of aqueous bromine to disodium dimethyl-maleate and -fumarate, *Org. Biomol. Chem.*, 5 (2007) 4001–4009.
174. J. G. Buchanan, G. D. Ruggiero, and I. H. Williams, Dyotropic rearrangement of α-lactone to β-lactone: A computational study of small-ring halolactonisation, *Org. Biomol. Chem.*, 6 (2008) 66–72.
175. M. J. Bonne, K. J. Edler, J. G. Buchanan, D. Wolverson, E. Psillakis, M. Helton, W. Thielemans, and F. Marken, Thin-film modified electrodes with reconstituted cellulose-PDDAC films for the accumulation and detection of triclosan, *J. Phys. Chem.*, 112 (2008) 2660–2666.
176. M. M. P. Morais, J. D. Mackay, S. K. Bhamra, J. G. Buchanan, T. D. James, J. S. Fossey, and J. M. H. van den Elsen, Analysis of protein glycation using phenylboronate acrylamide gel electrophoresis, *Proteomics*, 10 (2010) 48–58.

SAUL ROSEMAN

1921–2011

Professor Saul Roseman, a pioneer glycobiologist who made contributions to carbohydrate chemistry and biochemistry during more than six decades, passed away at the age of 90 years on July 2, 2011 in Pikesville, Maryland. He was survived by his wife of many years, Martha Roseman, a former dean of academic advising at the Johns Hopkins University, three children (Mark, Dorinda, and Cynthia, wife of the Glycobiologist Ron Schnaar), seven grandchildren, and 11 great-grandchildren. His wife Martha, who had been suffering from Parkinson's disease for a long time, passed away within a few weeks after Saul died.

Roseman was recognized for his many contributions in several areas, most notably his work on sugar structures and carbohydrate metabolism; the proposed pathway for biosynthesis of the GD1a ganglioside from ceramide in the brain (known as the Basu–Roseman pathway); bacterial sugar transport, and the phosphotransferase system (PTS); and the degradation of chitin, the skeletal polymer of crustaceans.

Roseman was the recipient of many national and international awards and honors, among them being his election, in 1971, to the US National Academy of Sciences and, in 1972, the degree *Doctor of Medicine Honoris Causa* from the University of Lund, Sweden. In 1973, he received the Sesquicentennial Award from the University of Michigan, and in 1974, the Rosenstiehl Award from the Brandeis University. Others include the T. Duckett Jones Memorial Award from the Helen Hay Whitney Foundation in 1978, the International Award from the Gairdner Foundation in 1981, and the Karl Meyer award from the Society of Glycobiology in 1993.

Among his extensive contributions in the carbohydrate field, he published 54 years ago the first correct structure of sialic acid, a nonulosonic acid occurring as the terminal sugar in many macromolecules that reside on the surface of cells. He devoted his research to studies on the biosynthesis of cell-surface macromolecules, glycoproteins (GPs), and glycosphingolipids (GSLs), and the sugar phospho-transport system (PTS) in bacterial cells. He was a charismatic figure in the American biochemical world and was nicknamed "Saul Serendipity" Roseman.

Roseman was born in Brooklyn, NY, on March 9, 1921; his immigrant father died when he was a boy and his family lived in poverty. His mother raised him by working hard in a tailoring shop. In school, he skipped grades; he was very bright and was often the youngest in his class. He received his bachelor's degree in chemistry, with minors in biology and physics, in 1941 from the City College of New York. He entered in the graduate school of the University of Wisconsin, and directly after he received his M.S. degree in chemistry in 1944, his studies were interrupted, as was common for young men of his generation, by his World War II military service in Europe, where he served in the infantry. After his discharge, he returned to the University of Wisconsin to complete his Ph.D. degree in 1951 under Professor Karl Paul Link, the discoverer of the anticoagulant, warfarin (Coumadin).[1] As a doctoral student he studied the metabolism of warfarin.[2] In 1951, his postdoctoral training, studying the biosynthesis of glycoconjugates (specifically proteoglycans), began in the laboratory of Professor Albert Dormfan[3] in the Department of Pediatrics at the University of Chicago Medical School. He joined the Biological Chemistry Department of the University of Michigan in Ann Arbor as a young assistant professor in 1954. His laboratory was a crowded place, because his research group was also sharing space in the Rackham Arthritis Research unit at the Medical School's Kresge building with at least two other clinical physicians, Drs. Giles Bole (who retired in 1986 as the Dean of the University of Michigan Medical School), David Castor, and Edward Heath (who was assistant professor of microbiology). There in 1958 he made his celebrated serendipitous discovery of the correct structure of sialic acid[4,5] (Scheme 1) with his first postdoctoral fellow, Donald Comb.[4,5]

He often told the story of how, by accident and careful observation, he and Dr. Comb arrived at that correct structure of sialic acid, a compound released through hydrolysis catalyzed by a newly discovered enzyme that was then termed nanaldolase.[4,5] They proposed the correct structure of sialic acid as resulting from the aldol

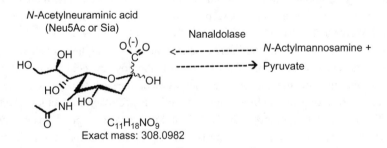

SCHEME 1. N-Acetylmannosamine from degradation of sialic acid (Neu5Ac) as catalyzed by "Nanaldolase" (N-acetylneuraminate lyase, EC 4.1.3.3), a reversible reaction.

condensation of pyruvate and N-acetylmannosamine 6-phosphate.[6] At that time, Gottschalk and others had reported a wrong structure in the literature, as they had taken clues from the chemical condensation of N-acetylglucosamine and pyruvate. Many years later, Roseman related his belief in serendipity, "finding the unexpected is just Nature's way of telling researchers where to look for the really interesting and important stuff, and serendipity is simply a tool of the trade. I consider serendipity to be a major tool in scientific investigation, and many important discoveries in biology and medicine have come through it." Roseman said, "take the human brain, for example. We think there are about 100,000 different types of enzymatic reactions that take place in the human brain. When you start from scratch looking for something in that, the chances that you are going to find what you are seeking are pretty low."

The story of the sialic acid structure was told to every one of his new students and postdoctoral fellows, and by the touch of that golden wand, many fortuitous discoveries took place in his laboratory. In the field of biosynthesis of glycoproteins and glycolipids, a series of glycosyltransferases[7-15] in the eukaryotic system was discovered between 1963 and 1970. The characterization of each of those new glycosyltransferases (GLTs) materialized not only through the work of the very active team he built at that time (1960–1970) at the University of Michigan but also by the touch of serendipity. Some of those stories are already documented.[16] The GLT (glycosyltransferase) group was composed of very able postdoctoral fellows, including William (Bill) Jourdian, Donald (Don) Carlson, Bernard (Bernie) Kaufman, Edward (Ed) Kean, Edward (Eddy) McGuire, and Harry Schachter (who took sabbatical leave for two years from the University of Toronto), and predoctoral fellows Jack (Jack) Distler, Subhash Basu (Basu), Allen (Al) Schultz, and Manju Basu (Manju). He always had more postdoctoral fellows than graduate students in his laboratory. A very productive group investigating the phosphotransferase system (PTS) in prokaryotes became established in his laboratory between 1965 and 1975 subsequent to an accidental discovery by Sudhamoy Ghosh (a postdoctoral fellow from India), who found in 1962 that a kinase in a bacterial system[17] catalyzed the transfer of phosphate from enolpyruvate phosphate (phosphoenol pyruvate, PEP) to sugars without the involvement of ATP. Later on, that PTS group was strengthened by the recruiting at the Johns Hopkins University between 1965 and 1975 of Freddie Kundig, Werner Kundig, Byron Anderson, Robert Simoni, Milton Sair, Teruko Nakazawa, Atsushi Nakazawa, Nancy Weigel, Norman Medow, Namat Keyhani, and others.

Roseman had conducted his postdoctoral research during four years (1951–1954) at the University of Chicago in the laboratory of the late Dr. Albert Dorfman, studying the structures of hyaluronic acid and chondroitin sulfates and the enzymes that catalyze their syntheses and degradation.[3] From Chicago he moved in 1954 to the

University of Michigan as a young assistant professor, spending 11 years there, being appointed as a Full Professor in 1960. In 1965 he moved his laboratory to the Department of Biology, McCollum–Pratt Institute, at the Homewood campus of the Johns Hopkins University in Baltimore, Maryland. Roseman remained there for the rest of his life, serving as Chairman of the Department of Biology (1969–1973) and Director of the McCollum–Pratt Institute (1988–1990), and holding the title Ralph S. O'Connor Chair, Professor of Biology. He remained active as a researcher and teacher throughout his years at the Homewood campus, publishing nearly 230 papers and over 30 reviews during his long and distinguished career. These publications included 15 papers in the last decade of his life, between 2001 and 2011 after he had reached the age of 80 years, in top-ranked journals: *J.Biol. Chem.* (7), *Proc.Natl. Acad. Sci. U.S.A.* (3), and *Biochemistry* (2). He was also an active member of the editorial board of the *Journal of Biological Chemistry* and published over 130 papers in that journal alone. Even as late as March 2011, he submitted a paper to that journal, and it was accepted in May shortly before he passed away on July 2, 2011.

His publications from the University of Michigan and the Johns Hopkins University may be divided into six distinct themes, as detailed here.

Biosynthesis and Metabolism of Simple Sugars

During his years (1955–1965) at the University of Michigan, he studied the biosynthesis of hexosamines[17–20] and nucleotide sugars (UDP-glucose, UDP-Gal, UDP-GalNAc, UDP-GlcNAc, and CMP-Neu5Ac[21]).

Biosynthesis of Mucin Glycoproteins and Brain Gangliosides

Roseman's group synthesized various nucleotide sugars having a ^{14}C radioactive label in the sugar residue by combined chemical[22] and enzymatic processes,[21] in collaboration[22] with Nobel laureate Har Gobind Khorana (in 1958, when Dr. Khorana was in Vancouver, BC, Canada). This collaboration[22] was a giant step in Roseman's research career. The availability of a range of radioactively labeled nucleotide sugars of high specific activity proved to be an ingenious tool in the Roseman laboratory, leading to the discovery (1963–1970) of several glycosyltransferases (enzymes in EC group 2.4.1) in the glycobiology field; later on, these became familiar as sialyltransferases (SATs),[7–13] galactosyltransferases (GalTs),[9,14,15] *N*-acetylgalactosaminyltransferases (GalNAcTs),[9,23] and *N*-acetylglucosaminyltransferases (GlcNAcTs).[13,24] Between 1963 and 1970, at least 14 new glycosyltransferases were characterized from the eukaryotic system in Roseman's laboratory, catalyzing the biosynthesis *in vitro* of several glycoproteins[13,24] and gangliosides.[9–13,23]

Discovery of the Sialyltransferase Catalyzing the *In Vitro* Biosynthesis of Sialyl-Lactose

Sialyl-lactose (SL) was first isolated from rat mammary gland by Trucco and Caputto in 1954. Later on (1958), Kuhn and Brossmer obtained from human milk two isomers of SL [α-NeuAc-(2→3)-β-Gal-(1→4)-Glc and α-NeuAc-(2→6)-β-Gal-(1→4)-Glc]. Jourdian, Carlson, and Roseman first reported[7] the enzymatic synthesis of the second sialyl-lactose [α-NeuAc-(2→3)-lactose] isomer by a particulate preparation from rat mammary gland. In addition, preliminary evidence showed that this preparation catalyzed the synthesis of a sialyl conjugate with lactosamine [β-Gal-(1→4)-GlcNAc]. In both cases, the donor used was CMP-^{14}C-NeuAc, prepared in Roseman's laboratory. However, the same particulate preparation from rat mammary tissue did not catalyze the synthesis of the GM3 ganglioside from lactosyl-ceramide and CMP-^{14}C-NeuAc. In the next two years, Basu and Kaufman[8-12] in the Roseman laboratory isolated a particulate preparation (buffy coat) from a 9-day-old embryonic chicken brain that catalyzed the biosynthesis *in vitro* of the GM3 ganglioside from lactosyl-ceramide and CMP-^{14}C-NeuAc in the presence of a detergent mixture: Triton-CF-54 and Tween-80.[9-12] Again, this was another chance discovery in Roseman's laboratory, as described later.[16] It is now established that, in the presence of neutral detergents, the water-insoluble glycolipid substrates form mixed micelles [25] and the membrane-bound glycosyltransferases are also solubilized.[26] During the discovery of these sialyltransferases (SATs),[8-12] it was not known whether this was due to the activity of one sialyltransferase (that is, one gene product) or three different sialyltransferases catalyzing three different reactions. The biosynthetic pathway for the GD1a ganglioside (Scheme 2), from lactosyl-ceramide, was proposed in 1966.[9,11-13]

At that time, individual enzymes were not solubilized or separated. Being a committed chemist, Roseman considered that each linkage is formed by catalysis of different[13,24] gene products. His intuitive conclusion came from the results of specificity studies obtained with different sources of particulate systems isolated from different animals. Later on, the concept crystallized into his theory of "One Enzyme for One Linkage." Nowadays about 13 different sialyltransferase genes have been characterized, and each gene product catalyzes different linkages or recognizes different terminal or penultimate sugars of the oligosaccharides. With the advent of gene cloning, Roseman's intuitive hypothesis was confirmed. During that time (1963–1970), Roseman's laboratory reported four different sialyltransferases for biosynthesis of glycoproteins (GP-SATs)[24] and four different sialyltransferases for gangliosides (GSL-GLTs)[16,27] (Scheme 2). With today's knowledge, it may be safely concluded that all these eight sialyltransferases (SATs) are different gene products.[27] Based on substrate-specificity

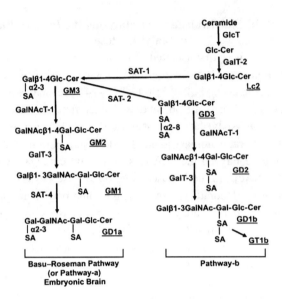

SCHEME 2. Proposed pathway for biosynthesis of the GD1a ganglioside from ceramide (Basu–Roseman pathway).[16,27]

studies on the mucin glycoprotein[24] and the GD1a ganglioside (Basu, Ph.D. thesis, 1966),[9] biosynthetic pathways were proposed by 1970 from the Roseman laboratory[16,24,27] that now appear in textbooks. Of course, behind the discovery of those pathways, there is more than one serendipity story.[16] Because the leader of the laboratory was a strong believer in scientific discovery originating through exploitation of chance observations, a person trained in that laboratory could not disregard the facts. Some of those stories have been published in later days.[16]

The biosynthetic pathway proposed in 1970 for ganglioside GD1a (starting from ceramide) is known as the Basu–Roseman pathway.[9,16,27–29] Workers elsewhere in the world established[30] much later the branching of that pathway for the biosynthesis of gangliosides GD2 or GT1a. In 1968, using buffy coat from embryonic chicken brain (ECB, now known to contain 60% of Golgi bodies),[29] Kaufman and Basu[12] from Roseman's laboratory first showed the biosynthesis of GD3 ganglioside [α-NeuAc-$(2 \rightarrow 8)$-α-NeuAc-$(2 \rightarrow 3)$-β-Gal-$(1 \rightarrow 4)$-β-Glc-ceramide] from GM3 ganglioside [α-NeuAc-$(2 \rightarrow 3)$-β-Gal-$(1 \rightarrow 4)$-β-Glc-ceramide]. Using a 50% Golgi-rich system from rat liver, Basu and his coworkers[27,29] and Sandhoff and Van Echten-Deckert[30] suggested that GD3 is the branching point for many other di-, tri-, and tetra-sialyl-gangliosides. Further biosynthetic work on glycoproteins[31,32] and blood group-active glycolipids (including GlcNAc-containing gangliosides)[16,27–29,33–40] was continued independently after 1970 in the laboratories of Schachter and Basu, respectively. The branching enzymes (five different GlcNAc-transferases) for the biosynthesis of di-, tri-, and tetra-antennary complex N-linked glycoproteins were proposed from Schachter's laboratory[32] at the University of Toronto. The biosynthetic steps for autoimmune antigen, Ii-glycolipids containing β-(1→6)- and β-(1→3)-GlcNAc-branched glycolipids, were established by Basu and Basu in rat lymphomas.[33] Subsequent to the "cloning decade" (1985–1995), we now have sufficient proof that all the GlcNAc-transferases (I–V) responsible for the biosynthesis of N-linked glycoproteins[32] and the GlcNAc-transferases (1–3) responsible for Ii-glycolipid[33] biosynthesis are of different gene origins. Other steps for the biosynthesis of blood-group H-active[34] and B-active glycosphingolipids were established in rabbit bone marrow,[35,36] embryonic chicken brains,[37] and bovine spleen[38] in the Basu laboratory (1970–1990) at Notre Dame. Biosynthesis of Le^X and sialyl Lewis-X { α-NeuAc-(2→3)-β-Gal-(1→3)-[Fuc-(1→4)]-β-Gal-(1→4)-β-GlcNAc-(1→3)-β-Gal-(1→4)-β-Glc-ceramide} was established in colon carcinoma cells by coworkers of both Basu[37] and Hakamori.[41] It is noteworthy that students originating from Roseman's laboratory flourished independently after 1970 in establishing different pathways for biosynthesis of glycoproteins and glycolipids, while Roseman's laboratory during the subsequent decade (1970–1980) at the Johns Hopkins University focused primarily on the bacterial PTS and chitin projects described next.

Biosynthesis of Oligoglycosyl-Glycoproteins Studied at the Johns Hopkins University (1965–1970)

The step-by-step synthesis of the oligoglycosyl moiety attached to the complex glycoproteins was established up to 1970 in the laboratory of Roseman. However, when Behrens and Leloir[42] discovered that the transfer of mannosyl-oligosaccharide to the protein backbone took place via its dolichol derivative, it became clear that two different pathways[24,43,44] synthesize the high-mannose-containing glycoproteins and the mucin-type glycoproteins. However, in the synthesis of the finally processed complex oligoglycosyl chains of the glycoproteins, the addition of terminal galactose, N-acetylglucosamine, and sialic acid takes place by stepwise addition of the sugars, as proposed by Roseman and his coworkers.[24] Branching of the oligoglycosyl moieties in glycoproteins or glycolipids occurs by the catalysis of specific glycosyltransferases present in the Golgi bodies, as revealed by Schachter in glycoprotein[45] and Basu in glycolipid[29,39,40] biosyntheses. In fact, except for one or two glycosyltransferases, the rest of them (at least 100 gene products have been characterized and reported) reside in the Golgi bodies. Later on, it was discovered that the sialyltransferases in the rat mammary gland or the buffy coat of embryonic chicken brain contained the Golgi bodies. However, an exact theory to account for the presence of these glycosyltransferases in the Golgi bodies and their regulation in the synthesis of oligosaccharides is not fully known and remains the subject of current research.[27,29,39] Most important, wide-ranging studies of the glycosyltransferases (a term introduced from the Roseman laboratory) is now the research subject in at least 200 laboratories around the world, and the literature has grown to a remarkable degree.

Bacterial Phosphotransferase System (PTS) Studied at the Johns Hopkins University (1965–1985)

Discovery of the PTS in prokaryotes started in the Roseman laboratory (1963) after the chance discovery by Sudhamoy Ghosh[17,46] of the phosphorylation of sugars and sugar amines by enolpyruvate phosphate (Scheme 3) in the presence of a cell-free bacterial extract.

The Roseman group moved to the Johns Hopkins University in the fall of 1965, and shortly thereafter, Roseman and Werner Kundig[46–48] purified the two enzymes, Enzyme I and Enzyme III, and another heat-sensitive protein, HPr.[49] Enzyme II is part of the bacterial-membrane particles of the PTS. It is involved in a series of sequential phosphotransfer reactions between proteins, and it simultaneously phosphorylates and

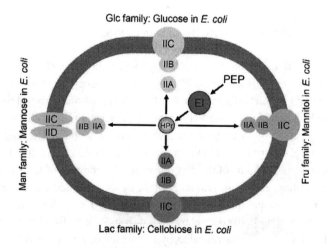

SCHEME 3. Sugar transport by the bacterial phosphotransferase system (PTS). (For color version of this figure, the reader is referred to the online version of this chapter.)

translocates its sugar substrates across the bacterial membranes (Scheme 3).[50–52] The source of the phosphate group is not ATP or another nucleoside triphosphate, but is strictly enolpyruvate phosphate (PEP). Transfer of the phosphate group from PEP was shown to occur first to the imidazole group of a histidine residue in the HPr. Later on, HPr was abbreviated as histidine-containing protein[49,50] and Enzyme II catalyzes its phosphorylation. Later on, at the Johns Hopkins University, with Robert Simoni and Milton Sair, Roseman proved that this novel phosphorylation was synchronized with sugar translocation across the bacterial membranes, as shown in the scheme.[52] However, no such PTS has been discovered in the sugar-transport system in eukaryotic cells.

During the 1965–1970 period, Roseman's laboratory was divided into three major groups. The workers on the glycosyltransferases (GTs) of eukaryotic systems involved in ganglioside biosynthesis were Bernard Kaufman, Subhash Basu,[53] and Helina Den[54]. Harry Schachter, Edward McGuire, and Manju Basu[55] were involved in the alpha-1 blood glycoprotein and mucin proteoglycan biosynthesis. Those working on bacterial PTS were Werner Kundig, Freddie D. Kundig, Robert Simoni, Milton Sair, Bob Hayes, Atsushi Nakazawa (visiting professor from Japan), Teruko Nakazawa (visiting professor from Japan), Jeff Stock, Byron Anderson, and Nancy Weigel (the last three were graduate students).

Intercellular Adhesion and the "Fish Theory" from the Johns Hopkins University (1966–1972)

In early 1966, Roseman's laboratory was engaged (with Charles Orr, Stephen Roth, and others) in the identification of specific intercellular-adhesion processes essential to normal embryonic development and morphogenesis. Perhaps the final identification of the molecules remains unpublished. The "intercellular adhesion" group (Charles W. Orr, Stephen Roth, S. Bozzaro, and others) believed that cell-surface glycosylation is translated to the intercellular adhesion. Based on their work, Roseman proposed two hypotheses, first, that there is intercellular hydrogen bonding between carbohydrate chains and, second, that intercellular adhesion results from binding to substrates and the cell-surface enzymes; this was termed as the "Fish theory" (Scheme 4).[44,56]

Later on, this was proved not to be the cause of intercellular adhesion. It is true that some glycosyltransferases of the Golgi bodies may exist on the plasma membrane of eukaryotic cells, but as proposed by Roth and Roseman,[44,56] intercellular adhesion is much more complex than just the lock and key concept for the interaction of an enzyme with a specific substrate. The principal criticism of that

I. *Hydrogen bonds (oligosaccharide chains)*

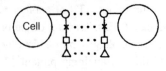

II. *Enzyme-substrate complex*

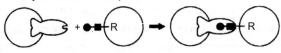

SCHEME 4. Cell-surface hydrogen bonding and enzyme–substrate interaction (Fish Theory).[44]

theory was that, in any enzymatic reaction, formation of the transition complex (enzyme–substrate) is momentary, whereas intercellular adhesion is stable and persistent. The intercellular-adhesion project turned into the attempted characterization of a "chicken factor" that caused cellular adhesion. Roseman spent a long period with at least three to four postdoctoral fellows to establish the structure of that factor, but was unable to establish a final structure. Even before he passed away (in 2011), he was interested in completing that project, and he sent the purified chicken factor (which was difficult to isolate) to major mass-spectrometry laboratories around the world (including the Chemistry and Biochemistry Department of the University of Notre Dame).

Degradation of Chitin and Hexosamine Metabolism in the Sea World (1985–2011)

During the last decades of his career, Roseman became interested in the enzymatic degradation of chitin, one of the most abundant and stable biopolymers in Nature. This venture could be termed the "Marine Chitin Degradation" project (Scheme 5). Back in his graduate classes at the University of Michigan in 1962, he lectured on the structure and properties of this chitin [β-(1 → 4)-linked polymer of N-acetylglucosamine]. Incoming graduate students were given a bottle of chitin and the protocol for isolation of the oligomers (chitobiose to chitodecaose) from the polymeric chitin.

From the middle of the 1980s until almost the end of his life (2011), Roseman was interested in the enzymatic degradation of chitin. Each year by some estimates 100 billion tons of discarded crustacean shells sink through world's oceans. A major

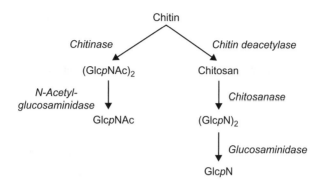

SCHEME 5. Degradation of chitin in the marine world.

component of these shells is chitin, the insoluble flexible polysaccharide that gives shells their toughness. Roseman and his later coworkers showed that chitin degrades by two-step processes: when a crustacean molts, it releases an enzyme, chitinase, to loosen the shell. This produces some oligosaccharides of GlcNAc, the building blocks of the large chitin polymer. The bacterium *Vibrio furnisii* can detect these short-chain oligomers on the shell and become attached when the bacteria begin to produce the enzymes needed to break and digest the chitin. Part of this process was unraveled in his Nieuwland lectures delivered at the University of Notre Dame in 1989. Later on, with his coworkers at the Johns Hopkins University (Chorley Yu, Bonnie Basler, Nemat Keyhani, O. Boudker, L. X. Wang, Alexi Fomenkov, and Z.-B. Li),[57–63] the complete story was told (Scheme 5).

We remember Roseman not only as an outstanding glycobiologist[63] but also as an outstanding teacher and lecturer. He used many slides, but always finished in the time allotted for his lectures. He insisted that his coworkers make lecture slides as simple as possible, with perhaps not more than seven lines on a slide. He preferred black and white slides to multicolored ones. He would jokingly say, "People use color slides when they are lacking colorful data." He always encouraged his laboratory personnel to work hard and to obtain positive results. To the incoming students and postdoctoral fellows he would say, "... do not waste my time by bringing negative results with hundreds of complaints or excuses" (his definition of "snow jobs"). He insisted on seeing triplicate results so that he could see the trend, and most of the time he was right in drawing conclusions from his intuitive thinking ability.

Saul Roseman was of solid build and kept himself physically active (until age 90 years); during the Baltimore period, he enjoyed sailing in his boat and swimming on most days, and he was a keen baseball and football fan. During the Ann Arbor days, he was always to be seen smoking his pipe, but he probably relinquished this habit in the mid-1980s. His son Mark continues the scientific tradition, being a professor of biochemistry at the Naval University in Washington, DC.

When he had time he loved to work shoulder to shoulder with those in the laboratory, clapping his hands and saying, "Let's go." This occurred quite often at the University of Michigan when his group was in the Arthritis Research Unit in a very limited space (only three rooms with a total of six benches of 6 ft. long shared by eight postdoctoral fellows, two graduate students (Jack and Basu), and at least four or five technicians). Two graduate students and a young technician, Ben Snyder, often worked the night shift. This was a trend Roseman tried to maintain in his larger laboratory at the Johns Hopkins University when he was not too busy with administration (1965–2010). Few could forget his famous, "Monday Night Research Group (MNRG) Meetings" from 8 PM until 10:30 or 11 PM. Each week one person in the laboratory was assigned to present his or her fresh data, and 12–15 pairs of eyes were used to judge the validity of those new results, along with Roseman's critical comments about proper "controls." When any one was going to give a paper at a Federation Meeting or other national or international meeting, the talk was first delivered at one of the MNRG meetings. This tradition was maintained by many of his ex-students and postdoctoral fellows in their own groups after they had left to establish independent laboratories. At every meeting, Roseman sat on the front row, and at the end of the talk, he would usually be the first person to raise a hand and ask relevant questions. He loved the search for new knowledge, and his passion for research was passed on to his students and coworkers with untiring enthusiasm.

In the fields of hexosamine metabolism, sialic acid structure and biosynthesis, characterization of at least 16 new glycosyltransferases (in glycoprotein and ganglioside biosynthesis), bacterial sugar transport by the PTS, intercellular adhesion, and chitin degradation mechanism in the sea world, Roseman left fundamental contributions for the future generations with his untiring dedication.

Subhash C. Basu

Acknowledgments

The picture of Dr. Roseman is from his portrait published in the Internet after his death. The schemes used in this chapter are redrawn from the sources as indicated and are not the exact duplicates of the originals. The author thanks Mrs. Dorisanne Nielsen and Mr. Eric Kuehner for their help during the preparation of this chapter and also thanks Dr. Arun Ghosh and Dr. Manuka Ghosh for assistance with the drawings.

References

1. M. A. Stahmann, C. F. Huebner, and P. K. Link, Studies on the hemorrhagic sweet clover disease, *J. Biol. Chem.*, 138 (1941) 513–527.
2. R. S. Overman, M. A. Stahmann, S. F. C. F. Hubener, W. R. Sullivan, J. Spero, L. Dohertey, D. G. Ikawa, L. Graf, S. Roseman, and K. P. Link, Studies on the hemorrhagic sweet clover disease: XIII. Anticoagulant activity and structure in the 4-hydroxycoumarin group. *J. Biol. Chem.*, 153 (1944) 5–24.
3. S. Roseman, F. E. Moses, J. Ludowig, and A. Dorfman, The biosynthesis of hyaluronic acid by group A *Streptococcus*. I. Utilization of 1-^{14}C-glucose, *J. Biol. Chem.*, 203 (1953) 213–225.
4. D. G. Comb and S. Roseman, Composition and enzymatic synthesis of *N*-acetylneuraminic acid (sialic acid), *J. Am. Chem. Soc.*, 80 (1958) 497–499.
5. D. G. Comb and S. Roseman, The sialic acids. I. The structure and enzymatic synthesis of *N*-acetylneuraminic acid, *J. Biol. Chem.*, 235 (1960) 2529–2537.
6. S. Roseman, G. W. Jourdian, D. Watson, and R. Wood, Enzymatic synthesis of sialic acid-9-phosphate, *Proc. Natl. Acad. Sci. U.S.A.*, 47 (1961) 958–961.
7. G. W. Jourdian, D. M. Carlson, and S. Roseman, Enzymatic synthesis of "sialyllactose", *Biochem. Biophys. Res. Commun.*, 10 (1963) 352–357.
8. S. Basu and B. Kaufman, Ganglioside biosynthesis in embryonic chicken brain, *Federation Proc.*, 24 (1965) 479.
9. S. Basu, Studies on the biosynthesis of gangliosides, Ph.D. Thesis (1966) University of Michigan, Ann Arbor, MI, [Micro film dissertation No. 6910].
10. B. Kaufman, S. Basu, and S. Roseman, Embryonic chicken brain sialyl-transferases, *Methods Enzymol.*, 8 (1966) 365–368.
11. B. Kaufman, S. Basu, and S. Roseman, Studies on the biosynthesis of gangliosides, In: S. M. Aronson and B. W. Volk, (Eds.), *Inborn Errors of Sphingolipid Metabolism,* Pergamon Press, New York, 1967, pp. 193–213.
12. B. Kaufman, S. Basu, and S. Roseman, Enzymatic synthesis of disialogangliosides from monosailogangliosides by sialyltransferases from embryonic chicken brain, *J. Biol. Chem.*, 243 (1968) 5804–5806.
13. S. Roseman, The synthesis of complex carbohydrates by multiglycosyltransferase systems and their potential function in intercellular adhesion, *Chem. Phys. Lipids*, 5 (1970) 270–297.
14. S. Basu, B. Kaufman, and S. Roseman, Conversion of Tay-Sachs ganglioside to monosialoganglioside by brain uridine diphosphate D-galactose: Glycolipid galactosyltransferase, *J. Biol. Chem.*, 240 (1965) 4114–4117.
15. S. Basu, B. Kaufman, and S. Roseman, Enzymatic synthesis of ceramide-glucose and ceramide lactose by glycosyltransferases from embryonic chicken brain, *J. Biol. Chem.*, 243 (1968) 5802–5804.
16. S. Basu, The serendipity of ganglioside biosynthesis: Pathway to CARS and HY-CARS glycosyltransferases, *Glycobiology*, 1 (1991) 469–475.
17. S. Ghosh and S. Roseman, Enzymatic phosphorylation of *N*-acetyl-D-mannosamine, *Proc. Natl. Acad. Sci. U.S.A.*, 47 (1961) 955–958.
18. S. Roseman and D. G. Comb, Chromogen formation and epimerization of *N*-acetylhexosamines, *J. Am. Chem. Soc.*, 80 (1958) 3166.
19. J. J. Distler, J. M. Merrick, and S. Roseman, Glucosamine metabolism. III. Preparation and *N*-acetylation of crystalline D-glucosamine- and D-galactosamine-6-phosphoric acids, *J. Biol. Chem.*, 230 (1958) 497–509.
20. C. T. Spivak and S. Roseman, Administration of N-acetyl-D-mannosamine to mammals, *J. Am. Chem. Soc.*, 81 (1959) 2403.

21. E. L. Kean and S. Roseman, The sialic acids. X. Purification and properties of cytidine-5'-monophosphosialic acid synthase, *J. Biol. Chem.*, 241 (1966) 5643–5650.
22. S. Roseman, J. J. Distler, J. G. Moffat, and H. Khorana, Synthesis of sugar nucleotides, *J. Am. Chem. Soc.*, 83 (1961) 659–663.
23. J. C. Steigerwald, S. Basu, B. Kaufman, and S. Roseman, Enzymatic synthesis of Tay-Sachs ganglioside, *J. Biol. Chem.*, 250 (1975) 6727–6734.
24. H. Schachter and S. Roseman, Mammalian glycosyltransferases: Their role in the synthesis and function of complex carbohydrates and glycolipids, In: W. Lennarz, (Ed.), *Biochemistry of Glycoproteins and Proteoglycans*, Plenum Press, New York, 1980, pp. 85–86.
25. M. Basu and S. Basu, Micelles and liposomes in metabolic enzymes and glycolipid glycosyltransferase assays, In: S. Basu, M. Basu, and J. M. Walker, (Eds.), *Liposomes: Methods and Protocols in Methods in Molecular Biology*, Humana Press, New York, 2002, pp. 107–130.
26. H. Higashi, M. Basu, and S. Basu, Biosynthesis in vitro of diasialosyl-neolactotetraosylceramide by a solubilized sialyltransferase from embryonic chicken brain, *J. Biol. Chem.*, 260 (1985) 824–828.
27. S. Basu, R. Ma, J. R. Moskal, and M. Basu, Ganglioside biosynthesis in developing brains and apoptotic cancer cells. X. Regulation of glyco-genes involved in GD3 and sialyl-Le x/a synthesis, *Neurochem. Res.*, 37 (2012) 1245–1255.
28. M. Basu, S. Basu, A. Stoffyn, and P. Stoffyn, Biosynthesis in vitro of sialyl alpha2-3neolactotetraosylceramide by a sialyltransferase from embryonic chicken brain, *J. Biol. Chem.*, 257 (1982) 12765–12769.
29. T. W. Keenan, J. D. Morre, and S. Basu, Ganglioside biosynthesis: Concentration of glycosphingolipid glycosyltransferase in Golgi apparatus from rat liver, *J. Biol. Chem.*, 249 (1974) 310–316.
30. G. V. Echten-Deckert and K. Sandhoff, Organization and topology of sphingolipid metabolism, In: D. H. R. Barton, K. Nakanishi, and O. Meth-Cohen, (Eds.), B. M. Pinto, (Ed.), *Comprehensive Natural Products Chemistry*, Vol. 3, Pergamon Press, New York, 1999, pp. 87–106.
31. M. Sarkar, E. Hull, Y. Nishikawa, R. J. Simpson, R. L. Moritz, R. Dunn, and H. Schachter, Molecular cloning and expression of cDNA encoding the enzyme that controls conversion of high-mannose to hybrid complex N-glycans: UDP-N-acetylglucosamine: 3-D-mannoside, 2-N-acetylglucosaminyltransferase I, *Proc. Natl. Acad. Sci. U.S.A.*, 88 (1991) 234–236.
32. H. Schachter, The "yellow brick road" to branched complex N-glycans glycoproteins: Carbohydrates to cloning, *Glycobiology*, 1 (1991) 453–461.
33. M. Basu and S. Basu, Biosynthesis in vitro of Ii-core glycolipids from neolactotetraosylceramide by beta1-3- and beta1-6-N-acetylglucosaminyltransferases from mouse T-lymphoma, *J. Biol. Chem.*, 259 (1984) 12557–12562.
34. S. Basu, M. Basu, and J. L. Chien, Enzymatic synthesis of blood group H-related glycosphingolipid by an alpha-fucosyltransferase from bovine spleen, *J. Biol. Chem.*, 250 (1975) 2950–2962.
35. M. Basu and S. Basu, Enzymatic synthesis of a tetraglycosylceramide by a galactosyltransferase from rabbit bone marrow, *J. Biol. Chem.*, 247 (1972) 1489–1495.
36. M. Basu and S. Basu, Enzymatic synthesis of a blood group B specific pentaglycosylceramide by an α-galactosyltransferase from rabbit bone marrow, *J. Biol. Chem.*, 248 (1973) 1700–1706.
37. M. Basu, J. W. Hawes, Z. Li, S. Ghosh, F. Khan, B. Zhang, and S. Basu, Biosynthesis in vitro of SA-Lex and SA-diLeX by alpha1-3-fucosyltransferases from colon carcinoma cells and embryonic brain tissues, *Glycobiology*, 1 (1991) 527–535.
38. K. A. Presper, M. Basu, and S. Basu, Biosynthesis in vitro of a blood group B-active fucose-containing hexaglycosylceramide from neolactosylceramide in bovine spleen, *J. Biol. Chem.*, 257 (1982) 169–173.
39. S. Basu, M. Basu, S. Dastgheib, and J. W. Hawes, Biosynthesis and regulation of glycosphingolipids, In: D. H. R. Barton, K. Nakanishi, and O. Meth-Cohen, (Eds.), B. M. Pinto, (Ed.), *Comprehensive Natural Products Chemistry*, Vol. 3, Pergamon Press, New York, 1999, pp. 107–128.

40. S. Basu, K. Das, and M. Basu, Glycosyltransferases in glycosphingolipid biosynthesis, In: B. Ernst, P. Sinaÿ, and G. Hart, (Eds.), *Oligosaccharides in Chemistry and Biology, a Comprehensive Handbook*, Wiley-VCH Verlag, GmbH, Germany, 2000, pp. 329–547.
41. M. R. Stroud, S. B. Levery, S. Mårtensson, M. E. Salyan, H. Clausen, and S. Hakomori, Human tumor-associated Le(a)-Le(x) hybrid carbohydrate antigen IV3(Gal beta 1→3[Fuc alpha 1→4] GlcNAc)III3FucnLc4 defined by monoclonal antibody 43-9F: Enzymatic synthesis, structural characterization, and comparative reactivity with various antibodies, *Biochemistry*, 33(35), (1994) 10672–10680.
42. N. Behrens and L. Leloir, Dolichol monophosphate glucose: An intermediate in glucose transfer in liver, *Proc. Natl. Acad. Sci. U.S.A.*, 66 (1970) 153–159.
43. A. Parodi, N. Behrens, L. Leloir, and H. Carminatti, The role of polyprenol-bound saccharides as intermediates in glycoprotein synthesis in liver, *Proc. Natl. Acad. Sci. U.S.A.*, 69 (1972) 3268–3272.
44. S. Roseman, Reflections in glycobiology, *J. Biol. Chem.*, 276 (2001) 41527–41542.
45. P. J. Letts, L. Pinteric, and H. Schachter, Localization of glycoprotein glycosyltransferases in the Golgi apparatus of rat and mouse testis, *Biochim. Biophys. Acta*, 372 (1974) 304–320.
46. W. Kundig, S. Ghosh, and S. Roseman, Phosphate bound to histidine in a protein as an intermediate in a novel phosphotransferase system, *Proc. Natl. Acad. Sci. U.S.A.*, 52 (1964) 1067.
47. W. Kundig, F. D. Kundig, B. Anderson, and S. Roseman, Restoration of active transport of glycosides in *Escherichia coli* by a component of a phosphotransferase system, *J. Biol. Chem.*, 241 (1966) 3243–3246.
48. W. Kundig and S. Roseman, "Sugar Transport" I. Isolation of a phosphotransferase system from *Escherichia coli*, *J. Biol. Chem.*, 246 (1971) 1393–1406.
49. B. Anderson, N. Weigel, W. Kundig, and S. Roseman, Sugar Transport. III. Purification and properties of a phosphotransferase protein (HPr) of the phosphoenol pyruvate-dependent phosphotransferase system of *Escherichia coli*, *J. Biol. Chem.*, 246 (1971) 7023–7033.
50. R. D. Simoni, S. Roseman, and M. H. Saier, Suppression of defects in cyclic adenosine 3'-5' monophosphate metabolism in *Escherichia coli*, *J. Biol. Chem.*, 251 (1976) 6584–6597.
51. N. Weigel, D. A. Powers, and S. Roseman, Sugar transport by the bacterial phosphotransferase system. Primary structure and active site of a general phosphocarrier protein (HPr) from *Salmonella typhimurium*, *J. Biol. Chem.*, 257 (1982) 14499–14508.
52. N. D. Medow and S. Roseman, Sugar transport by the bacterial phosphotransferase system: Isolation and characterization of a glucose-specific protein (IIIGlc) from *Salmonella typhimurium*, *J. Biol. Chem.*, 257 (1982) 14526–14537.
53. S. Basu, B. Kaufman, and S. Roseman, Enzymatic synthesis of glucocerebroside by a glucosyltransferase from embryonic chicken brain, *J. Biol. Chem.*, 248 (1973) 1388–1394.
54. H. Den, B.-A. Sela, S. Roseman, and L. Sachs, Blocks in ganglioside synthesis in transformed hamster cells and their revertants, *J. Biol. Chem.*, 249 (1974) 659–661.
55. S. Basu, A. Schultz, M. Basu, and S. Roseman, Enzymatic synthesis of galactocerebroside by a galactosyltransferase from embryonic chicken brain, *J. Biol. Chem.*, 243 (1971) 4272–4279.
56. S. Roth, E. J. McGuire, and S. Roseman, Evidence for cell-surface glycosyltransferases. Their potential role in cellular recognition, *J. Cell Biol.*, 51 (1971) 536–547.
57. S. Bozzaro and S. Roseman, Adhesion of *Dictyostelium discoideum* cells by to carbohydrates immobilized in polyacrylamide gels. I. Evidence for three sugar specific cell surface receptors, *J. Biol. Chem.*, 258 (1983) 13882–13889.
58. S. Bozzaro and S. Roseman, Adhesion of *Dictyostelium discoideum* cells to carbohydrates immobilized in polyacrylamide gels. II. Effect of D-glucose derivatives on development, *J. Biol. Chem.*, 256 (1983) 13890–13897.
59. N. O. Keyhani, X.-B. Li, and S. Roseman, Chitin catabolism in the marine bacterium *Vibrio furnisii*. Identification and molecular cloning of a chitoporin, *J. Biol. Chem.*, 275 (2000) 33068–33076.

60. N. O. Keyhani, L.-X. Wang, Y. C. Lee, and S. Roseman, The chitin disaccharide, N, N'-diacetylchitobiose, is catabolized by *Escherichia coli* and is transported/phosphorylated by the phosphoenolpyruvate: Glycose phosphotransferase system, *J. Biol. Chem.*, 275 (2000) 33084–33090.
61. N. O. Keyhani, O. Boudker, and S. Roseman, Isolation and characterization of IIAChb, a soluble protein of the enzyme II complex required for the transport/phosphorylation of N, N'-diacetylchitobiose in *Escherichia coli*, *J. Biol. Chem.*, 275 (2000) 33091–33101.
62. N. O. Keyhani, K. Bacia, and S. Roseman, The transport/phosphorylation of N, N'-diacetylchitobiose in Escherichia coli. Characterization of phospho-IIChb and of a potential transition state analogue in the phosphotransfer reaction between the proteins IIAChb and II Chb*, *J. Biol. Chem.*, 275 (2000) 33102–33109.
63. X.-B. Li and S. Roseman, The Chitinolytic cascade in Vibrios is regulated by chitin oligosaccharides and a two-component chitin catabolic sensor/kinase, *Proc. Natl. Acad. Sci. U.S.A.*, 101 (2003) 627–631.
64. N. Kresge, R. Simoni, and R. l. Hill, Hexosamine metabolism, sialic acids, and the phosphotransferase systems: Saul Roseman's contribution to glycobiology, *J. Biol. Chem.*, 281 (2006) e1–e5 (electronic publication).

DE NOVO ASYMMETRIC SYNTHESIS OF THE PYRANOSES: FROM MONOSACCHARIDES TO OLIGOSACCHARIDES

ALHANOUF Z. ALJAHDALI, PEI SHI, YASHAN ZHONG, and GEORGE A. O'DOHERTY

Department of Chemistry and Chemical Biology, Northeastern University,
Boston, Massachusetts, USA

I. Introduction	57
1. Background	57
II. Masamune–Sharpless Approach to the Hexoses	58
III. Danishefsky Hetero-Diels–Alder Approach to Various Pyranoses	60
IV. MacMillan Iterative Aldol Approach to Various Pyranoses	67
V. Asymmetric Oxidative Biocatalytic Aldol Approach to Various Pyranoses	71
VI. Sharpless Dihydroxylation/Enzymatic Aldol Approach 2-Ketoses	73
VII. Non-*De Novo* Asymmetric Approaches to Pyranoses	74
1. Dondoni Thiazole Approach	74
2. Seeberger Approaches	76
3. Reissig Approaches	80
VIII. O'Doherty *De Novo* Approach to Pyranoses	80
1. *De Novo* Use of the Achmatowicz Approach to Pyranoses	83
2. Iterative Dihydroxylation of Dienoates	84
3. Palladium-Catalyzed Glycosylation	88
4. Applications to Synthesis and Medicinal Chemistry	89
Acknowledgments	115
References	115

ABBREVIATIONS

2-LTT, 2-lithiothiazole; 2-TST, 2-(trimethylsilyl)thiazole; AD-mix-α, a reagent system for asymmetric dihydroxylation of alkene; Bacillosamine, 2,4-diamino-2,4,6-trideoxy-D-glucose; Backval oxidation, oxidation of 1,3-dienes to 1,4-diacetoxyalk-2-enes; Boc$_2$O, di-*tert*-butyl dicarbonate; cat. Pd(PPh$_3$)$_2$, palladium bistriphenylphosphine (the catalytic intermediate in Pd(0/II) cross-coupling reactions); CbzCl, Benzyl chloroformate; CSA, Camphorsulfonic acid; Danishefsky diene, *trans*-1-methoxy-3-trimethylsilyloxy-1,3-butadiene; DBU, 1,8-diazabicycloundec-7-ene; DCC, N,N'-dicyclohexylcarbodiimide; DDQ, 2,3-dichloro-5,6-dicyano-1,4-benzoquinone; DEAD, diethyl azodicarboxylate; DHAP, 3-hydroxy-2-oxopropyl phosphate; DIBAL-H, diisobutylaluminum hydride; Diimide, the reactive intermediate from *o*-nitrophenylsulfonyl hydrazide/triethylamine; (+)-DIPT, (+)-diidopropyltartrate; DMAP, 4-dimethylaminopyridine; DMP/2,2-DMP, 2,2-dimethoxypropane; D-DKH, 2-amino-2,3,6-trideoxy-D-*xylo*-hexos-4-ulose; Evans aldol, an asymmetric aldol reaction that uses a chiral auxiliary; FDP adolose, an enzyme that catalyzes the aldol reaction; Felkin/Felkin–Ahn addition, a transition-state model for the stereoselective addition to carbonyl compounds; Fod, 6,6,7,7,8,8,8-heptafluoro-2,2-dimethyl-3,5-octanedionato; HBTU, O-(Benzotriazol-1-yl)-N,N,N',N'-tetramethyluronium hexafluorophosphate; Henry Rxn, a base-catalyzed addition for nitroalkanes to ketones and aldehydes; HWE olefination, Horner–Wadsworth–Emmons olefination reaction; IBX, 2-iodoxybenzoic-acid; Johnson–Lemieux reaction, oxidative cleavage of an alkene with $NaIO_4$–OsO_4; IPA, Isopropyl alcohol; Kdo, 3-deoxy-D-*manno*-oct-2-ulosonic acid; LAH, lithium aluminum hydride; Ley-spiroketal, a bis-spiroketal-containing protecting group for 1,2-diols; Luche, $NaBH_4$ reduction of ketones to alcohols with lanthanide chlorides; *m*-CPBA, *m*-chloroperoxybenzoic acid; MTT, dimethyldiphenylthiazolyltetrazolium salt, a type of tetrazole; Mukaiyama aldol, a Lewis acid-catalyzed aldol; Myers 1,3-reductive rearrangement, a 1,3-reductive rearrangement of allylic alcohols to alkenes; Nap, methylnaphthyl; NBS, *N*-bromosuccinimide; NBSH, 2-nitrobenzenesulfonyl-hydrazide; NIS, *N*-Iodosuccinimide; NMM, *N*-methylmorpholine; NMO, *N*-methylmorpholine *N*-oxide; (R,R)-Noyori, (R)-Ru(η^6-mesitylene)-(R,R)-TsDPEN; Payne rearrangement, a based-catalyzed rearrangement of epoxy allylic alcohols; $Pd_2(dba)_3$, Tris(dibenzylideneacetone)dipalladium(0); Petersen olefination, the conversion of aldehydes and ketones to alkenes via a beta-hydroxy silane; PMB, *p*-methoxybenzyl; PMBOH, *p*-methoxybenzyl alcohol; Pummerer reaction, conversion of an alkyl sulfoxide into an α-acyloxythioether; RedAl, sodium bis(2-methoxyethoxy) aluminum hydride; RSK, ribosomal s6 kinase; Rubottom oxidation, oxidation of silylenol ethers to α-hydroxyketones with *m*-chloroperoxybenzoic acid; SAR, structure–activity relationship; Saegusa oxidation, oxidation of an enol ether to α,β-unsaturated ketone with $Pd(OAc)_2$; Sharpless asymmetric epoxidation (SAE), condition for the asymmetric epoxidation of allylic alcohols; TBAF, tetra-*n*-butylammonium fluoride; TBDPSCl, *tert*-Butyl(chloro) diphenyl silane; TBS, *tert*-butyldimethylsilyl; TBSCI, *tert*-butyldimethylsilyl chloride; TFA, trifluoroacetic acid; THF, tetrahydrofuran; TIPSCl, triisopropylsily chloride; TMEDA, tetramethylethylenediamine; TMS, trimethylsilyl; TMSOTF, trimethylsilyl trifluoromethanesulfonate; Upjohn dihydroxylation, oxidation of alkenes to vicinal diols with catalytic OsO_4 and NMO

I. INTRODUCTION

1. Background

The synthesis of carbohydrates has held a key niche in the realm of organic synthesis for as long as the discipline of organic chemistry has been in existence. At one point in the evolution of carbohydrate chemistry, the synthesis of carbohydrates has diverged from what has become considered as "traditional" organic synthesis. It may be wondered whether Emil Fischer would have recognized these boundaries. Nonetheless, these borders have become more closely defined as the field of carbohydrate synthesis has begun to tackle increasingly complex targets associated with carbohydrate chemistry and biology. In contrast, the broader synthetic organic community has remained dominated by the synthesis of "traditional" natural products. Of course, carbohydrates are themselves natural products and this differentiation is obviously incorrect.

From a practical point of view, this schism can be seen in the choice of starting materials. For instance, the synthetic carbohydrate community has long seen the monosaccharide sugars as the obvious starting materials for the practical synthesis of more-complex carbohydrates. The logic behind this preference is most readily apparent when it comes to the synthesis of oligosaccharides, where the enantiomeric purity of the monosaccharide building blocks is of ultimate importance. In contrast, in the synthesis of "traditional" natural products, the monosaccharides are viewed as one of many sources of chiral starting materials for synthesis.

Interestingly, when it comes to the synthesis of the monosaccharide sugars, the two worlds converge again. Dating back to the seminal work of Fischer and of Kiliani, the synthetic interconversions of sugars has played an integral role in determining their stereochemistry.[1,2] In later days, the monosaccharides have served as stereochemically rich molecules that can be used to test both the diastereoselectivity and enantioselectivity of a reaction. For instance, the sugars have proved to be ideal models for exploring the balance between substrate and reagent control in diastereoselective reactions. As the power of these reactions develops, the synthetic complexity for which they can be used increases commensurately.

As the synthetic utility of these methods has advanced, the goal of these efforts has evolved from the development of basic synthetic methods to addressing important needs of medicinal chemistry. Thus much interest now comes from the medicinal-chemistry community, as these new routes often provides access to unnatural sugars that could be of use in structure–activity relationship (SAR) studies. In addition, the synthesis of monosaccharides, and in particular, hexoses, has served as a synthetic challenge and a measuring stick to the synthetic organic community.

Of particular interest are those routes to hexoses that start from achiral starting materials where asymmetric catalysis is used to install the stereocenters. These routes to sugars are described as "*de novo*" or "*de novo* asymmetric" routes in the synthetic organic community. It should be noted that in the carbohydrate synthesis community the term *de novo* has taken on very different meanings. While these approaches are empowered by asymmetric catalysis, which is the new tool of modern asymmetric synthesis, the approaches are founded in the fundamental principles of carbohydrate chemistry dating back to the earliest work of Fischer.[1,2] These seminal discoveries can be seen in various review articles in accounts over the years, for example, the 1946 *Advances* review by Lespieau, which describes the early work toward building up of the polyol portion of the hexose and pentose polyenols,[3] the 1974 *Advances* survey by Mizuno and Weiss, which documents the "formose" sugars (C_2 to C_8 sugars) produced from formaldehyde,[4] and the 1977 *Advances* review by Černý and Staněk, which outlines the synthesis and chemistry of anhydro sugars.[5] Finally, this chapter builds upon the 1982 contribution from Zamojski, Banaszek, and Grynkiewicz, which comprehensively reviews the then-known (as of 1982) synthetic work toward sugars from noncarbohydrate substrates; this coincidently marks the beginning of the practical use of asymmetric catalysis in total synthesis.[6] For the purposes of this chapter, the term *de novo* asymmetric synthesis refers to the use of catalysis for the asymmetric synthesis of carbohydrates from achiral compounds.[7–9]

II. Masamune–Sharpless Approach to the Hexoses

The challenge of a *de novo* synthetic approach to carbohydrates has been addressed by several groups during recent decades. Of these approaches, only the iterative epoxidation strategy of Masamune and Sharpless provides access to all eight hexoses. Their work utilized iterative asymmetric epoxidation of allylic alcohols to obtain the eight possible L-hexoses. The route is noteworthy in that it is the first approach of its kind since the work of Fischer. Because both the Fischer approach and that of Sharpless and Masamune set out to synthesize all of the hexoses, their routes have also been the least practical (that is, requiring the most steps and protecting groups). Decreasing the number of steps and the avoidance of protecting groups has guided the development of alternative synthetic endeavors (vide infra).[10,11]

The route to establish the four C-2 to C-5 stereocenters of a given hexose involves two net antiadditions of two hydroxyl groups across two alkenes. To accomplish this, the antiaddition is accompanied by an epoxidation/ring-opening reaction sequence, where the ring opening occurs in a Payne-type rearrangement. Stereochemical variability is installed by the incorporation of reagent-controlled epoxidation and *cis*- to *trans*-epimerization of an isopropylidene acetal.

The Masamune–Sharpless synthesis begins with the Sharpless asymmetric epoxidation of (*E*)-4-diphenylmethoxybut-2-en-1-ol (**2**), with the (+)-DIPT ligand system (SAE: Sharpless asymmetric epoxidation) to give epoxide **3**. The first of the two key Payne rearrangements then gives the diol **4** (Scheme 1). In this reaction, NaOH is used to equilibrate epoxides **3** and diol **4** as well as generate NaSPh, which selectively reacts with epoxide **8** in an S_N2 reaction to give diol **4**, the product of net 1,2-*anti*-dihydroxylation and sulfide substitution of the precursor **2**. The resulting *anti*-diol was then acetonated, the sulfide selectively oxidized to a sulfoxide with *m*-CPBA, and a subsequent Pummerer rearrangement transferred that oxidation state to the carbon atom, as in **7**. In this process, the requisite stereochemistry has been installed in compound **7**, which in turn is ready for further manipulation at the sulfide α- and β-carbons.[12,13]

Thus, the four-carbon backbone in compound **1** corresponds to C-3 to C-6 of the eight L-hexoses, while C-3 in **1** specifically corresponds to C-5 in the L-hexoses. Compound **7** can be synthesized with an overall yield of 93% through a three-step

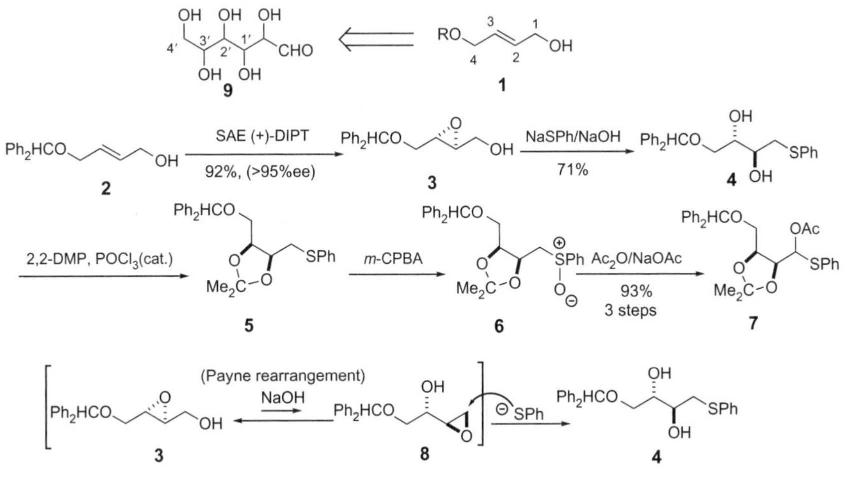

SCHEME 1. Synthesis of a key tetrose fragment.

sequence: protection, oxidation of the thioether to the sulfoxide, and subsequent Pummerer reaction.

With the two stereocenters in epoxide **7** established, the reaction sequence progresses in a stereodivergent fashion. In the first direction, it begins with the hydrolysis of **7** using DIBAL-H as the nucleophile to give aldehyde **10** (L-*erythro*) without any epimerization (91%) (Scheme 2). In contrast, treatment of **7** with K_2CO_3/MeOH combines a C-2 epimerization with the hydrolysis to form **11** having L-*threo* stereochemistry. A two-carbon extension of **10** and **11** is achieved through a Wittig reaction followed by reduction of an intermediate aldehyde to furnish the corresponding allylic alcohols **12** and **13**. The asymmetric epoxidation procedure is used once again to generate four new diastereomeric epoxides **14–17**, all with high selectivity (>20:1). Importantly, these epoxidations are reagent-controlled reactions. Thus, the stereoselectivity comes from the titanium tartrate reagent and not from the preexisting stereocenters in allylic alcohols **12** and **13**. Once again, subjecting these four diastereoisomers to the NaSPh/Payne rearrangement conditions affords, after acetonation, the sulfides **18**, **19**, **20**, and **21**.

With the four required stereocenters installed, the four diastereomeric sulfides **18**, **19**, **20**, and **21** are converted, respectively, into L-allose, L-altrose, L-mannose, L-glucose, L-gulose, L-idose, L-talose, and L-galactose in a sequence analogous to the previous one (Scheme 3). For example, with compound **18**, the process begins with oxidation of the sulfide, followed by a Pummerer rearrangement to gives **22** (90%). Once again, at compound **22**, the route can diverge into two diastereomeric directions. The first route uses DIBAL-H to form aldehyde **23** (81%) directly, and this, after acid-catalyzed deacetonation (TFA) and hydrogenolysis (H_2, Pd/C) of the benzylidene groups, gave L-allose (**25**). In an alternative route, L-altrose (**26**) can also be produced from **22**. The alternative route employs potassium carbonate in methanol to induce a C-2 epimerization, and after hydrolysis, it furnished compound **24** (48%). Then, subjecting aldehyde **23** under the same reaction conditions to hydrolyze the protecting groups gives L-allose (**25**) and L-altrose (**26**). Correspondingly, L-mannose (**27**) and L-glucose (**28**) are synthesized from **19**, L-gulose (**29**) and L-idose (**30**) arise from **20**, while **21** gives L-talose (**31**) and L-galactose (**32**).

III. Danishefsky Hetereo-Diels–Alder Approach to Various Pyranoses

Concurrent with the development of the Sharpless epoxidation chemistry, Danishefsky was demonstrating the utility of 1,3-dialkoxydienes (commonly termed "Danishefsky dienes") in Diels–Alder cycloaddition chemistry for synthesis.

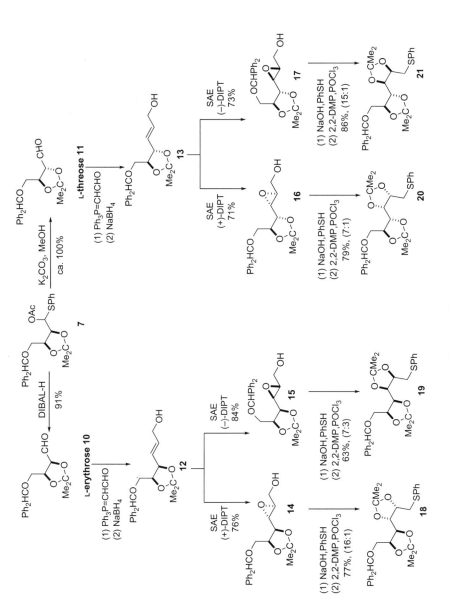

SCHEME 2. Synthesis of four hexose precursors.

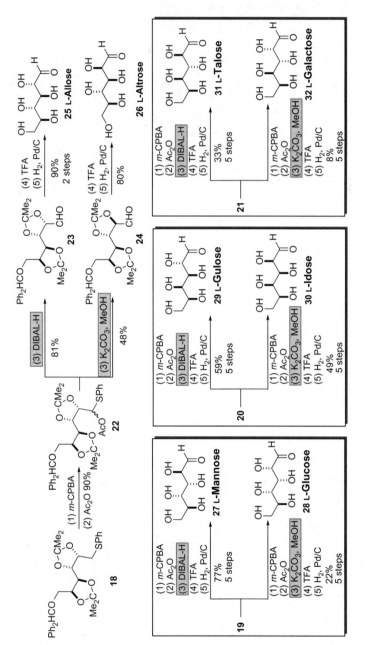

SCHEME 3. Synthesis of the eight L-hexoses.

One particularly powerful application is their use in hetero-Diels–Alder reactions for the synthesis of several glycals (Scheme 4). This in turn led to the development of a flexible asymmetric approach to various hexoses. The success of these approaches eventually inspired further studies by the Danishefsky group toward oligosaccharide synthesis, although the glycals used in these oligosaccharide assemblies are mostly derived from chiral carbohydrate starting materials.[14,15]

This approach begins with the synthesis of either a 1-alkoxy-3-silyloxydiene (**33**) or a L-methoxy-3-((trimethylsilyl)oxy)-1,3-butadiene **38** (also termed "Danishefsky dienes"), which in turn were allowed to react with aldehydes in a Lewis acid-catalyzed hetereo-Diels–Alder cycloaddition/elimination to give the *cis*- and *trans*-dihydropyrans **34** and **39**. After the cycloaddition with diene **38**, a Mn(OAc)$_3$-promoted α-acetoxylation was used to install the C-4 acetoxy group with *trans*-stereochemistry. Typical Lewis acids used were BF$_3$, ZnCl$_2$, and Yb(fod)$_3$, where the fod ligands serve as chiral ligands that worked well in combination with a chiral auxiliary on the diene. A Luche reduction of the enone functionality installed the C-3 equatorial stereochemistry, via an axial addition of hydrogen. Dihydroxylation of the double bond in the glycal provides access to four hexopyranosides with *talo*-, *galacto*-, *manno*-, or *gluco*-stereochemistry. This was accomplished via an epoxidation/ring opening or dihydroxylation procedure. The hydroxyl-directed epoxidation of **35a** gave the *talo*-sugar derivative **36**. Acylation of **35a** to give **35b**, followed by reaction with *m*-CPBA, afforded diastereomer **37** having the *galacto* stereochemistry. A similar hydroxyl-directed epoxidation of **40a** gave the mannopyranose derivative **41**, whereas dihydroxylation of **40b** with osmium tetraoxide gave the *gluco* diastereoisomer **42**.[16]

A particularly powerful application of this hetero-Diels–Alder approach to pyranoses is in the synthesis of rare sugars. Excellent examples of this can be seen in Danishefsky's synthesis of the Kdo framework and the sugar component of the antifungal agent papulacandin (Schemes 5 and 6).[17] The approach to Kdo begins with the synthesis of the trialkoxydiene **45** (as a mixture of *E,Z*-isomers) from 2-acetylfuran (**43**). A BF$_3$-catalyzed cycloaddition of **45** and aldehyde **47** (as a chiral acrolein equivalent), followed by TFA-promoted elimination of methanol, Luche reduction, and formation of a TMS ether provided the glycal derivative **48**. Alcoholysis of the double bond in the glycal with benzyl alcohol, followed by selenide oxidation/elimination, affords the intermediate **49**. A diastereoselective dihydroxylation is then used to install the C-7/8 hydroxyl groups to give, after acylation, compound **50**. Finally, the electron-rich furan ring can be oxidatively cleaved to a carboxylic acid, which is then esterified with diazomethane. The anomeric hydroxyl group was removed under hydrogenolysis conditions, followed by acylation to give the Kdo derivative **51**.

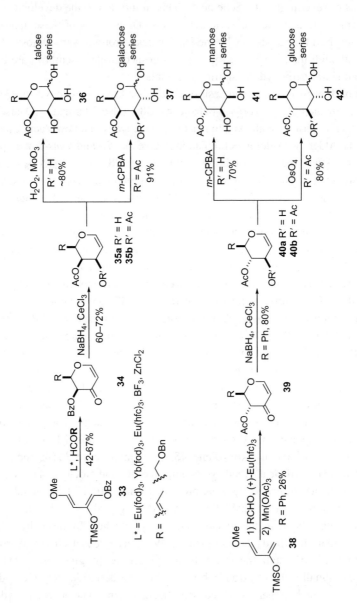

SCHEME 4. Asymmetric stoichiometric chiral auxiliary.

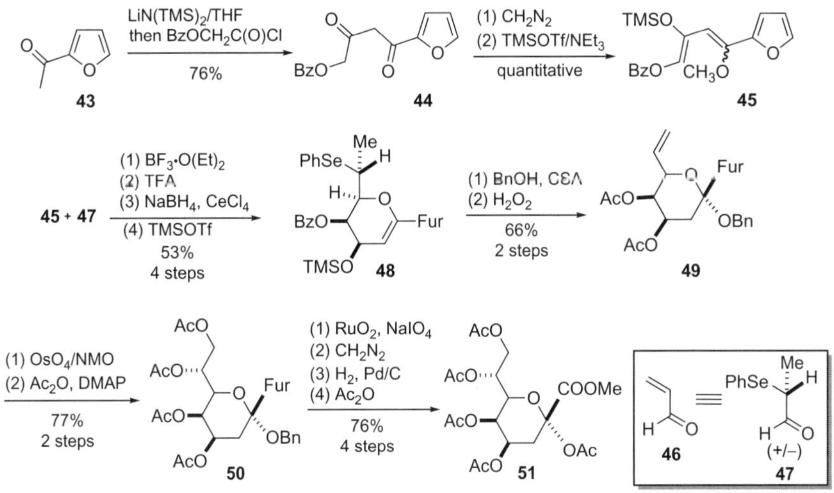

SCHEME 5. Synthesis of a (+/−)-3-deoxy-*manno*-oct-2-ulopyranosonate (Kdo) derivative.

The Danishefsky approach to the papulacandin spiroketal ring system begins with a Yb(fod)$_3$-catalyzed hetero-Diels–Alder reaction between diene **38** and aldehyde **52** to give, after loss of methanol, the glycal derivative **53**. The C-6 carbon atom can be diastereoselectively introduced by a vinyl-cuprate addition to give compound **54**. The vinyl group in **54** is then converted into a benzoate-protected hydroxymethyl group by a three-step Johnson–Lemieux oxidative cleavage, aldehyde reduction, and a benzoylation sequence to provide the intermediate **55**. Sequential enol ether generation, Rubottom oxidation, and benzoylation provided compound **56** having the C-4 stereocenter in place. Subsequent generation of an enol ether and Saegusa oxidation provided enone **57**, which was reduced with DIBAL-H and acylated to give glycal derivative **58**. Epoxidation and ring opening in methanol gave the *gluco*-configured sugar derivative **59**, which upon successive benzoylation, spiroketal formation, debenzylation, and peracylation gave the papulacandin ring system shown in **61**.[18]

While these approaches to various sugars by the Danishefsky group always had the potential to constitute *de novo* asymmetric syntheses, in practice, they were actually only racemic synthesis, as the Danishefsky hetero-Diels–Alder reaction was inevitably a difficult one to make asymmetric on a practical scale, as it would routinely require both a chiral catalyst and a chiral auxiliary. This all changed in 1999 with the discovery by Jacobsen that related chromium(III) complexes could effectively catalyze the Danishefsky hetero-Diels–Alder/elimination reaction (**38** to **63**) in excellent

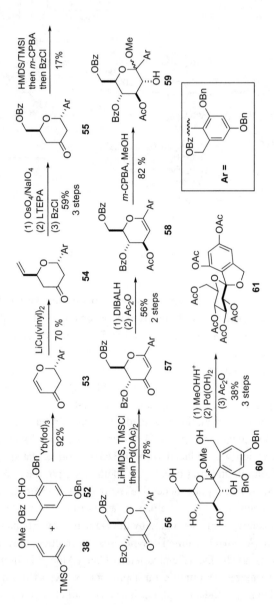

SCHEME 6. Synthesis of the ring system of papulacandin D.

yields and enantiomeric excesses (all-*cis* configuration) (Scheme 7). This work by Jacobsen and others transforms the early synthetic approach by Danishefsky into formal *de novo* asymmetric syntheses.[19]

IV. MacMillan Iterative Aldol Approach to Various Pyranoses

The challenge of developing a practical *de novo* asymmetric synthesis of the hexoses has more recently been taken up by MacMillan, who along with others had been studying the use of chiral amines as catalysts for the aldol reaction. Ultimately, his group developed an iterative aldol approach (a proline-catalyzed aldol reaction followed by a subsequent diastereoselective aldol reaction) to produce various hexoses (**64** to **70**) (Scheme 8). In this regard, they found conditions wherein L-proline can be used for catalytic dimerization of aldehydes. The reaction, an enamine–aldol process, occurs via the intermediacy of an enamine to give aldol products having *threo* stereochemistry (such as **67** from two molecules of aldehyde **64**, via

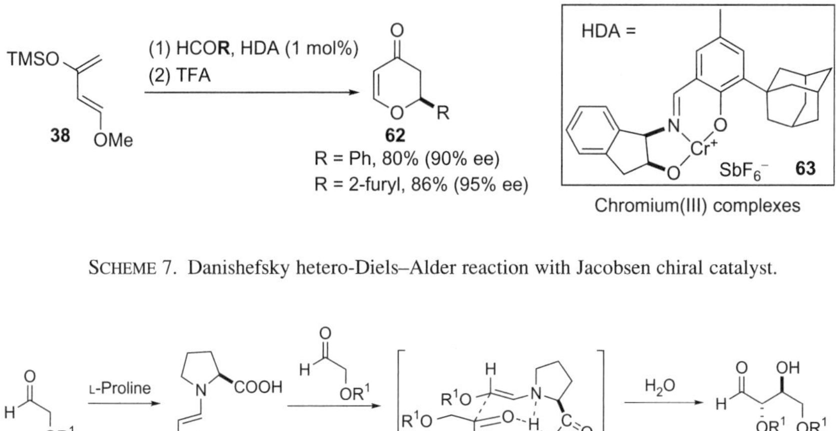

SCHEME 7. Danishefsky hetero-Diels–Alder reaction with Jacobsen chiral catalyst.

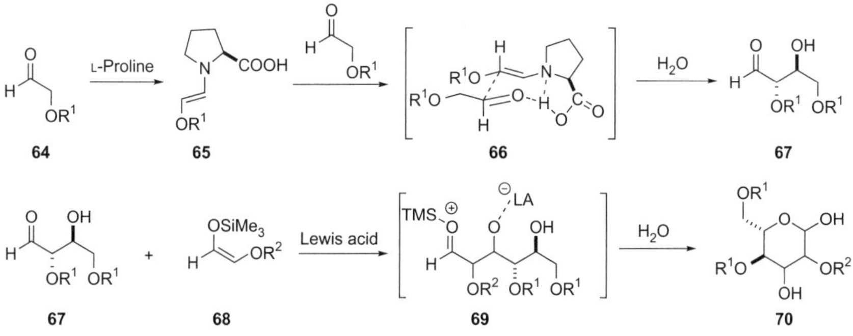

SCHEME 8. Two-step aldol reactions to synthesize sugars.

compound **66**). Key to the success of this approach is the low reactivity of the aldol product **67** toward enamine **65** in comparison to the relatively reactive aldehyde **64**. Thus, a second Mukaiyama-type Lewis acid-catalyzed aldol reaction (**67**+**68**) is required to convert the C_4 sugar derivative **67** into the hexopyranose **70**. Because various aldol reaction mechanisms are accessible depending on the Lewis acid used, various C-2/3 hexose diastereomers can be accessed.

In practice, various protected α-hydroxy aldehydes (**72a–d**) can be produced in two steps from (Z)-2-buten-1,4-diol (**71**) (Scheme 9). Conversion into **72** and exposure to catalytic amounts (10 mol%) of L-proline produced the anti-aldol products **73** in excellent yield and enantiomeric purity. The required silylenol ether **68** for the second aldol coupling can also be produced from diol **71** via the protected α-hydroxyaldehyde **72d**. Depending on the conditions and the Lewis acid used, compounds **68** and **73c** can be combined diastereoselectively to give three hexose aldol products. Thus, using the β-hydroxyaldehyde-chelating Lewis acid $TiCl_4$, the Felkin aldol product **76** having *allo* stereochemistry can be produced (Scheme 10). Switching the Lewis acid to $MgBr_2$ changes the reaction to an anti-Felkin aldol reaction where, when Et_2O is used as the solvent, the *gluco*-hexose product **74** is formed, whereas when CH_2Cl_2 is used as solvent, the *manno* hexose product **75** is produced. Thus, three different hexose diastereomers can be selectively produced in four linear steps (7 total steps). The relatively large amount of proline used in the aldol reactions is not a drawback, as L-proline at least is a readily available natural product.[20]

The utility of the MacMillan iterative aldol approach can be seen by its application in the synthesis of several sugar-containing natural products. For comparative purposes, this is probably best exemplified by Chandrasekhar's synthesis of the

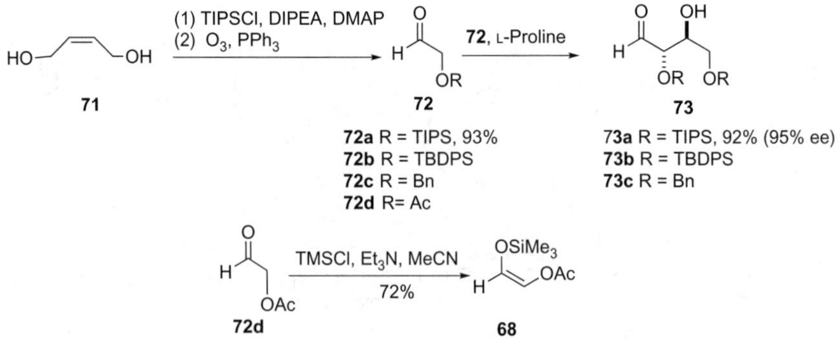

SCHEME 9. Synthetic intermediates **73** and **68**.

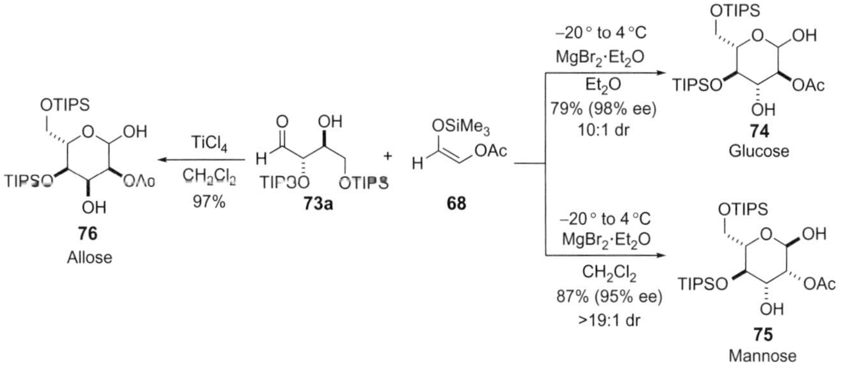

SCHEME 10. Synthesis of glucose, mannose, and allose derivatives.

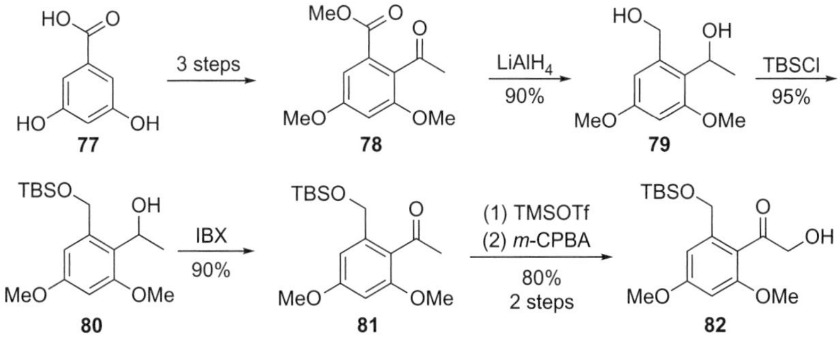

SCHEME 11. Synthesis of a hydroxyacetophenone derivative.

papulacandin ring system. The Chandrasekhar route is designed to be stereodivergent, as it targets three different stereoisomers of the papulacandin ring system.

The route begins with the synthesis of the required substituted 3,5-dihydroxyacetophenones **81** and **82** (Scheme 11).[21,22] Thus, in three steps, the keto ester **78** was prepared from 3,5-dihydroxybenzoic acid (**77**). Concurrent reduction of the ester and keto groups using LiAH$_4$ gave diol **79**. Selective TBS protection, followed by oxidation with 2-iodoxybenzoic acid, led to the acetophenone derivative **81**. Rubottom oxidation converted **81** into the hydroxyacetophenone derivative **82** for further aldol reaction.

With the required enolate precursor in hand, compound **82** could be coupled with the D-proline-mediated *erythro* aldol dimer **83** (Scheme 12). Thus exposing **82** and **83** to the titanium-promoted enolate aldol conditions developed by MacMillan gave **84**,

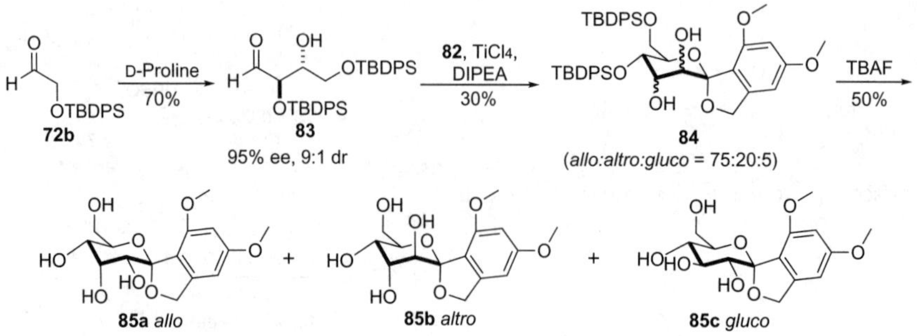

SCHEME 12. Synthesis of three stereoisomers of the papulacandin sugar framework.

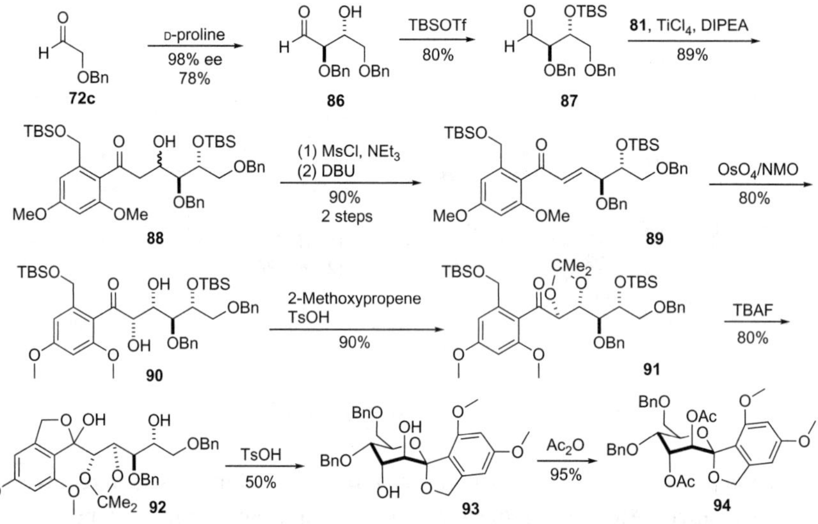

SCHEME 13. Synthesis of the *altro*-papulacandin.

the product of tandem aldol and spiroketal formation, as a mixture of hexose diastereomers. Desilylation of compound **84** with an excess of tetrabutylammonium fluoride gave a 15:4:1 mixture of *allo*-, *altro*-, and *gluco* isomers of the papulacandin core sugar.[23,24] While this route produced the papulacandin ring system in a relatively short route, it suffered from poor stereocontrol.

In a second-generation approach, Chandrasekhar targeted a stereoselective synthesis of the *altro* diasteromer of the papulacandin structure (Scheme 13). The revised route begins with the proline-catalyzed dimerization of benzyloxyacetaldehyde (**72c**) to form the C_4 aldehyde **86**, and TBS protection of the β-hydroxyl group gave **87**. Once again, a titanium-mediated aldol reaction between **81** and **87** afforded compound **88**, which upon dehydration gave enone derivative **89**. Substrate-controlled diastereoselective (98:2 *anti/syn*) dihydroxylation of **89** gave **90**, acetonation of which gave **91** and subsequent desilylation gave **92**. Acid-catalyzed deacetonation and ketal formation gave the *altro* stereoisomer **93** of papulacandin, which upon acetylation gave the product **94**.

V. Asymmetric Oxidative Biocatalytic Aldol Approach to Various Pyranoses

An alternative *de novo* asymmetric approach to the hexoses is one that uses biocatalysis instead of traditional catalysis. Examples of this can be seen with Hudlicky's use of genetically engineered microorganisms, the use of lipases by Johnson for the desymmetrization of meso-diols, and the Sharpless–Wong iterative asymmetric dihydroxylation/enzyme aldol reaction approach to hexoses (Schemes 14–18).

Possibly one of the best examples of the simplicity of this approach is the Hudlicky and Johnson use of microorganisms for the asymmetric oxidation of substituted benzenes. Hudlicky first reported the application of an enzyme-catalyzed general protocol to the synthesis of mannose derivatives.[25] Chlorobenzene (**95**) was subjected to whole-cell oxidation by *Pseudomonas putida* 39/D[26,27] or *Escherichia coli* JM109 (pDTG601).[28] In this example, chlorobenzene can be oxidized to form diol **96** in excellent enantiomeric selectivity. Acetonation followed by diastereoselective dihydroxylation affords **97**. Finally, ozonolysis of the chloroalkene in methanol to the aldehyde/methyl ester, followed by reduction with $NaBH_4$ and peracetylation, provided compound **98**, albeit in diminished overall yield.[29]

In a related organism-free approach, Johnson demonstrated the power of an enzymatic desymmetrization approach to both enantiomers of 2,4-dideoxy-*erythro*-hexoses

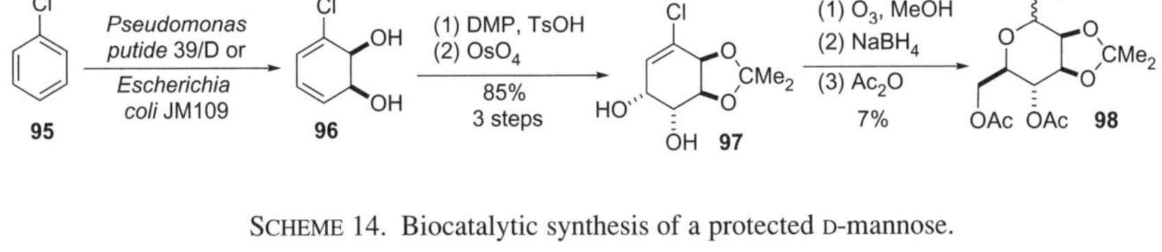

SCHEME 14. Biocatalytic synthesis of a protected D-mannose.

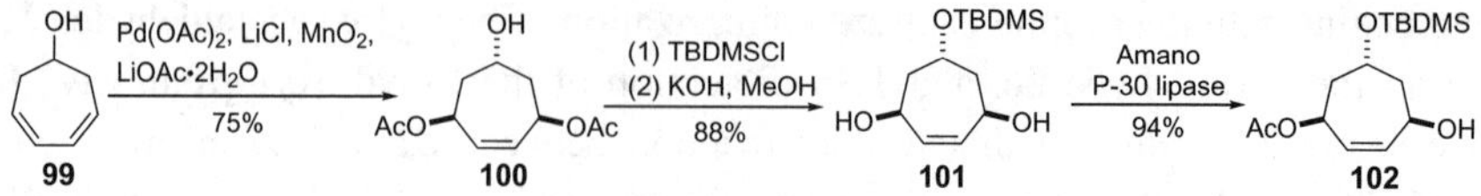

SCHEME 15. Synthesis of enantiomerically pure monoacetate **102**.

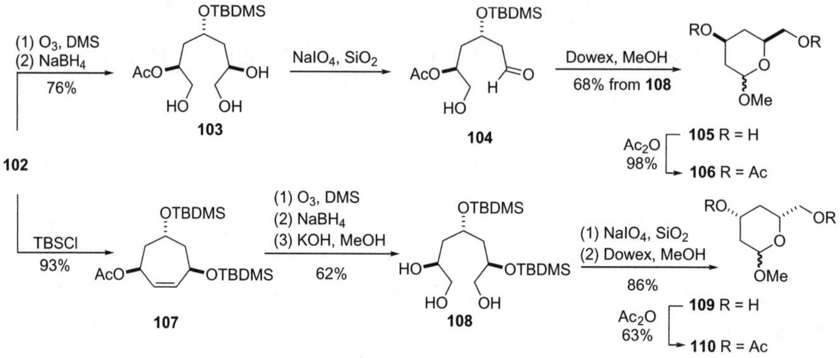

SCHEME 16. Synthesis of 2,4-dideoxyhexoses.

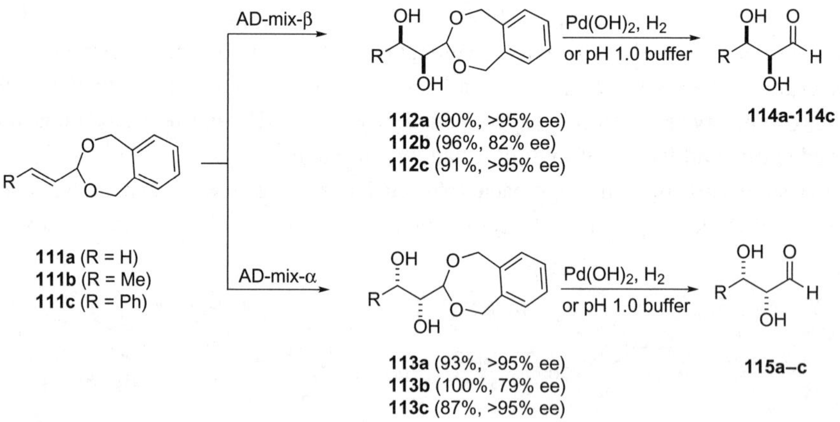

SCHEME 17. Synthesis intermediates **114** and **115**.

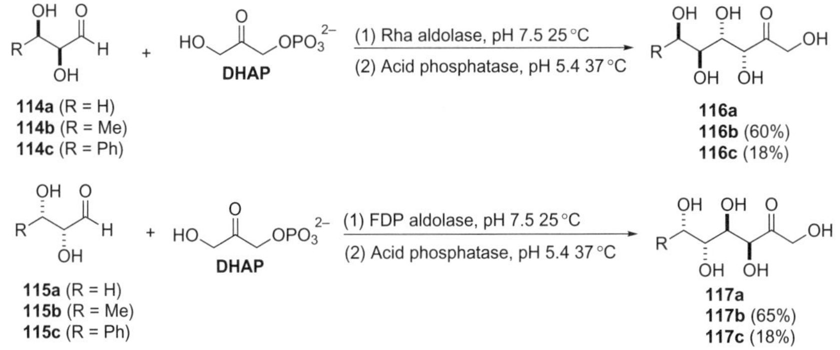

DE NOVO ASYMMETRIC SYNTHESIS OF THE PYRANOSES

SCHEME 18. Synthesis of sugar derivatives **116** and **117**.

(Schemes 15 and 16). The Johnson approach begins with the diastereoselective synthesis of the C_S symmetric triol diacetate **100** via a highly stereoselective palladium-catalyzed Backval oxidation of diene **99**.[30] Protecting-group adjustments converted **100** into the C_S symmetric diol **101**. Enzymatic desymmetrization of **101** with Amano P-30 lipase in isopropenyl acetate for 5 days produces acetate **102**.

This enantio-divergent route starts with the ozonolysis of **102** to form triol **103**. Periodate cleavage of **103** to hydroxyaldehyde **104**, which in acidic methanol cyclizes to give 2,4-dideoxy-D-*erythro*-hexopyranose (**105**) and, upon peracetylation, **106**, as 5:1 mixtures of α and β anomers. The enantiomeric sugar **110** can be prepared in a similar route from **102**. The alternative route begins with TBS protection of **102** to give **107**. The alkene in **107** is then ozonolyzed and reduced, followed by hydrolysis of the acetate group to provide triol **108**. Periodate cleavage of **108** followed by treatment with acidic methanol forms 2,4-dideoxy-L-*erythro*-hexopyranose derivatives **109** and **110** upon peracetylation.

VI. SHARPLESS DIHYDROXYLATION/ENZYMATIC ALDOL APPROACH 2-KETOSES

A decade later, Sharpless returned his attention to the *de novo* synthesis of sugars. In this second-generation effort, Sharpless teamed up with his colleague Wong. This alternative approach combines the use of the subsequently developed Sharpless asymmetric alkene dihydroxylation with Wong's expertise in enzyme-catalyzed reactions in organic synthesis. In particular, this employs enzyme-catalyzed aldol reactions for the synthesis of several 2-keto-hexoses.[31]

Key to the success of this approach is the use of a 1,5-dihydrobenzo[1,3]dioxepine group as an aldehyde-protecting group that is removable under hydrogenolysis conditions. The enantio-divergent route begins with the asymmetric dihydroxylation of alkenes **111a–c** with either the AD-mix-β or the AD-mix-α reagent system to give the diol **112a–c** or **113a–c**, respectively. The key hydrogenolytic, acetal-protection reaction is used to generate aqueous solutions of aldehydes **114a–c** and **115a–c**. This acetal deprotection strategy is important, as aldehydes **114a** and **115a** were not stable under conditions of simple acid hydrolysis.

With the glyceraldehyde derivatives in hand, asymmetric aldol reactions between these aldehydes and DHAP are catalyzed under Rha or FDP aldolases to produce sugar derivatives **116a–c** and **117a–c** without difficulty. The two aldolases used in this case are specific for DHAP as the nucleophile, while the aldehyde substrate can be changed.

VII. NON-*DE NOVO* ASYMMETRIC APPROACHES TO PYRANOSES

For the purposes of this chapter, the term *de novo* asymmetric synthesis refers to the use of catalysis for the asymmetric synthesis of carbohydrates from achiral precursors. This term thus precludes consideration of those *de novo* processes that produce sugars from molecules having preexisting chiral centers. To highlight these differences, we present here the work of the groups of Dondoni, Seeberger, and Reissig.[32–36] These efforts are nicely representative, as they constitute three different approaches with divergent aims.

1. Dondoni Thiazole Approach

The Dondoni thiazole approach to the hexopyranoses is reminiscent of the original work of Fischer, in that it homologates simpler sugars to more-complex ones. The major differences it has with the Fischer approach is the use of thiazole anions instead of cyanide and the focus on diastereoselectivity. An example of this approach is seen in the conversion of D-arabinose **118** into *N*-acetyl-D-mannosamine (**123**) (Scheme 19).

The route begins with a bis-acetonation of D-arabinose (**118**) with 2,2-dimethoxypropane to provide the free aldehyde **119**, which is then condensed with *N*-benzylhydroxylamine to form the *N*-benzylnitrone **120**. A highly diastereoselective addition of 2-lithiothiazole to **120** proceeds with antiselectivity to form the *N*-benzylhydroxylamine derivative **121**. This hydroxyamine can then be reduced

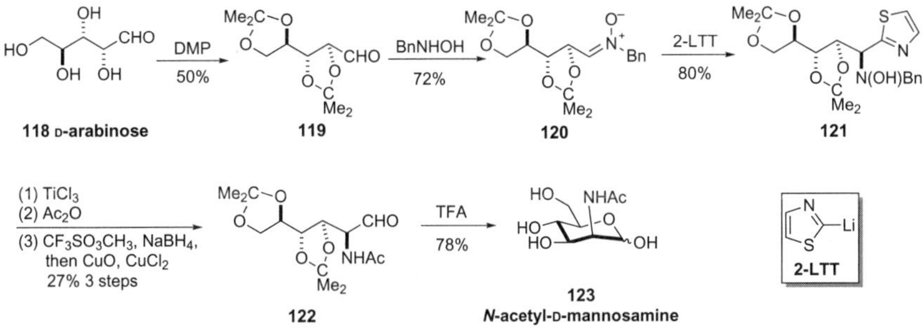

SCHEME 19. Dondoni's synthesis of *N*-acetyl-D-mannosamine.

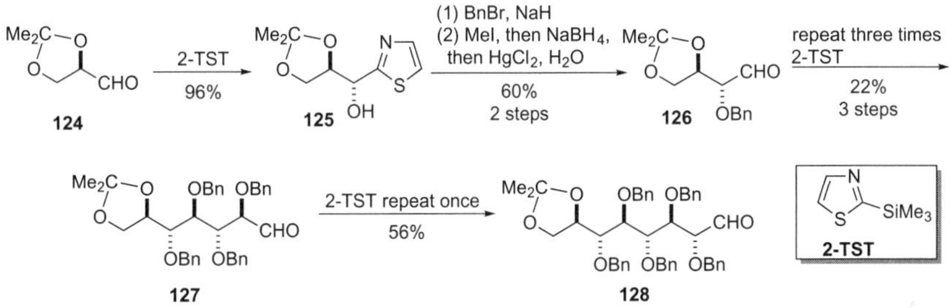

SCHEME 20. Homologation of D-glyceraldehyde to higher sugars.

with TiCl$_3$ and acetylated. The thiazole is then reduced and the product hydrolyzed to form aldehyde **122**. Deacetonation of the latter with trifluoroacetic acid then gives the desired *N*-acetyl-D-mannosamine (**123**) in good overall yield.[37]

Possibly the closest relationship between Dondoni's work and that of Fischer is Dondoni's 1985 report of the first thiazole-based synthesis of sugars of various chain lengths from 2,3-*O*-isopropylidene-D-glyceraldehyde (**124**) (Scheme 20). The synthesis begins with the diastereoselective addition of 2-(trimethylsilyl)thiazole (2-TST) to **124**. Alcohol protection as a benzyl ether followed by reductive hydrolysis of the thiazole gives the homologated tetrose derivative **126**. Repeating this two-step sequence three more times affords the heptose derivative **127**. One more iteration of the Dondoni protocol converts **127** into the octose derivative **128**. Thus, from the protected C$_3$ *aldehydo* sugar **124**, the same coupling–elaboration–unmasking sequence can be repeated to give a series of one-carbon higher homologues all the way up to the octose derivative **128**.[38]

The Dondoni homologation strategy was used most effectively for synthesis of the naturally occurring C_7 and C_8 sugars lincosamine (**133**) and destomic acid (**136**) (Scheme 21). Both routes begin with 1,2:3,4-di-*O*-isopropylidene-α-D-*galacto*-hexodialdopyranose (**129**). The aldehyde **129** is first condensed with a hydroxyamine to form the *N*-benzylnitrone **130**. A chelated and a nonchelated addition of 2-lithiothiazole to nitrone **130** provides the two required diastereomeric thiazole adducts **131** and **134**. The hydroxyamine group of **131** can then be successively reduced with $TiCl_3$, acetylated, the thiazole reduced, and then hydrolyzed to form aldehyde **132**, which can be converted into lincosamine (**133**). In a similar manner, the hydroxyamine group of **134** can be reduced by $TiCl_3$, the amine benzyloxycarbonylated, the thiazole reduced, and the product hydrolyzed to form alcohol **135**, which can be converted into destomic acid (**136**).[39]

2. Seeberger Approaches

Seeberger's group has also been interested in the *de novo* synthesis of sugars from chiral starting materials (Schemes 22–27). This includes the synthesis of deoxy and aminodeoxy sugars, as exemplified in syntheses of the aminodeoxy sugars 2,4-diamino-2,4,6-trideoxy-D-glucose (D-bacillosamine, **145**) and 2-amino-2,6-dideoxy-D-*xylo*-hexopyranos-4-ulose ("DKH," **146**), as well as related diastereomeric amino sugars (Schemes 22 and 23). The route starts with the commercially available L-Garner aldehyde **137**. Chelation-controlled addition of propynylmagnesium bromide converts aldehyde **137** into the alkyne **138**. An *E*-alkene-selective reduction with RedAl, along with protection by a 2-naphthylmethyl group, affords **139**. Deacetonation gives the primary alcohol **140** which after Dess–Martin oxidation forms aldehyde **141a**. Simple exchange of a benzyl group for the naphthyl group affords

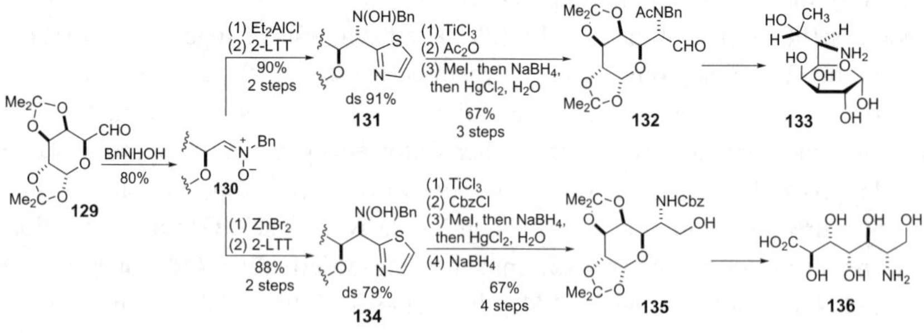

SCHEME 21. Synthesis of destomic acid and lincosamine.

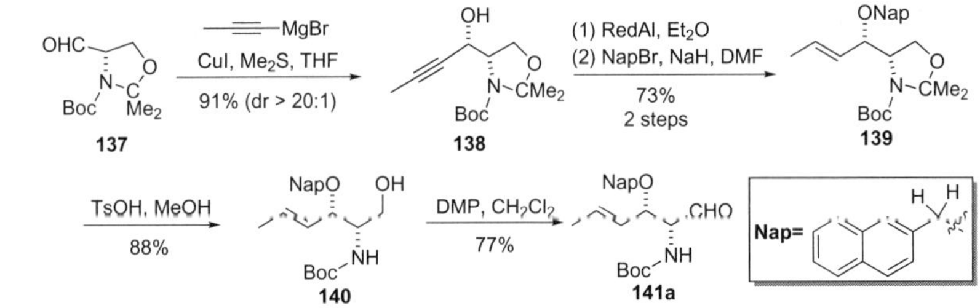

SCHEME 22. Synthesis of precursor **141a** for a monosaccharide building block.

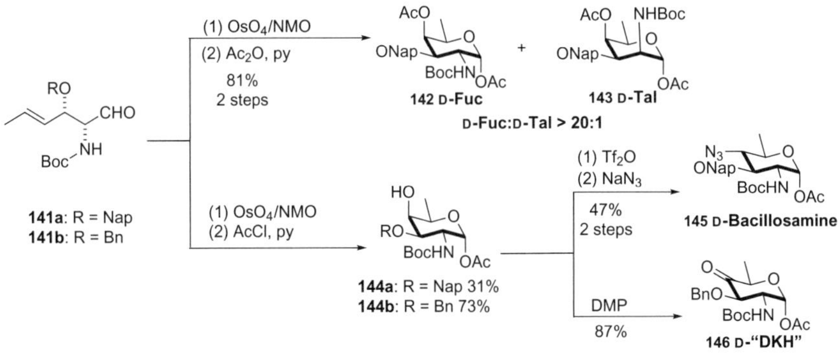

SCHEME 23. Synthesis of monosaccharide building blocks **142**, **143**, **145**, and **146**.

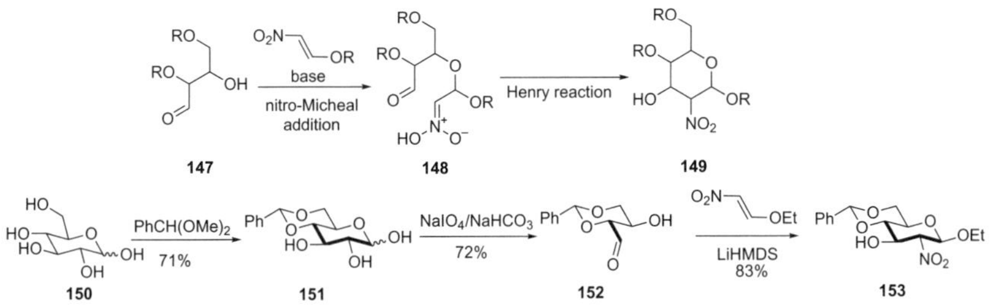

SCHEME 24. Synthesis of a 2-deoxy-2-nitro-hexopyranoside **153**.

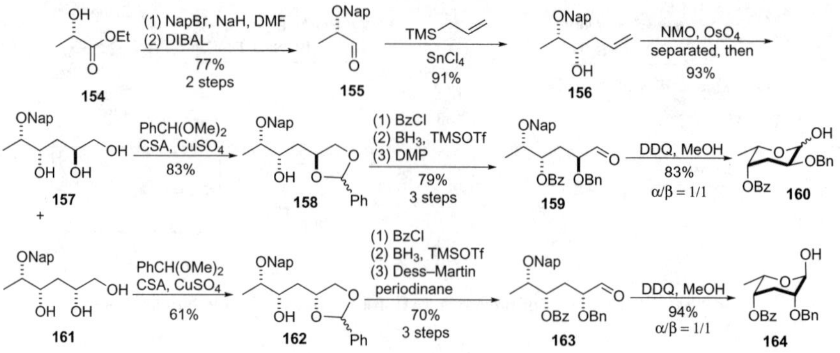

SCHEME 25. Synthesis of building blocks for L-colitose and 2-*epi*-colitose.

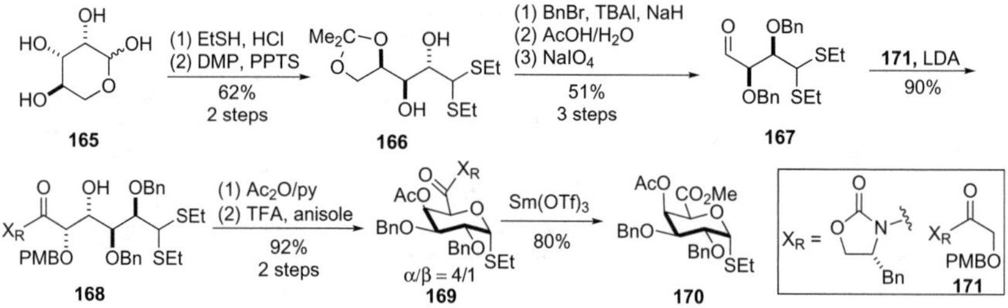

SCHEME 26. Synthesis of a D-galacturonic acid derivative.

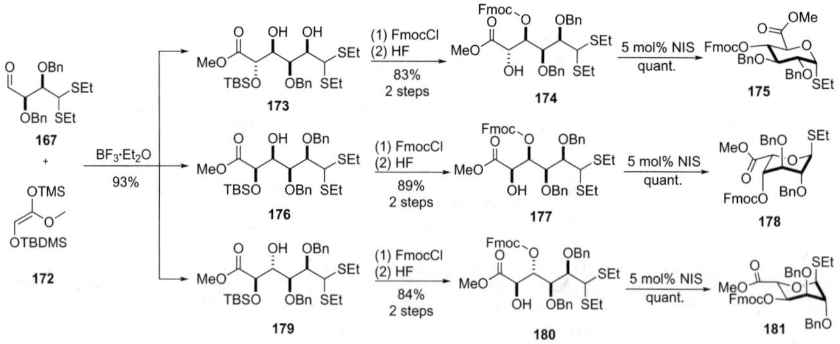

SCHEME 27. Synthesis of L-glucuronic, L-iduronic, and L-altruronic acid building blocks.

141b. Use of the Dess–Martin reagent is critical, as other conditions gave significant amount of epimerization.[40,41]

A diastereoselective Upjohn-type dihydroxylation of alkene **141a** provides a pyranose derivative, which after peracetylation with acetic anhydride, gives the D-fucose derivative **142** with only a minimal amount of the C-2 epimer **143** (D-*galacto*/D-*talo*, >20:1). Simply switching the acylating reagent from acetic anhydride to acetyl chloride gives selective acylation at the anomeric position, affording intermediates **144a/b**. This step allows for selective functionalization at C-4 position as an azide group, by triflation and azide displacement to give the desired building block **145** for the bacillosamine derivative, 2,4-diamino-2,4,6-trideoxy-D-glucose. Similarly, a Dess–Martin oxidation of the 4-hydroxyl group in **144b** produces the building block **146** for 2-amino-2,6-dideoxy-D-*xylo*-hexopyranos-4-ulose (D-"DKH").

Seeberger's group has also developed a counterintuitive homologation strategy to a 2-deoxy-2-nitro derivative (**153**) of D-glucose from D-glucose itself (**150**) via an erythrose derivative **152**. The route uses a reversible 1,4-nitroalkene addition/Henry reaction to establish the all-equatorial stereochemistry in **153**. The approach to compound **153** begins with the 4,6-benzylidene acetal **151** of D-glucose, periodate cleavage of which gives the protected erythrose derivative **152**. Exposure of aldehyde **152** to 2-ethoxynitroalkene and base diastereoselectively gave the 2-nitro derivative **153**, a potential precursor to D-glucosamine.[42–44]

The Seeberger group has also developed *de novo* asymmetric approaches to the deoxy sugars L-colitose and its 2-epimer from ethyl L-lactate (**154**). The route begins with Nap-protection, followed by reduction with DIBAL-H to give **155**, and a Cram-chelated controlled allylation to form compound **156**. A nonselective dihydroxylation of **156** gives a diastereomeric mixture of diols **157** and **161**. Benzylidene protection of the terminal diol in **157** gives **158**, which after benzoylation, reduction of the benzylidene group, and Dess–Martin oxidation gave aldehyde **159**. Exposure of **159** to DDQ selectively removed the naphthyl group to afford the 2-*epi*-colitose derivative **160**. Applying an identical procedure to **161** converts it into the colitose derivative **164** in good overall yield.[45]

The Seeberger group has also explored the use of the Evans aldol reaction for the synthesis of rare sugars (Schemes 26 and 27). The aldol route starts with a protected threose derivative **167**, which was prepared from L-arabinose (**165**) in five steps. An Evans aldol reaction between **167** and the chiral amide **171** affords **168** having *galacto* stereochemistry. Acylation and PMB deprotection converted **168** into **169**, from which the chiral auxiliary was removed with $Sm(OTf)_3$ to give the D-galacturonic acid derivative **170**.[46]

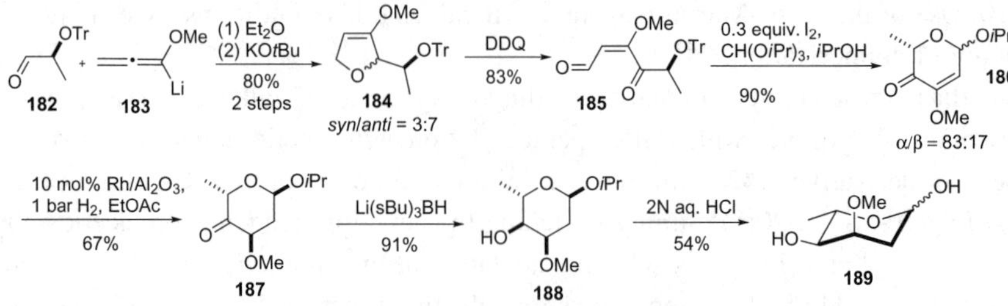

SCHEME 28. Synthesis of L-cymarose.

By removing the chiral auxiliary and performing a Mukaiyama-type aldol reaction on **167**, Seeberger's group was also able to produce the additional glucuronic acid precursors **175**, **178**, and **181**. Thus, performing the aldol reaction between compounds **167** and **172** with BF_3 as the Lewis acid gives a mixture of three diastereomeric aldol products, **173**, **176**, and **179**, which upon Fmoc protection and TBS deprotection gives compounds **174**, **177**, and **180**. Finally, the pyranose ring is formed upon exposure to a catalytic amount of NIS to form selectively the derivatized L-glucuronic (**175**), L-iduronic (**178**), and L-altruronic (**181**) acids.

3. Reissig Approaches

Finally, Reissig's group has also explored the *de novo* synthesis of various hexoses from simpler sugars.[47] Their approach begins with the chiral aldehyde **182**, which is readily prepared from L-lactic acid in three steps. Addition of the lithiated methoxyallene **183** to aldehyde **182** gave the dihydrofuran derivative **184** (Scheme 28). Oxidation of the furan ring in **184** with DDQ gives keto-aldehyde **185**, which upon exposure to an acidified solution of 2-propanol provides the pyranone **186** without loss of enantiomeric purity. Reduction under 1 bar of hydrogen pressure and 10 mol% rhodium on aluminum oxide led to the diastereomeric glycoside **187**. Reduction of **187** with L-selectride gave the alcohol **188**. Finally, acid hydrolysis of **188** yielded 2,6-dideoxy-3-*O*-methyl-L-*ribo*-hexose (L-cymarose) as the free sugar in 10 steps from lactic acid.[48]

VIII. O'DOHERTY DE NOVO APPROACH TO PYRANOSES

The O'Doherty group has also been involved in the development of several *de novo* approaches to the hexoses, and more generally to oligosaccharides, with the aim of using these methods to address questions associated with medicinal chemistry (Schemes 29–63). These approaches can be categorized by two general methods for the *de novo* synthesis of hexopyranolactones having variable substitutions at C-6. The first of these approaches involves a *de novo* asymmetric synthesis of a furan alcohol

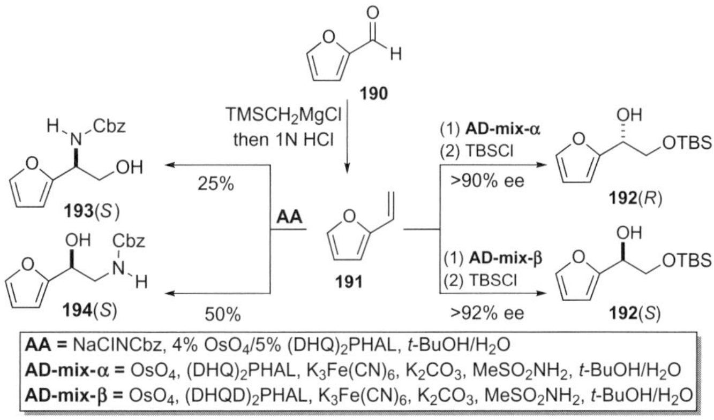

SCHEME 29. *De novo* asymmetric oxidation approaches to chiral furan alcohols.

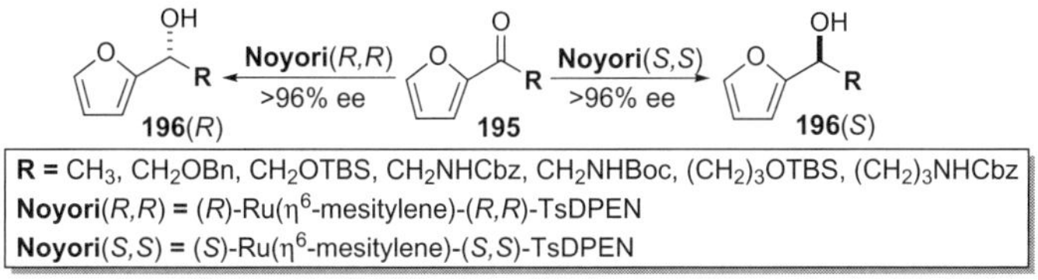

SCHEME 30. Asymmetric-reduction approaches to chiral furan alcohols.

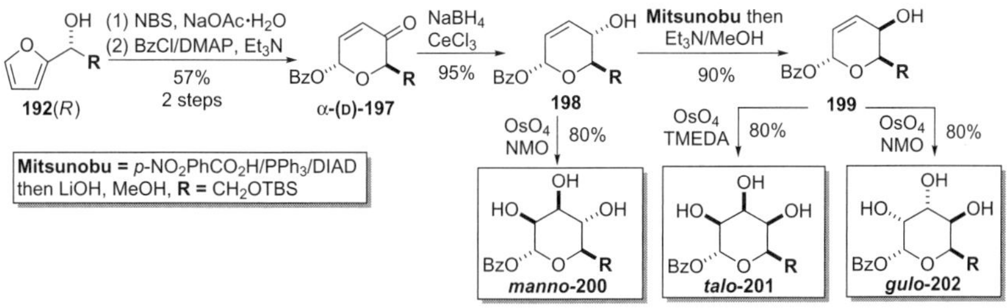

SCHEME 31. Syntheses of the D-enantiomers of *manno*-, *gulo*-, and *talo*-pyranoses.

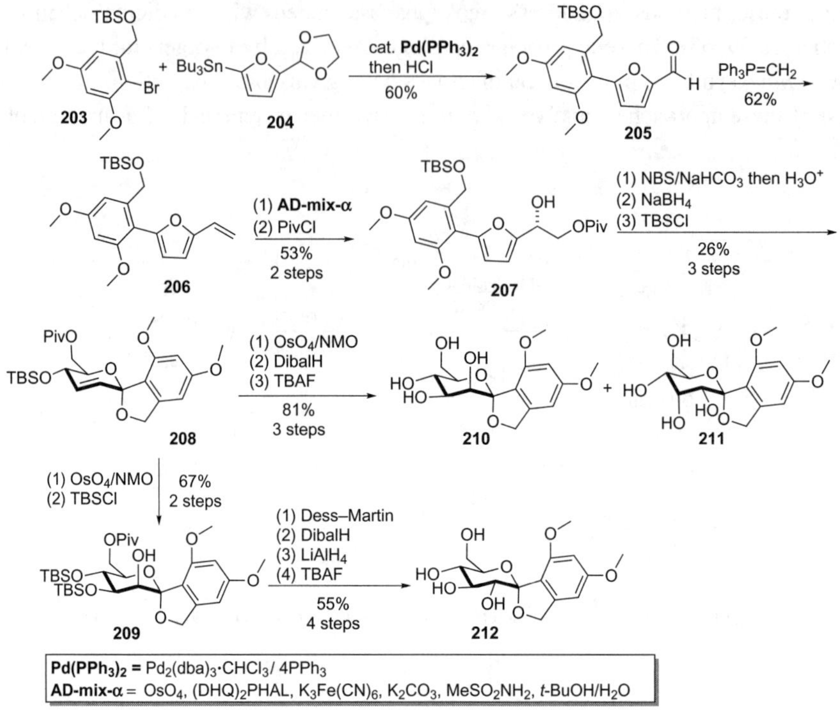

SCHEME 32. Enantioselective synthesis of the papulacandin ring system.

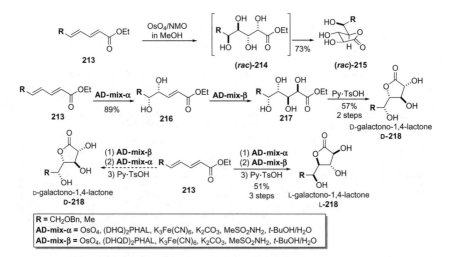

SCHEME 33. Enantioselective syntheses of alactono-1,4-lactones and deoxy sugars.

coupled with an Achmatowicz reaction (Scheme 31).[49,50] The second approach involves an iterative dihydroxylation strategy to various hexonolactones (Scheme 33).[51–54]

Both of these efforts result in the discovery of two orthogonal approaches to hexopyranoses having variable substitutions at C-6. Of the two, the iterative asymmetric dihydroxylation of dienoates (Scheme 33) is the most efficient in terms of steps, in that it can produce a sugar molecule in one step for racemic and three steps for enantiomerically pure sugars. On the other hand, the Achmatowicz approach is significantly broader in terms of synthetic scope. The approaches' potential can be seen in the highly efficient *de novo* routes to various mono-, di-, tri-, tetra-, and heptasaccharide motifs, where these routes in turn enable biological and medicinal structure–activity studies. A recurring theme to these approaches is the minimal use of protecting groups and the facility for producing various sugar diastereomers having variable substitutions at C-6.

1. *De Novo* Use of the Achmatowicz Approach to Pyranoses

An important aspect of this Achmatowicz approach is the facile and practical access to furan alcohols in enantiomerically pure form from achiral furans (such as **190** and **195**). The synthesis begins with the asymmetric synthesis of various furan alcohols (for instance, **192** and **194**) and amines (such as compound **193**) by the use of asymmetric catalysis. The first iteration of this synthesis involves the asymmetric oxidation of vinylfuran (**191**), which in turn can be prepared from furfural **190** (Scheme 29). The inherent volatility and instability of vinylfuran require its generation *in situ* via the Petersen olefination. The solutions of vinylfuran generated where compatible with both the Sharpless asymmetric dihydroxylation (**191** to **192**) and the amino-hydroxylation (**191** to **193** and **194**) procedures.[55,56]

The second-generation furan alcohol synthesis uses a Noyori hydrogen-transfer reaction for the asymmetric reduction of acylfurans (**195** to **196**). The two routes are easily adapted to 100-g scale synthesis and use readily available reagents. The route that uses the asymmetric Sharpless dihydroxylation is the most amenable to the synthesis of hexoses having a hydroxyl group at C-6, whereas the Noyori route is more flexible toward a wide range of substitutions at the C-6 position.[57–59]

Access to such enantiomerically pure furan alcohols as **192**, enables a very practical de novo approach to mannose derivatives when combined with four additional steps. The route is also amenable to the *talo* and *gulo* diastereomers. The route begins with the NBS-promoted oxidative hydration of the chiral furan alcohol **192** to substituted 6-hydroxy-2*H*-pyran-3(6*H*)-ones which, after an α-selective benzoylation affords benzoyl-protected compound **197**. A highly diastereoselective Luche reduction (NaBH$_4$/CeCl$_3$) converts **197** into the allylic alcohol **198**. A subsequent Upjohn dihydroxylation (OsO$_{4(cat)}$/NMO) gives the D-mannopyranose derivative **200** as a single diastereomer. By incorporating a Mitsunobu/hydrolysis step (**198** to **199**) into the sequence, the route can be expanded to the synthesis of D-talopyranose and D-gulopyranose derivatives. Exposure of the diastereomeric allylic alcohol **199** to the same Upjohn dihydroxylation (OsO$_4$/NMO) sequence provides the gulopyranose structure **202**, whereas a hydroxyl-directed dihydroxylation (OsO$_4$/TMEDA) affords the diastereomeric talopyranose structure **201**. Other deoxy and aminodeoxy sugars were also prepared by using this approach. An interesting feature of this procedure is the use of C=C and C=O π-bond functionality in the triol precursor or protecting group.[60,61]

As with the other *de novo* approaches (Schemes 6 and 11–13), the Achmatowicz approach to the hexoses can also be applied to the *de novo* synthesis of the papulacandin ring system (Scheme 32). The papulacandins are an important group of microbial antifungal antibiotics,[62] which have been shown to inhibit (1→3)-β-D-glucan synthase. The O'Doherty approach to this class of glycosylarene natural products is outlined in Scheme 32. The route begins with synthesis of the 5-arylfurfural **205** via a Pd-catalyzed coupling of bromoarene **203** and furfural derivative **204**. Wittig olefination of **205** provides the vinyl derivative **206**, which upon Sharpless dihydroxylation gives, after pivaloylation of the primary alcohol group, the furan alcohol **207**. Achmatowicz oxidation of **207** followed by spiroketalization, Luche reduction, and TBS protection provided spiroketal **208**. Upjohn dihydroxylation of **208** followed by removal of the pivaloyl and TBS groups provided the *manno-* and *allo-*papulacandin derivatives, **210** and **211**, respectively, in 4:1 ratio. Incorporating a protection step after the dihydroxylation provides the *manno-*papulacandin isomer **209**. The C-2 axial alcohol of **209** can be inverted by successive oxidation (DMP), reduction (DIBAL-H), and deprotection (LAH/TBAF) to give the *gluco-*papulacandin **212**.[63–65]

2. Iterative Dihydroxylation of Dienoates

In addition to the Achmatowicz approach, O'Doherty has also developed an alternative *de novo* approach to the hexoses. This second approach involves an iterative dihydroxylation of dienoates to produce various sugar lactones (Scheme 33). In its most efficient form, it requires only one to three steps for sugars in either racemic or enantiomerically pure form. For instance, such dienoates as ethyl sorbate (**213**, R=CH$_3$) react under the Upjohn conditions (OsO$_4$/NMO)[66] to give racemic galactono-1,4-lactones (**215**) in a single flask. The reaction occurs via a sequential bis-asymmetric dihydroxylation [that is, **213** to (*rac*)-**215** via (*rac*)-**214**]. Key to this discovery is the recognition that the more electron-rich γ,δ-double bond of dienoate **213** reacts first to form diol **216**, which once formed, reacts again in a diastereoselective fashion (>4:1) to give the tetrol **214**. When tetrol **214** is formed in a polar protic solvent such as methanol, it undergoes a base-catalyzed (NMM) lactonization to give the galactono-1,4-lactone (*rac*)-**215**. Alternatively, the initial dihydroxylation can be performed with the Sharpless reagent to give diol **216**, which when dihydroxylated with OsO$_4$/NMO in MeOH gives the galactonolactone **218** in high enantiomeric excess. By performing the second dihydroxylation with the enantiomeric Sharpless reagent, this allows for a highly stereoselective three-step synthesis of the galactono-1,4-lactones **218** with diverse substitution at C-6 and near-perfect enantio- and diastereocontrol.[67]

By taking advantage of the ability to form a palladium π-allyl intermediate (**221**), this approach is also adaptable for the synthesis of 1,5-lactones (Scheme 34). Thus,

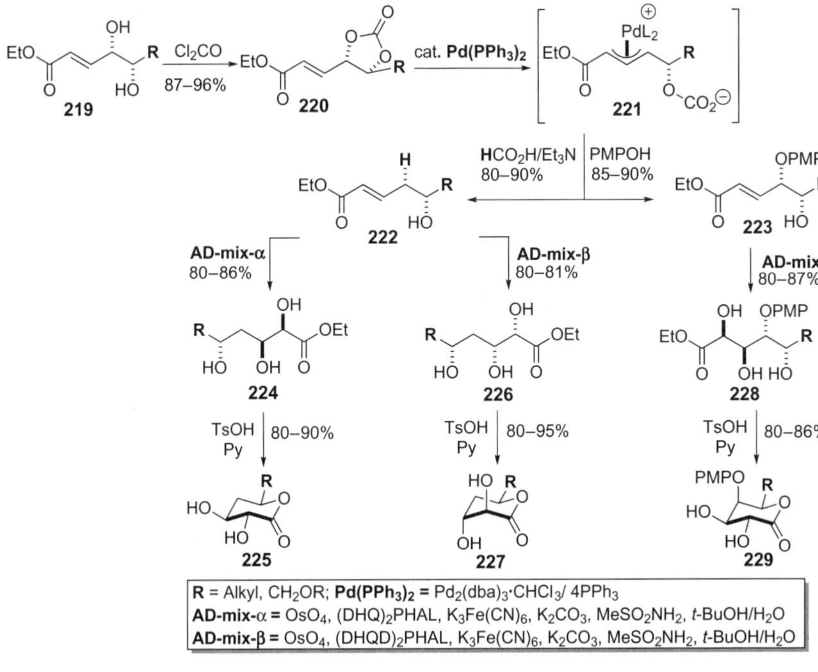

SCHEME 34. Synthesis of galactono-1,5-lactones.

the hydroxyl group at C-4 can be selectively removed or protected with a *p*-methoxyphenyl group, which enforces lactonization to a six-membered ring. This can be accomplished by converting the initial diol **219** into the cyclic carbonate **220**. Treatment of **220** with Pd(0) generates the π-allyl intermediate **221**, which can be generated and trapped by various nucleophiles (for instance, H, OAr). When the C-4 hydroxyl group is replaced by an aryl ether (giving compound **223**), increased diastereoselectivity is seen in the dihydroxylation under the Upjohn conditions (OsO$_4$/NMO), leading to compound **228**, which can be lactonized to produce the 1,5-lactone **229**. In contrast, when the hydroxyl group is replaced by hydrogen to form the unsaturated ester **222**, no diastereoselectivity in the dihydroxylation is observed. This loss of stereocontrol can be circumvented by use of the Sharpless reagents (AD-mix-α or -β), which can provide selectively either of the 4-deoxy-1,5-lactones (**225** and **227**).[68–72]

A similar variant of this approach can also be applied for the synthesis of a 4-deoxy-4-fluoroaldono-1,5-lactone (Scheme 35). The route begins with the regioselective S$_N$2-type displacement of the 4-hydroxyl group in **219** by a fluoride anion to give after hydrolysis compound **230**. A second reagent-controlled dihydroxylation with the AD-mix-α provides the L-glucono-1,5-lactone **232** having fluoride at C-4.[73]

Interestingly, by switching the stereochemistry of double bonds in the dienoate from *E,E*- as in **219** to the *Z,E*-dienoate **233**, the iterative dihydroxylation can also be applied for a two-step synthesis of talonolactones (Scheme 36). Exposure of dienoate **233** to the Sharpless dihydroxylation conditions provides the 1,4-lactone **234**. A second Upjohn dihydroxylation of **234** then produce the talono-1,4-lactone **235** diastereoselectively.[74]

This iterative dihydroxylation approach to sugars can also be applied to the papulacandins (Scheme 37). In this case, the synthesis provides the *galacto*-papulacandin framework in both the pyranoid (**241**) and furanoid (**242**) forms. This second-generation route begins with the synthesis of dienone **238** via a Horner–Wadsworth–Emmons olefination between intermediates **236** and **237**.

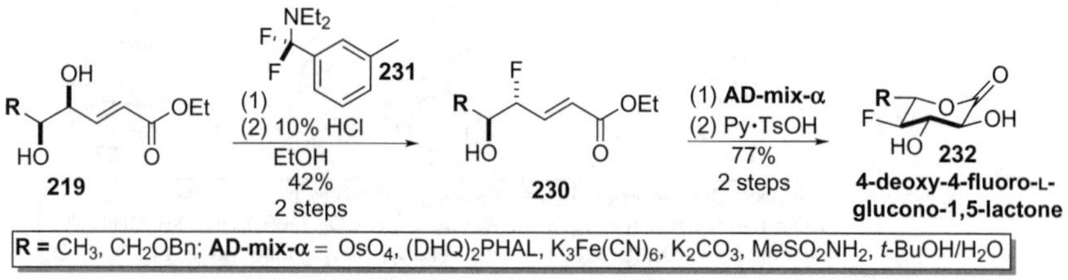

SCHEME 35. Asymmetric syntheses of a 4-substituted aldono-1,5-lactone.

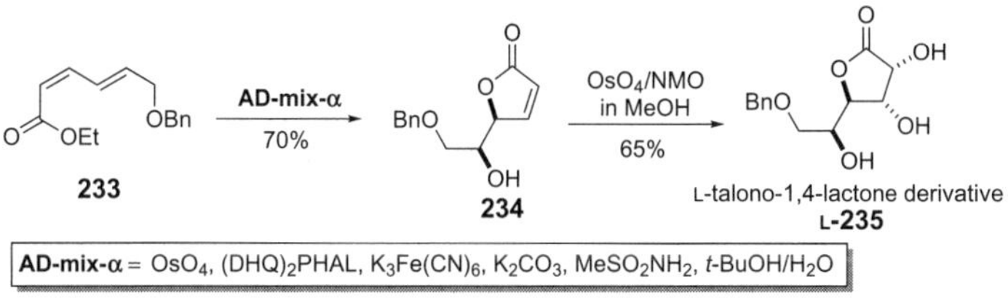

SCHEME 36. Asymmetric syntheses of an L-talono-1,4-lactone derivative.

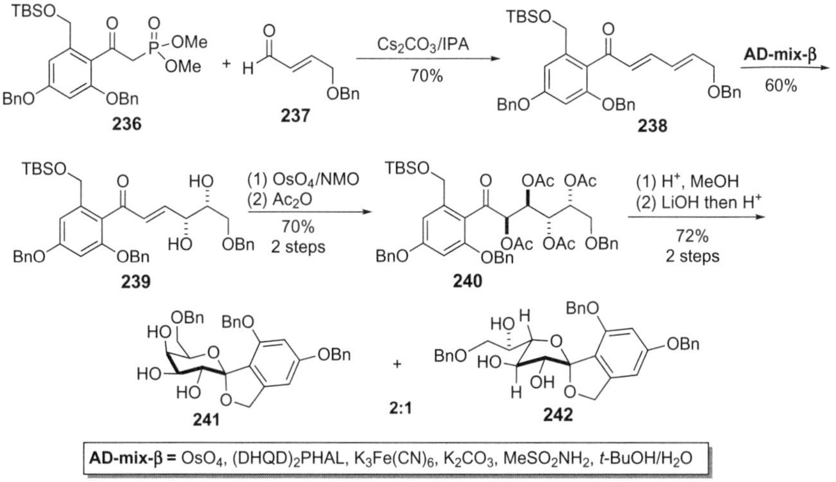

SCHEME 37. *De novo* asymmetric synthesis of galacto-papulacandin analogues.

A regioselective asymmetric dihydroxylation of dienone **238** provides the diol **239**. A second substrate-controlled dihydroxylation leads, after peracetylation, to the tetraacetate **240**. A one-pot deprotection/spiroketalization sequence provides a 2:1 mixture of the *galacto*-papulacandins with the sugar group in both the pyranose and furanose forms (**241** and **242**, respectively).[75]

This Achmatowicz can also be applied to produce imino sugars (Scheme 38). The approach returns to the benzyloxycarbonyl-protected furan amine **243**. In this instance, the aza-Achmatowicz reaction is best accomplished by using *m*-CPBA to form the piperidine derivative **244**. The hemiacetal hydroxyl group can be protected as an ethyl glycoside and the enone reduced to form the allylic alcohol **246**. Dihydroxylation of **246** selectively gave **247**, which after hydrogenolysis produced the

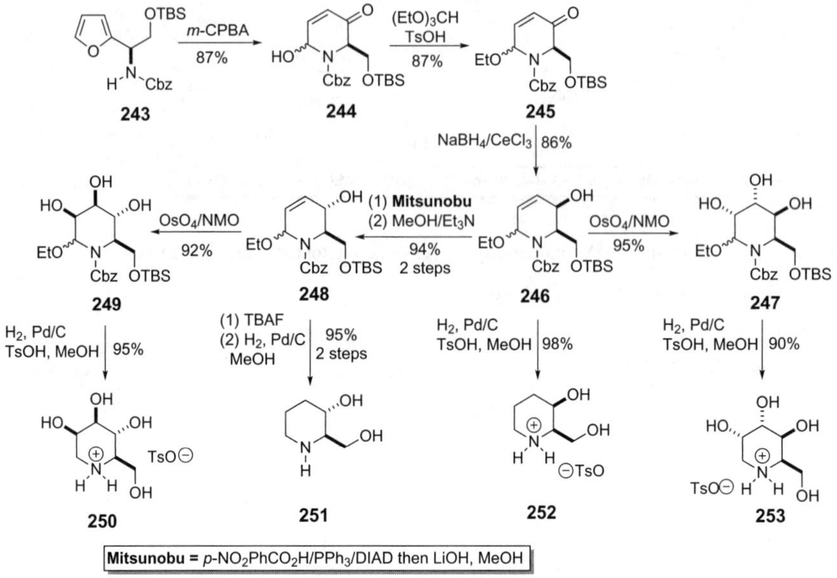

SCHEME 38. Synthesis of D- and L-deoxymannojirimycin.

gulo analogue (**253**) of deoxynojirimycin. As with the pyranose case, incorporating a Mitsunobu/hydrolysis sequence (**246** to **248**) into the procedure allows the route to be expanded to the synthesis of deoxymannojirimycin (**250**).[76]

3. Palladium-Catalyzed Glycosylation

It is difficult to compare these various *de novo* routes in terms of number of steps, availability of starting materials, and/or atom economy. Clearly the best metric for evaluation is to compare the routes in terms of utility for synthesis and biology. The Achmatowicz approach, in this regard, distinguishes itself from the other approaches. This is especially true when it comes to practical application for rare sugars, medicinal chemistry, and more particularly oligosaccharides. These advantages result from the Achmatowicz's compatibility with the Pd–π-allyl-catalyzed glycosylation, which allows for stereospecific formation of the glycosidic bond.[77–79]

As outlined in Scheme 39, the Pd(0)-catalyzed glycosylation reaction is stereospecific and quite general.[80,81] The reaction affords high yields for both the α-**254** to α-**255** and β-**254** to β-**255** systems. When using the $Pd_2(dba)_3 \cdot CHCl_3$ as the Pd(0) source and triphenylphospine as the ligand, in a 1:2 Pd/PPh_3 ratio, it occurs rapidly at

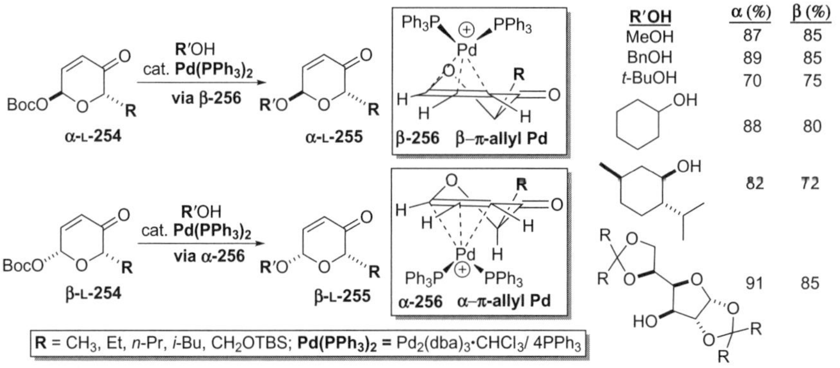

SCHEME 39. Stereospecific palladium-catalyzed glycosylation.

room temperature. While the reaction does not at first glance appear to be a typical glycosylation process, the enone resulting in these glycosylation products can be converted into mono-, di-, and triol products in postglycosylation reactions that transform them into more-traditional glycosylation products (see later). Critical to the success of these approaches is the chemoselective use of C—C and C—O π-bond functionality, as atom-less protecting groups (namely, the enone as a protected triol), as well as an anomeric directing group (via a Pd–π-allyl conjugate).[82]

4. Applications to Synthesis and Medicinal Chemistry

An excellent example of the power of this Pd-catalyzed glycosylation can be seen in its application to the synthesis of daumone (Scheme 40), a rare ascarylose-containing pheromone associated with the induction of a dauer state (an enduring and nonaging stage) in the nematode *Caenorhabditis elegans*. The de novo route to daumone features a convergent route, with the Pd-glycosylation step at the point of convergence (**259** + **260** to **261**). Because both components are chiral, the convergence requires the asymmetric synthesis of both halves. This is accomplished with two Noyori reduction steps, the first in the synthesis of pyranone **260** and the other with the reduction of ynone **257** to ynol **258**. Hydrogenation of **258** gave the saturated alcohol **259**, which coupled cleanly with enone **260** to form the α-glycoside **261**. After glycosylation, the stereochemistry for the ascarylose (3,6-dideoxy-D-*arabino*-hexose) component of daumone is introduced via a two-step epoxidation, reduction, and ring-opening sequence (**261** to **263** via **262**). With the sugar functionality in place, the

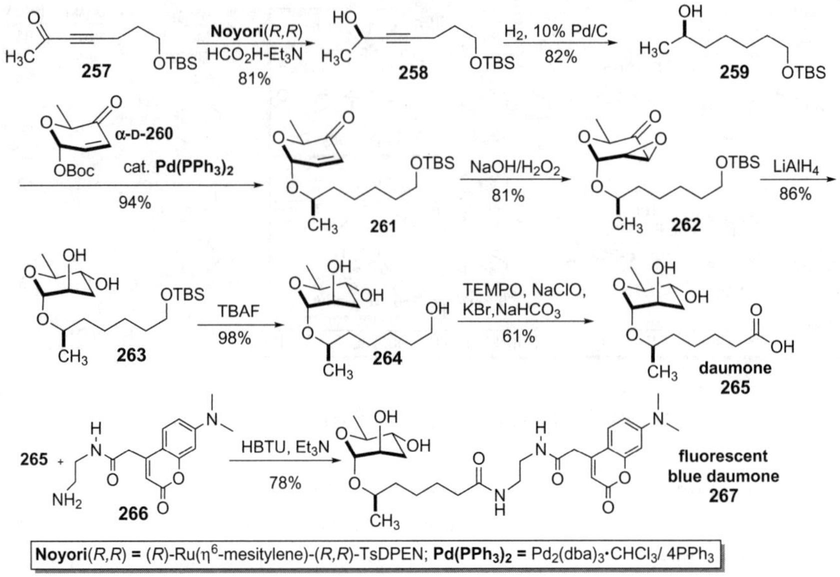

SCHEME 40. Asymmetric synthesis of daumone and analogues.

synthesis is completed by removal of the TBS group and selective oxidation of the terminal hydroxyl group to the carboxylic acid (**263** to **265** via **264**). As daumone is obtainable from natural sources in only minute quantities, this more abundant synthetic daumone was used for the synthesis of fluorescent analogues (such as **267**), which were used in mode-of-action studies.[83,84]

The Pd-catalyzed glycosylation can also be used for the synthesis of *N*-glycosyl products. This is best exemplified with the *de novo* synthesis of analogues of adenosine and 2′-deoxyadenosine ("L-homo-adenosine," **277**, and "L-homo-deoxyadenosine," **278**), having 2-deoxy-β-L-*arabino*-hexopyranosyl and 2,3-dideoxy-β-L-*erythro*-hexopyranosyl as the glycosyl groups (Scheme 41). The synthesis begins with the Pd-catalyzed glycosylation of 6-chloro-9*H*-purine (**269**) with the β-glycosyl donor β-L-**268** to form β-glycoside **270**. The C-4 ketone can be reduced with NaBH$_4$ to form allylic alcohol **271**, which can in a divergent fashion be converted into either the homo-adenosine **277** or the homo-deoxyadenosine **278**. The route to **277** starts with a Myers reductive rearrangement of the allylic alcohol to form **272**, which can be diastereoselectively dihydroxylated to form the diol **273**. With the desired stereochemistry installed in **273**, the 6-chloro group is displaced with ammonia (adenine derivative **276**) and the TBS-protection group is removed with TBAF to form the

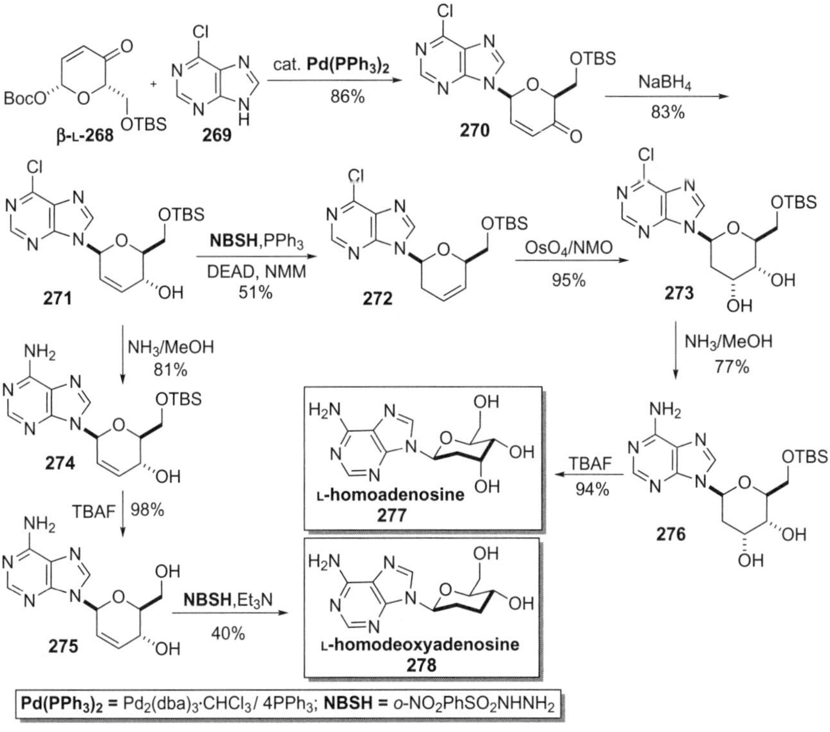

SCHEME 41. Asymmetric synthesis of homo-adenosine.

homo-adenosine **277**. Alternatively, the chloro group of **271** can be displaced with ammonia to form **274**, the TBS group can be removed and the alkene can be reduced with diimide to form the homo-deoxyadenosine **278**.[85]

Another application of the Achmatowicz approach is the synthesis of polyhydroxylated indolizidine alkaloids. The most well-known member of this class of natural products is D-swainsonine (**292**, Scheme 42).[86–90] The biological activity of swainsonine is attributable to its potent inhibitory activity toward both lysosomal α-mannosidase[91] and mannosidase II.[92–94] Interestingly, its enantiomer (L-swainsonine) has been shown to be a selective inhibitor of the enzyme narginase (L-rhamnosidase, $K_i = 0.45$ μM).[95] As a result of these activities, both D- and L-swainsonine have become important targets for synthesis. This is similarly true for the synthesis of their epimers as well as other analogues.[96–108] The O'Doherty *de novo* asymmetric synthesis of both enantiomers of swainsonine are outlined in Scheme 42.[109–111]

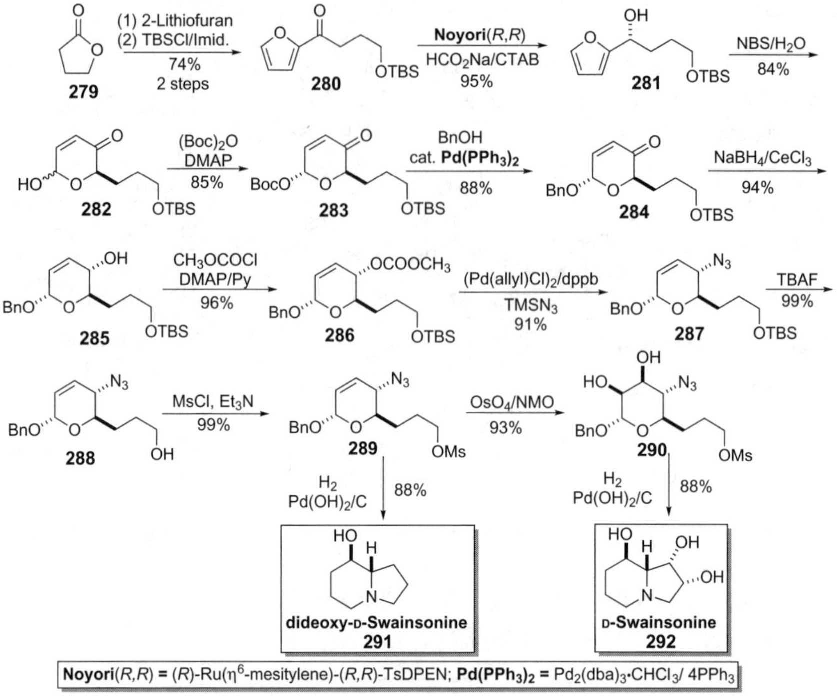

SCHEME 42. Syntheses of dideoxy-D-swainsonine and D-swainsonine.

The synthesis begins with the synthesis of an acylfuran (**280**) that has all the carbon atoms required for swainsonine, albeit none of the stereochemistry. Acylfuran **280** could be prepared in two steps from 2-lithiofuran and butanolactone (**279**). A Noyori asymmetric reduction (**280** to **281**) can be used to install the asymmetry, and then an Achmatowicz reaction and Boc protection convert **281** into the pyranone **282**. A Pd-catalyzed glycosylation of benzyl alcohol with **282** installs the required C-1 benzyl protecting group to give **284**. A combination of Luche reduction (**284** to **285**), carbonate formation (**285** to **286**), and Pd–π-allyl allylic azide displacement is used to install the C-4 nitrogen functionality in azide **287**. The TBS ether is then converted into a good leaving group (**287** to **289**) and then the 2,3 double bond in **287** is dihydroxylated. Finally D-swainsonine is cleanly produced in enantiomerically pure form, by a diastereoselective dihydroxylation (**289** to **290**) and exhaustive hydrogenolysis (**290** to **292**). This *de novo* route, in only 13 steps, provides either D- or L-swainsonine from achiral stating materials. The simpler dideoxy-D-swainsonine can also be prepared by exhaustive hydrogenolysis of azide **289**.

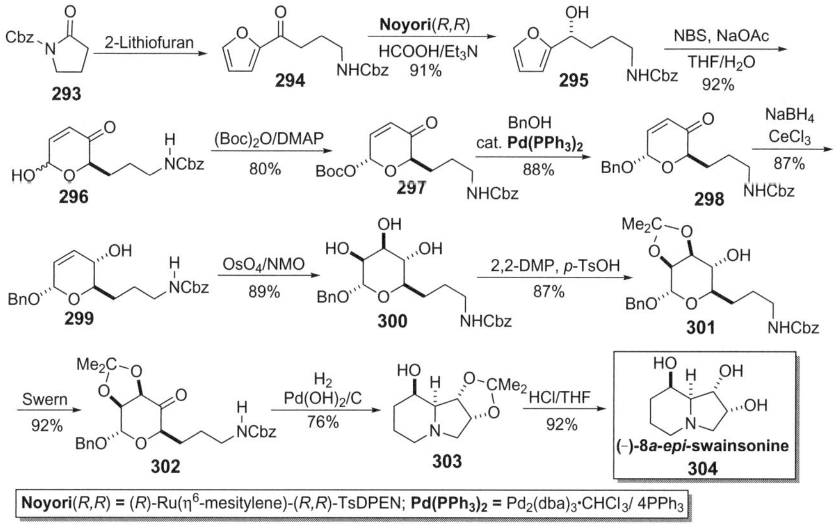

SCHEME 43. Synthesis of 8*a-epi*-swainsonine.

In a related route, the 8*a*-epimer (**304**) of D-swainsonine can also be prepared (Scheme 43). Because the route installs the nitrogen atom at the very beginning and the molecule cyclizes via a reductive amination, it saves two steps over the preceding route to swainsonine. The route starts with the addition of furanyllithium to amide **293** to form ketone **294**. Once again, a Noyori asymmetric reduction (**294** to **295**) and an Achmatowicz reaction (**295** to **296**) and Boc protection can be used to convert compound **294** into pyranone **297**. A Pd-catalyzed glycosylation between benzyl alcohol and **297** installs the required C-1 benzyl protecting group in **298**, and subsequent Luche reduction affords compound **299**. Dihydroxylation, acetonation (**300** to **301**), and Swern oxidation convert alcohol **301** into the 4-ketone **302**. Finally, a highly diastereoselective exhaustive hydrogenolysis of ketone **302** cleanly converts it into the indolizidine derivative **303**, which upon deacetonation provides (−)-8*a-epi*-swainsonine in enantiomerically pure form.[112,113]

The Achmatowicz route can also be applied toward the synthesis of disaccharide targets. An example of this is shown in its application to the synthesis of a bis-D-*manno* analogue of α,α-trehalose (Scheme 44). The route involves the iterative glycosylation of a water molecule to prepare the (1 ↔ 1)-disaccharide core **307**. The route begins from furan alcohol **192**, which can be prepared in enantiomerically pure form via a Noyori-type reduction of the acylfuran **305**. An Achmatowicz reaction

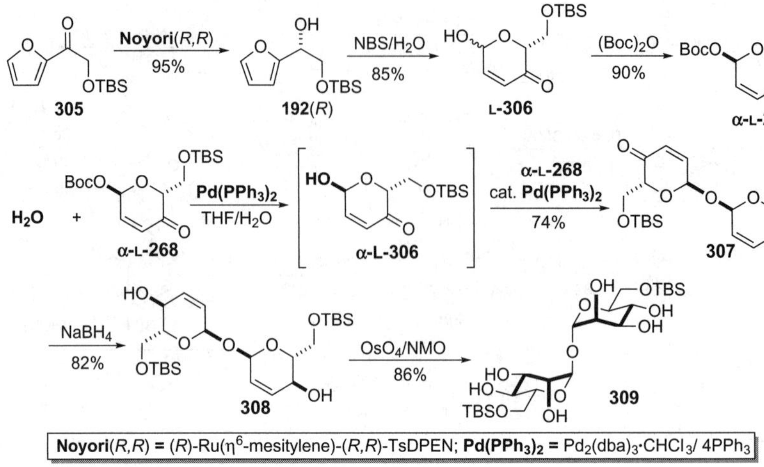

SCHEME 44. *De novo* synthesis of trehalose analogues.

(**192** to **306**) and Boc protection prepare the key glycosyl donor, pyranone **268**. The glycosylation of water with α-pyranone **268** stereoselectively affords α-**306** as a single diastereomer, which upon standing slowly epimerizes into a 1:1 α/β-mixture of diastereomer **306**. The crude alcohol α-**306** can be glycosylated with a second equivalent of α-pyranone **268** to give the (1 ↔ 1)-disaccharide precursor **307**. Finally, a bidirectional application of the postglycosylation modification (borohydride reduction and Upjohn dihydroxylation) provides the bis-*manno* α,α-trehalose isomer **309**.[114]

The Achmatowicz route is also applicable for the synthesis of such aryl glycosides as SL0101, the serine/threonine protein kinase (RSK) inhibitor (Scheme 45). Synthesis of this L-rhamnoside natural product began with Pd-catalyzed glycosylation of the benzyl-protected aglycon **310** with the rhamnosyl donor L-**260**. The 4-acetoxy group is introduced by borohydride reduction (to **312**) and acetylation to form the hex-2-enopyranoside **313**. An Upjohn-type dihydroxylation installs the 2,3-diol group (giving **314**), and this intermediate, using orthoester chemistry, undergoes regioselective acylation at the C-2 position to give compound **315**. Base catalysis (DBU) causes the axial acetyl group at C-2 to migrate to the equatorial C-3 position (**316**). Finally, the benzyl ether groups are removed by catalytic hydrogenolysis to generate the natural inhibitor, kaempferol L-rhamnopyranoside 3,4-diacetate (SL0101, **317**).[115,116]

In addition to the synthesis of pyranoses, the O'Doherty group has also expanded their methodology to permit access to 5*a*-carba sugars (formally cyclitols). While the carba sugars (with replacement of the sugar ring oxygen atom by carbon) have been

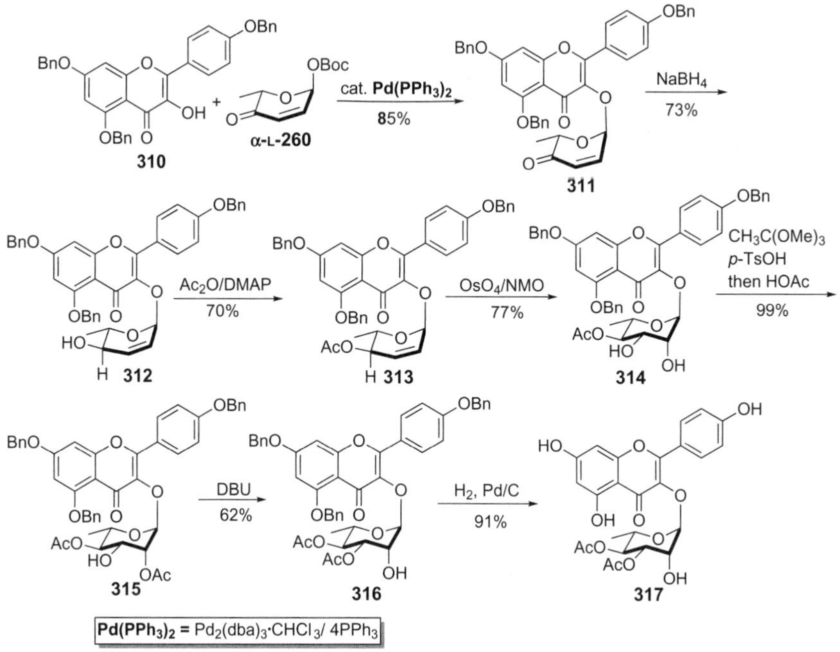

SCHEME 45. Synthesis of the RSK inhibitor SL0101.

studied both synthetically and in natural products, a unified strategy for their synthesis was lacking. The route relies on the synthesis of the enantiomerically pure enone **324**, which serves as an equivalent to compound **260** (see Scheme 40). The unifying theme for the two methodologies is the use of a Pd–π-allyl intermediate. Thus, in the approach to 5*a*-carba pyranoses (alternatively termed cyclitol derivatives), the route involves the use of a Pd-catalyzed conversion of Boc-ester **324** into benzyl ether **325**. The route begins with the asymmetric synthesis of enone **323** from quinic acid **318**, which requires 11 steps. In comparison, the synthesis of pyranone **260** from 2-acetylfuran **192** occurs in three steps. In contrast to the synthesis, there are mostly similarities between the glycosylation and cyclitolization, as well as the postglycosylation/cyclitolization sequences. For instance, the axial allylic alcohol **323** is converted into a *tert*-butyl carbonate (a good Pd-leaving group) with Boc$_2$O (giving **324**). Instead of using NaBH$_4$, the ketone **425** is stereoselectively reduced with LiAlH$_4$ (**325** to **326**), and the 2,3-alkene of **326** can be stereoselectively dihydroxylated using the Upjohn conditions, to afford the 5*a*-carba-D-mannopyranoside derivative **327** (formally a cyclitol derivative) (Scheme 46).

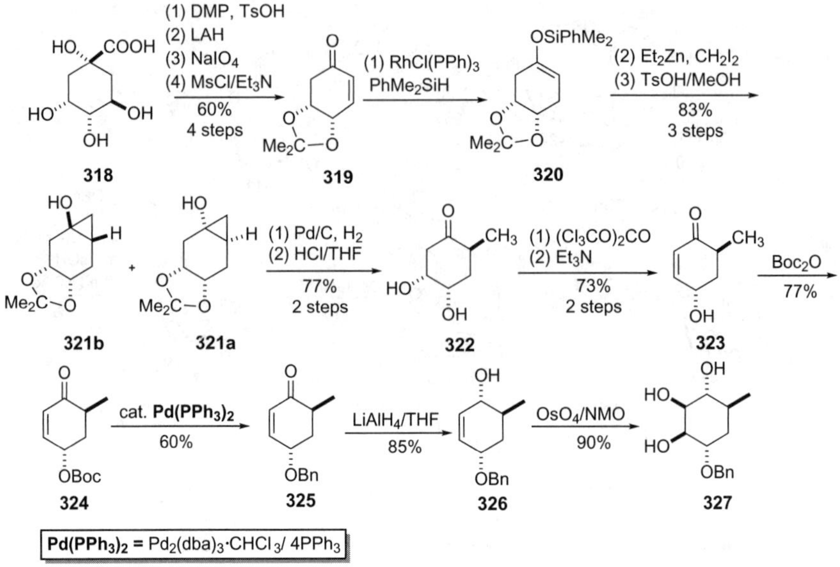

SCHEME 46. Synthesis of cyclitols (5a-carba pyranoses).

This installation methodology for a *manno*-carba sugar has great potential for applications in medicinal chemistry. The substitution of a carba sugar imparts stability toward acid and enzymatic hydrolysis, and thus provides substantially enhanced biostability. An example of this application is its use in the synthesis of carba sugar analogues of the inhibitor SL0101 **332** (Scheme 47). The synthesis of this cyclitol analogue of SL0101 (**332**) begins with a Pd-catalyzed cyclitolization of the benzyl-protected "aglycon" **310** with the cyclitol donor, (*ent*)-**324**. The C-4 acetate is introduced by LiAlH$_4$ reduction followed by acetylation to form **329**. An Upjohn dihydroxylation installs the C-2, C-3 diol (**330**), and this by using orthoester chemistry can be regioselectively acylated at the C-3 position to give **331**. Hydrogenolytic debenzylation of **331** produces compound **332**, the cyclitol analogue of the natural product.[117,118]

As part of a larger effort to enhance the activity of SL0101, this methodology was used to prepare analogues of SL0101 having the C-6 methyl group replaced by other groups (Scheme 48). This need requires access to pyranones **336b–e**. The synthesis starts with the addition of 2-lithiofuran to various carboxylic acids to afford **334b–e**. A Noyori asymmetric reduction of ketones **334b–e** provides alcohols **335b–e**. Exposure of **335b–e** to the Achmatowicz procedure, followed by reaction with Boc$_2$O,

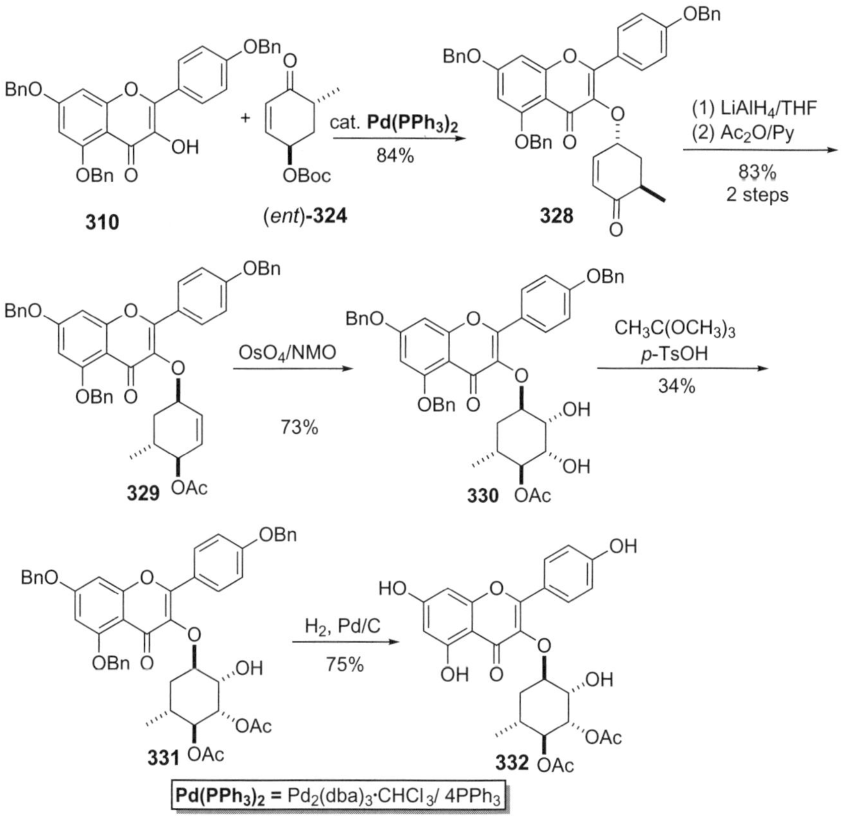

SCHEME 47. Synthesis of a carba sugar analogue of inhibitor SL0101.

affords pyranones **336b–e**. A Pd-catalyzed glycosylation of the benzyl-protected aglycon **310** with the C-6-modified rhamnosyl donors **336b–e** affords **337b–d**, which upon borohydride reduction yields allylic alcohols **338b–d**. Successive acylation at C-4, Upjohn dihydroxylation, followed by debenzylation, affords **339**. This effort led to the discovery of the more-potent RSK inhibitor **339** (R = n-Pr).[119]

In addition to monosaccharide targets, the *de novo* Achmatowicz methodology is also ideal for conducting medicinal-chemistry studies on a substituted disaccharide fragment of the cyclic glycopeptide antibiotics mannopeptimycins. In particular, it enables SAR studies on the cyclic hexapeptide mannopeptimycin-ε, which was isolated from *Streptomyces hygroscopicus* LL-AC98.[120] The key structural feature is the O-glycosylated tyrosine derivative having an α-(1→4-linked)-D-mannobiose

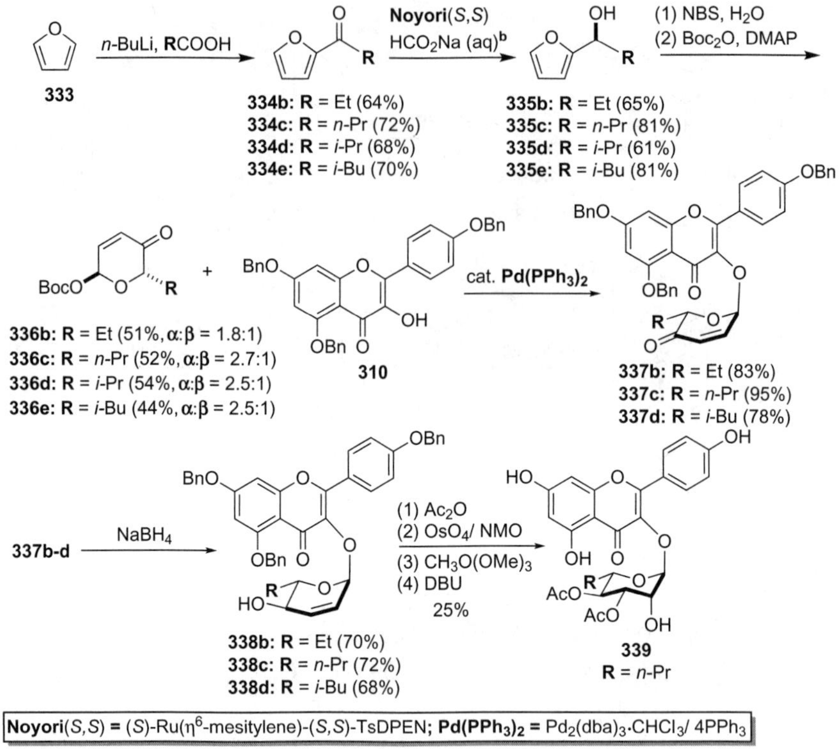

SCHEME 48. Synthesis of C-6-substituted analogues of inhibitor SL0101.

component bearing a 4'-O-(3-methylbutanoyl) group (**345**) which, via a *de novo* asymmetric Achmatowicz approach, was prepared, along with a C-4' 3-methylbutanoyl (isovaleryl) amide analogue **347**. The linear route involved application of the iterative bis-glycosylation (**340** to **341**), bis-reduction, bis-dihydroxylation, and acylation with 3-methylbutanoic acid of a protected tyrosine, in only seven steps (Scheme 49). To further study the structure–activity relationship (SAR) of the antibiotic mannopeptimycin, access to the C-4 amide analogues **347** is desired.[121] To accomplish this, the Pd–π-allyl-catalyzed allylic azide alkylation (compare Scheme 42) was envisioned as being used in conjunction with an azide reduction step and subsequent acylation. Specifically the allylic alcohol **343** is converted into an allylic carbonate, which is converted into the allylic azide **344**. The azide can be selectively reduced and the corresponding amine **346** acylated and the

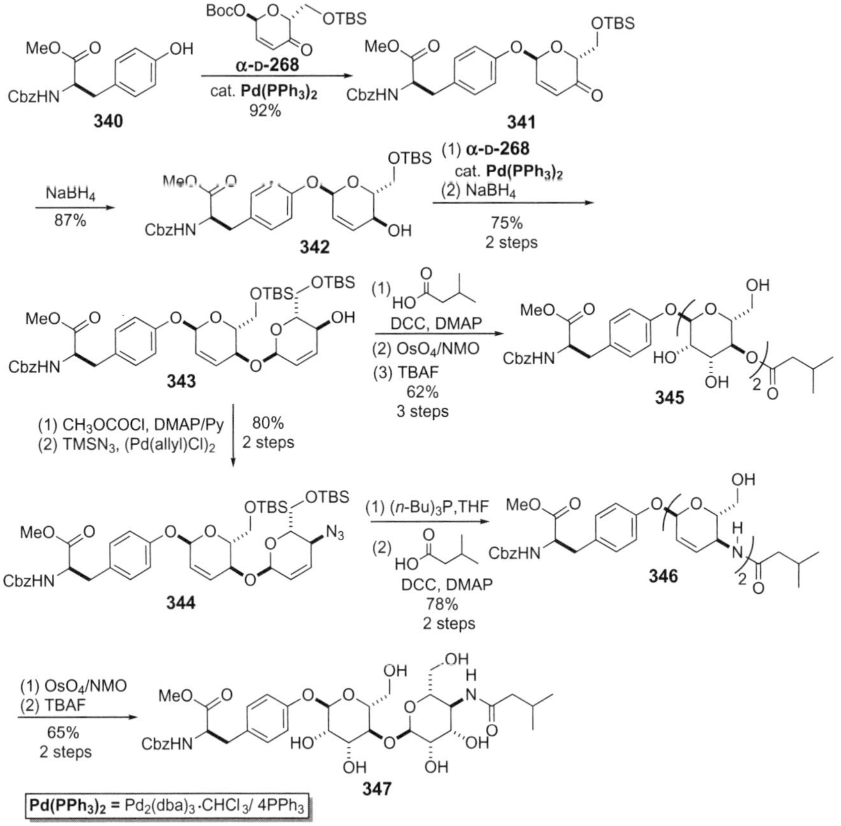

SCHEME 49. Synthesis of the glycosylated tyrosine portion of mannopeptimycin-ε **345** and its aminated analogue **347**.

remaining double bonds of **346** can be dihydroxylated, followed by removal of the TBS groups to provide the desired target tyrosine disaccharide **347**.[122]

The same chemistry for introduction of C-4 azide functionality can be used to make analogues of methymycin (Scheme 50), a glycosylated 12-membered-ring macrolide antibiotic isolated from *Streptomyces venezuelae* ATCC 15439. Like most macrolide antibiotics, the rare aminodeoxy sugar portion [desosamine, 3,4,6-trideoxy-3-(dimethylamino)-D-*xylo*-hexose] of methymycin is important for the bioactivity, and its modifications hold promise for the preparation of new macrolide antibiotics having enhanced antimicrobial activity. The *de novo* Achmatowicz approach to a library of methymycin analogues is outlined in Scheme 50. To accomplish this, the macrolide

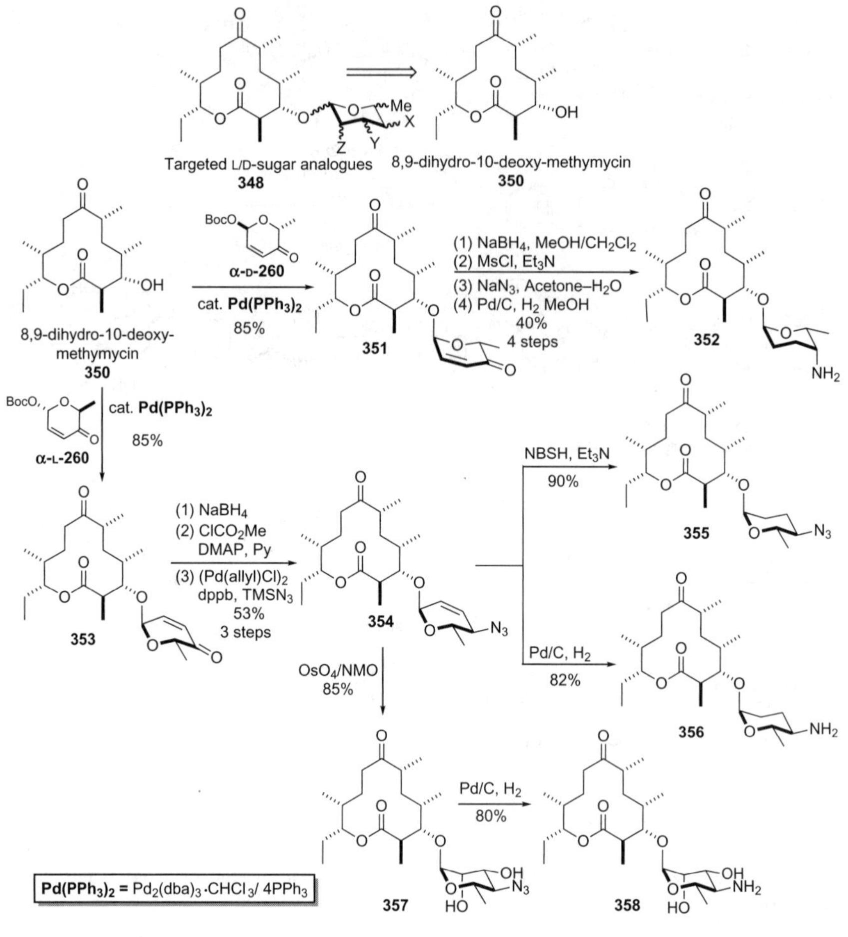

SCHEME 50. Synthesis of glycosylated methymycin analogues.

aglycon, 10-deoxymethynolide, is glycosylated in a stereodivergent manner (with D- or L-Boc pyranones **260**) to give the α-D-glycoside **351**, whereas when its enantiomer (***ent***)-**260** is used, the α-L-glycoside **353** is obtained. A subsequent postglycosylation transformation can be used to provide various unnatural aminodeoxy sugar congeners and stereoisomers. In practice, the installation of amino functional groups at the C-4 position was readily accomplished. For example, the α-D-pyranone ring in the methymycin analogue **351** is converted into a 4-amino-2,3,6-trideoxyhexose (as in **352**) having α-D-*threo*-stereochemistry in four steps. This occurs via a reduction, activation of the resulting alcohol, azide inversion, and reduction strategy. The approach also works for the installation of equatorial amino groups at the C-4 position, via a net retention of stereochemistry in the substitution reaction at C-4. The conversion begins with a Pd-catalyzed π-allyl reaction with trimethylsilyl azide (TMSN$_3$) to give allylic azide **354** from the α-L-methymycin analogue **353**. In turn, azide **354** is converted into 4-azido-/4-aminohexose analogues of methymycin having the α-L-*manno*- (**357** and **358**) and α-L-*threo*-configurations (**355** and **356**).[123]

For centuries, such cardiac glycosides as digitoxin and digoxin have been recognized for their ability to treat patients suffering from congestive heart failure. More recently, several cardiac glycosides have been found to possess potent anticancer activity.[124,125] *In vitro* studies show that a sub-cardiotoxic dose of digitoxin has quite a strong anticancer effect.[126] Detailed mechanism of action studies shows that pronounced apoptosis is induced in several different cell lines.[127] All these cardiac glycosides are more active than their corresponding aglycons, underlining the importance of the carbohydrate component to cytotoxic activity. More recently, Thorson and coworkers have prepared a library of digitoxin analogues (neoglycosides)[128–130] which includes examples having increased anticancer activity. The O'Doherty *de novo* asymmetric use of the Achmatowicz chemistry for the installation of rare sugars can also be used for the synthesis and medicinal evaluation of these cardiac glycoside analogues (Schemes 51–54). These efforts began with the preparation of the digitoxin mono- (**362**), di- (**366**), and trisaccharide (**367**) conjugates (Scheme 51); the neoglycoside analogues **374** (Schemes 52); and various other analogues (Schemes 53 and 54).[131,132]

This effort begins with the glycosylation of such alcohols as digitoxigenin with the β-D-glycosyl donor **260** to afford the β-D-pyranone **359**. In contrast to the α-pyranones, the borohydride reduction of the ketone group in **359** is rather unselective,[133] affording a mixture of allylic alcohols **360**. Fortunately, both diastereomers could be used in the subsequent Myers reductive rearrangement of **360**, cleanly providing alkene **361** as a single diastereomer. Finally, subjecting alkene **361** to the Upjohn conditions[66] (OsO$_4$/NMO), gives exclusively the *ribo*-diol **362** in good yield and as a single diastereomer.

The synthesis of **362** demonstrates the ability of this approach to glycosylate secondary alcohols selectively over tertiary ones as well as the compatibility of the postglycosylation chemistry with the butenolide functionality in digitoxigenin. Thus, the synthetic effort can be expanded toward the disaccharide and trisaccharide structures **366** and **367**. The installation of the second sugar group requires the regioselective acylation of the *syn*-3,4-diol in **362** to afford **363**, which was accomplished by using the regioselective orthoester.[134,135] The equatorial alcohol in **363** reacts cleanly under the Pd-catalyzed glycosylation and subsequent

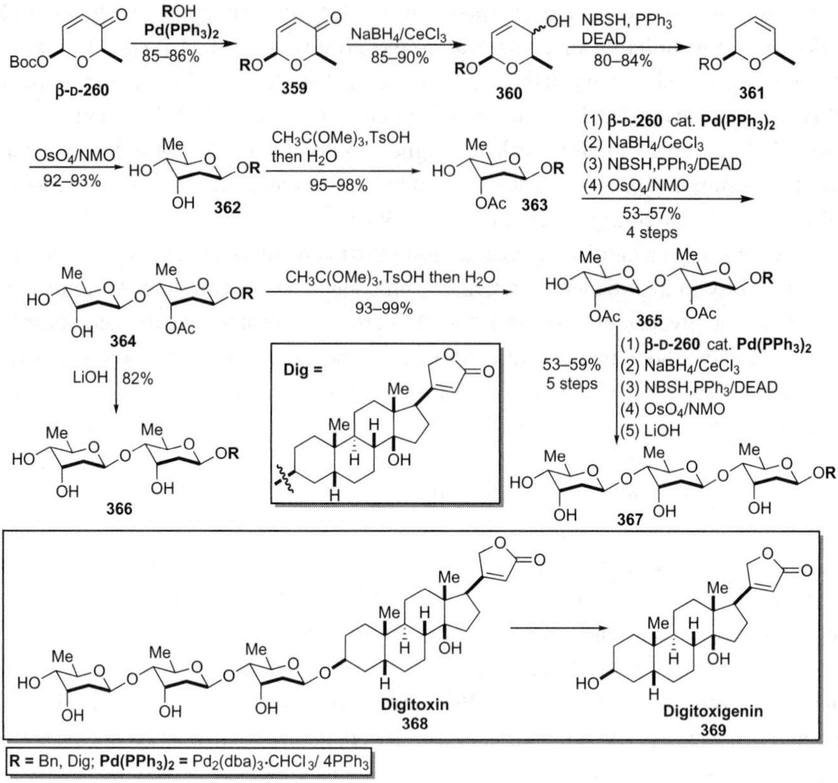

SCHEME 51. Syntheses of digitoxin and digitoxigenin mono- and di-digitoxosides.

postglycosylations to give first the disaccharides **364** and then **365**, and these in turn were hydrolyzed to give the disaccharide **366** and trisaccharide **367** derivatives, respectively. The synthesis proceeds smoothly, whether the aglycon was a benzyl group, digitoxigenin, or a digitoxigenin analogue (**373**) in which the OH group is replaced by a methoxyamino group to give the analogue **374**. For those products where the aglycon is a benzyl group, the free sugars (**370**, **371**, and **372**) can also be made (Scheme 52), as well as analogues. Biological evaluation of three O-glycosides (**362**, **366**, **367**) and three neoglycosides (**374**, $n = 1–3$) shows that O-glycosides are more active than neoglycosides and the monoglycosyldigitoxigenin is more active as an anticancer agent than the corresponding di- and trisaccharide derivatives.[136,137]

This *de novo* approach for the installation of unnatural pyranones is ideally suited for SAR studies of a myriad of substituted monosaccharides of varied substitution

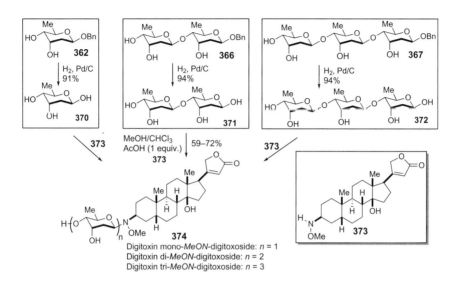

SCHEME 52. Syntheses of digitoxose mono-, di-, and trisaccharides and digitoxin methoxyamino neoglycosides.

patterns and stereochemistry. In terms of stereochemistry, these SAR studies include the synthesis and biological evaluation of a range of carbohydrate stereoisomers having both α- or β-, and D- or L-, configurations. To accomplish these for medicinal-chemistry studies, new postglycosylation methodologies are required. Examples of these efforts are outlined in Scheme 53, which include nucleophilic epoxidation (**375** to **376**) and opening (**376** to **379**), Wharton rearrangement (**376** to **377**), dihydroxylation (**377** to **378**), and diimide reduction (**377** to **380**). These studies have shown that unnatural digitoxin monoglycoside analogues having α-D-stereochemistry display enhanced anticancer activity.[138–143]

Antitumor evaluation of these various O-linked monosaccharide glycosides indicates that analogues having the rhamnose (6-deoxy-L-galactose) and amicetose (2,3,6-trideoxy-L-*erythro*-hexose) substitution mode and stereochemistry have the highest activity. This synthetic approach can be used to further explore this structure–activity relationship, and it has evolved into the synthesis and biological evaluation of mono- (**385a–d** and **386a–e**), di- (**387** and **388**), and trisaccharide (**389** and **390**) analogues having various substitution modes at C-6. The results of these two studies are outlined in tables I and II. From these studies we can conclude that α-L-rhamnosyl and α-L-amicetosyl analogues with just a simple methyl group at C-6 have the highest activity. The route to the (1→4)-linked trisaccharides starts with the glycosylation

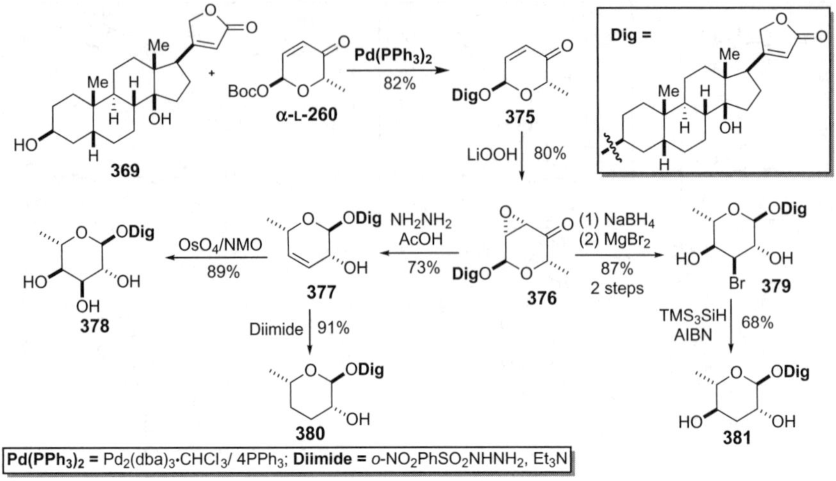

SCHEME 53. Digitoxin monosaccharide analogues: a survey of substitution modes and stereochemistry.

TABLE I
Cytotoxicity Evaluation of Digitoxin Analogues on NCI-H460 Human Lung Cancer Epithelial Cells

Compounds	IC_{50} (nM)[a]	GI_{50} (nM)[b]	Compounds	IC_{50} (nM)[a]	GI_{50} (nM)[b]
386a	52	2	385a	52	2
386b	57	2	385b	87	2
386c	116	5	385c	533	8
386d	130	6	385d	458	8
386e	180	13			

[a] The IC_{50} value was measured by a 12-h treatment in Hoechst and propidium iodide stain assays.
[b] The GI_{50} value was measured by a 48-h treatment in an MTT assay. All values represent the mean value of three independent experiments with duplicate determinations.

of digitoxigenin with enone **336a–d**, followed by borohydride reduction, which provides compounds **382a–e** having the 4-hydroxyl group ready for glycosylation to form disaccharide **383** and trisaccharide **384** (Scheme 54). Finally, a per-diimide reduction or dihydroxylation provides the desired target molecules **385** to **390**. Key to the success of this sequence is the bidirectional use of the diimide reduction and the highly stereoselective dihydroxylation reactions, which occur on a range of substrates without detrimental effects on yield and, for the dihydroxylation steps, any loss of stereoselectivity (thus installation of six stereocenters from **384** to **390**).[144–146]

This Achmatowicz approach can also be used for the synthesis of the trisaccharide portion of the glycosylated angucycline antibiotic natural products PI-080 and

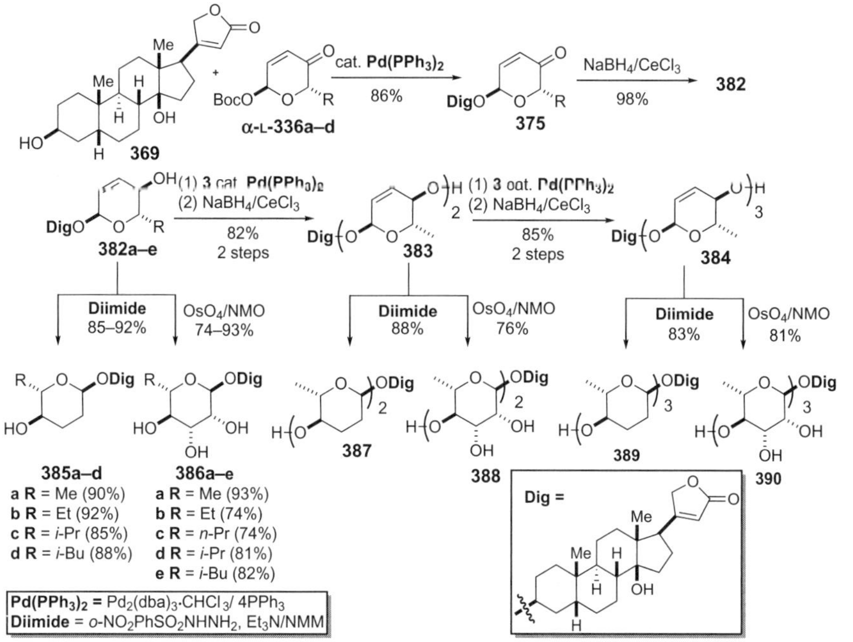

SCHEME 54. Synthesis of various C-5′-alkyl side-chains and oligodigitoxin analogues.

TABLE II
Cytotoxicity Evaluation of Digitoxin Analogues on NCI-H460 Human Lung Cancer Epithelial Cells

Compounds	IC$_{50}$ (nM)[a]	Compounds	IC$_{50}$ (nM)[a]
385a	48	386a	47
387	510	388	365
389	3963	390	1347

[a] The IC$_{50}$ values were measured by a 12-h treatment in Hoechst and propidium iodide stain assays. All values represent the mean value of three independent experiments with duplicate determinations.

landomycin E (Schemes 55 and 56). This approach, which uses the same postglycosylation strategy that installed the β-D-digitoxose sugar stereochemistry, can also be used to install the first β-D-olivose (2,6-dideoxy-D-*arabino*-hexose) component of the trisaccharide **398**, which consists of a β-D-olivose, α-L-rhodinose, and α-L-aculose (2,3,6-trideoxy-α-L-*glycero*-hex-2-enos-4-ulose) sequence. Thus, the same four-step synthesis that converted **359** into **362** was used for the conversion of β-D-**260** into **393**. Key to the success of this approach was the discovery of a highly regioselective

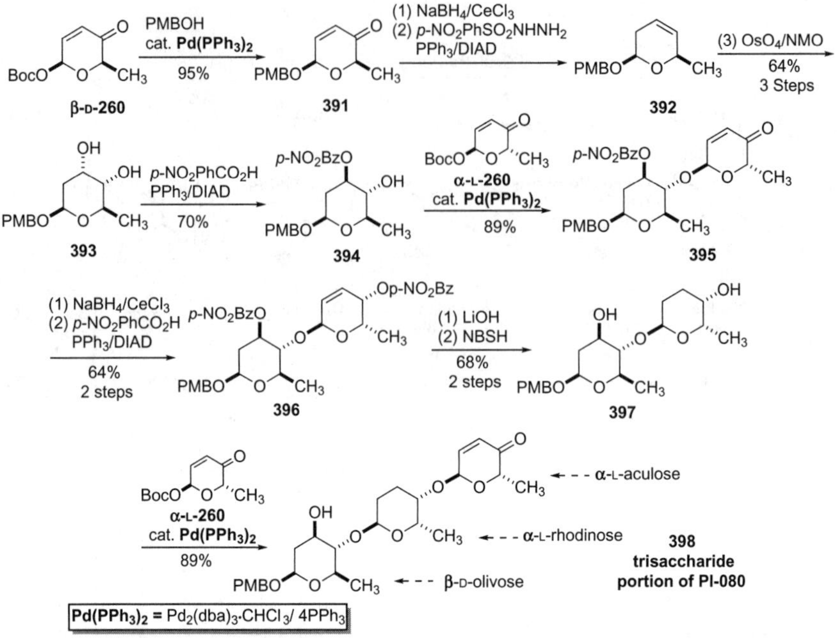

SCHEME 55. Synthesis of the trisaccharide portion of PI-080.

Mitsunobu-like inversion of the axial alcohol of a *cis*-1,2-diol (**393** to **394**). It is important to note that this transformation, in combination with a highly diastereoselective dihydroxylation reaction, becomes an excellent solution to the problem of 1,2-*trans*-diequatorial addition to a cyclohexene. In addition to a regioselective inversion, the Mitsunobu chemistry also differentially protects the diol for further glycosylation with enone α-L-**260** to give the coupled product **395**. 1,2-Reduction of the enone, Mitsunobu inversion (**395** to **396**), ester hydrolysis, and diimide reduction install the rhodinose (2,4,6-trideoxy-L-*threo*-hexose) stereochemistry of the second sugar in disaccharide **397**. Finally, a Pd-catalyzed glycosylation with α-L-**260** gives the PI-080 trisaccharide **398**. With the trisaccharide **398** in hand, it could be evaluated for anticancer activity. These studies found that **398** has wide-ranging cytotoxicity, displaying significant growth inhibition (GI_{50} from 0.1 to 11 µM) and cytotoxicity (LC_{50} from 5.1 to 100 µM) against various tumor cell lines.[147,148]

An approach similar to the PI-080 synthesis can also be used for synthesis of the trisaccharide portion of landomycin E. The synthesis begins with the conversion of β-D-pyranone **399** into the 2,6-dideoxyhexose **400** having β-D-*ribo* (digitoxose)

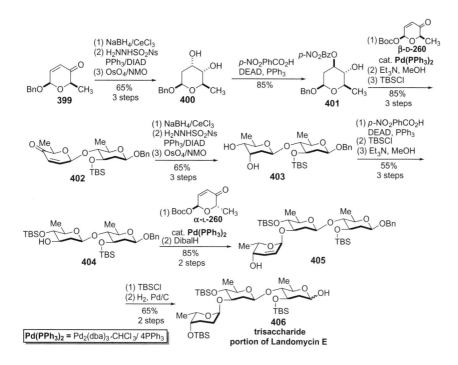

SCHEME 56. Synthesis of the trisaccharide portion of landomycin E.

stereochemistry. Once again the regioselective Mitsunobu inversion and protection can be used to prepare the 2,6-dideoxyhexose **401** having the β-D-*arabino* (olivose) stereochemistry. An additional nine-step sequence can be used to install a second β-D-olivose component to give **404** (via a β-D-digitoxose precursor **403**). Finally a Pd-glycosylation reduction/inversion strategy installs the terminal 2,4,6-trideoxy-L-*threo*-hexose (α-L-rhodinose) component in the landomycin E trisaccharide **406**.[149,150]

The O'Doherty group also examined the use of their Achmatowicz approach for the synthesis of analogues of the related angucycline quinonoid antibiotic jadomycin B, a compound that possesses a 2,6-dideoxy-L-*ribo*-hexose (α-L-digitoxose) component.[151] This effort involves the synthesis of a cyclitol analogue **417** of jadomycin B. The synthesis starts with enone **407**, which is reduced and reductively rearranged and dihydroxylated to install the β-L-*ribo* stereochemistry in **411**. Because the direct Pd-catalyzed glycosylation and cyclitolization failed (**418**+**419**), a Mitsunobu glycosylation of **412** with **411** was used to install the α-L-*ribo*-configured cyclitol.

Finally, a stepwise acid-catalyzed deprotection, oxazolidone formation, and deprotection provide the desired 5a-carba sugar analogue **417** of jadomycin B (Scheme 57).[152,153]

This Achmatowicz approach can also be used for the synthesis of oligosaccharide structural motifs that are even more complex. The power of this methodology is displayed nicely in its application to the dodecanyl tri- and tetrasaccharide glycosides, cleistriosides (**426, 427**) and cleistetrosides (**425** and **428** to **432**). Isolated from the African tree, *Cleistophilus patens*, these compounds are reported to possess significant antibacterial and possibly antimalarial properties.[154–156] The approach to these two classes of plant glycosides is outlined in Schemes 58 and 59. The *de novo* synthesis of these oligosaccharides, while clarifying various structural issues, can

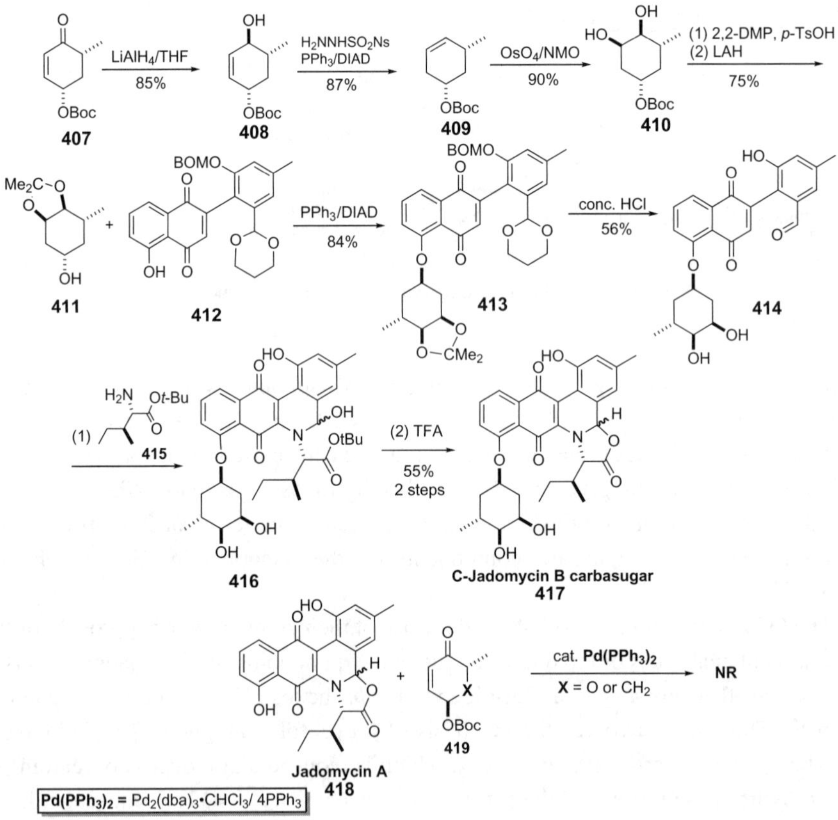

SCHEME 57. Synthesis of a carba sugar analogue of jadomycin B.

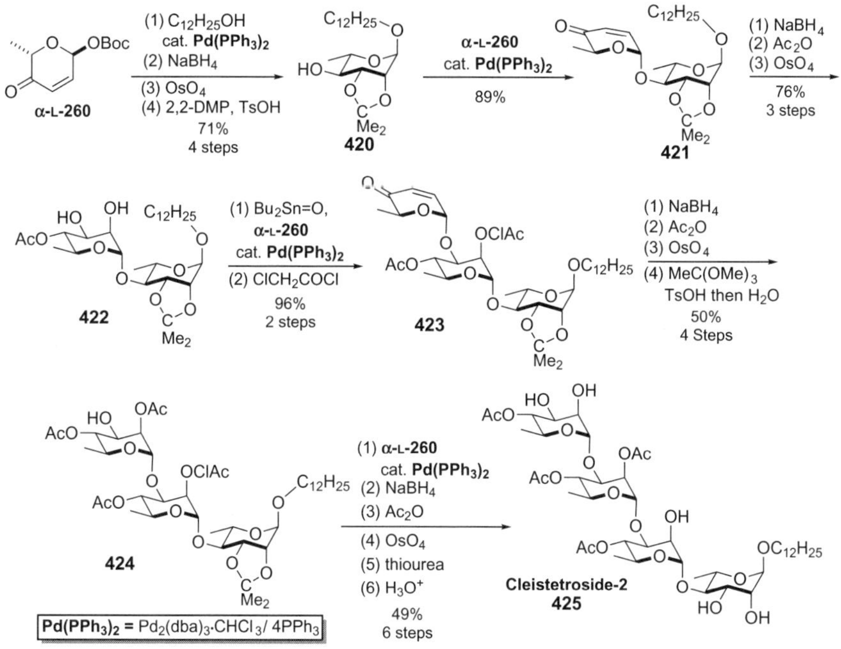

SCHEME 58. Syntheses of cleistetroside-2.

also be used to supply sufficient material for SAR-type studies as well as to serve as a test of the viability of this synthetic methodology.

The route begins with pyranone α-L-**260**, which in four steps (glycosylation, reduction, dihydroxylation, and acetonation) can be converted into the protected L-rhamnose **420**. A subsequent glycosylation (**420** to **421**), reduction, acylation, and dihydroxylation afford disaccharide **422**. A tin-directed regioselective (7:1) glycosylation of the diol in **422** installs a sugar at the C-3 position, which after chloroacetylation gives trisaccharide derivative **423**. Further elaborations of the pyranone component install the required L-rhamnose structure in **424** for both the cleistriosides and cleistetrosides. For example, in only six steps, compound **424** can be converted into **425** in a 49% overall yield. It should be noted that in terms of total steps, the *de novo* route to any of the cleistetrosides is comparable to the two previously reported routes to cleistetroside-2. In contrast to these two traditional routes, this *de novo* approach is more functionally divergent, which is desired for medicinal-chemistry studies.[157,158]

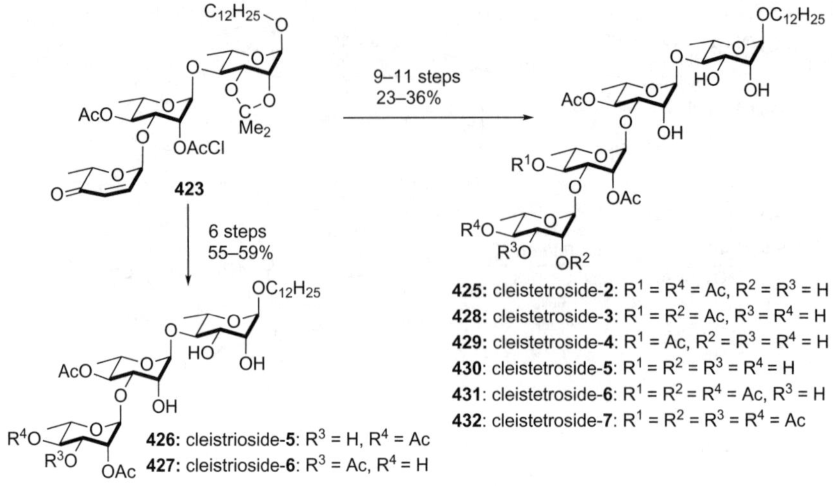

SCHEME 59. Syntheses of cleistrioside and cleistetroside families.

425: cleistetroside-2: $R^1 = R^4 = Ac, R^2 = R^3 = H$
428: cleistetroside-3: $R^1 = R^2 = Ac, R^3 = R^4 = H$
429: cleistetroside-4: $R^1 = Ac, R^2 = R^3 = R^4 = H$
430: cleistetroside-5: $R^1 = R^2 = R^3 = R^4 = H$
431: cleistetroside-6: $R^1 = R^2 = R^4 = Ac, R^3 = H$
432: cleistetroside-7: $R^1 = R^2 = R^3 = R^4 = Ac$

426: cleistrioside-5: $R^3 = H, R^4 = Ac$
427: cleistrioside-6: $R^3 = Ac, R^4 = H$

TABLE III
MIC(μM) Values for Cleistrioside and Cleistetroside Derivatives

Organism		426	427	425	428	429	430	431	432
E. coli	MG1655	>64	>64	>64	>64	>64	>64	>64	>64
	imp	>64	32	>64	>64	>64	16	>64	>64
B. subtilis	JH642	32	8	4	8	8	8	4	>64

Antibacterial activity was assessed by MIC method, which was performed by the broth dilution method described by the National Committee for Clinical Laboratory Standard Methods (M7-A6, 2003).

For instance, the pyranone ring in trisaccharide **423** is perfectly situated for further elaboration into the final rhamnose ring for cleistrioside-5 and -6 as well as the bis-rhamnose rings for cleistetrosides-2, -3, -4, -5, -6, and -7 (Scheme 59). This was accomplished in a range of 9–11 steps and 23–36% overall yields. This divergent approach is particularly valuable for medicinal-chemistry studies. The antibacterial activity of both the cleistriosides and cleistetrosides is outlined in Table III, while their anticancer activity is detailed in Table IV. It is worth noting that this combined synthetic/medicinal-chemistry effort was the first to disclose the anticancer activity for this class of natural products.

TABLE IV
Cancer Cytotoxicity of Cleistrioside and Cleistetroside Derivatives on NCI-H460 Human Lung Cancer Epithelial Cells

Compounds	IC_{50} (µM)	Compounds	IC_{50} (µM)
426	90.9	429	12.2
427	12.5	430	15.4
425	9.1	431	7.5
428	12.8	432	16.4

The IC_{50} value was measured by a 48-h treatment in an MTT assay. The results are an average of at least three independent experiments.

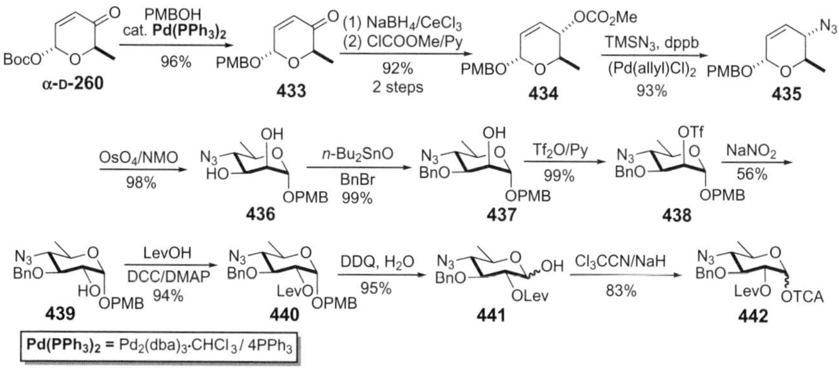

SCHEME 60. Synthesis of an anthrose glycosyl donor.

The related oligorhamnan, the anthrax tetrasaccharide **455** (an oligosaccharide expressed on the outermost surface of *Bacillus anthracis* spores), is probably the most structurally complex synthetic application to date of this *de novo* Achmatowicz approach to oligosaccharides. The approach merges the Pd-allylation chemistry to install amino groups at C-4 in sugars (Scheme 60) with the synthesis of rhamnose-containing oligosaccharides (Schemes 61 and 62).[159–161] Anthrax is the disease that is caused by the spore-forming bacterium *B. anthracis*.[162,163] As part of a larger effort to find a unique structural motif associated with the bacterium, the anthrax tetrasaccharide **455** was discovered. The novel tetrasaccharide consists of three L-rhamnose residues and a rare sugar, D-anthrose [4,6-dideoxy-4-(3-hydroxy-3-methylbutanamido)-2-*O*-methyl-β-D-glucose]. Several carbohydrate-based approaches to the anthrax tetrasaccharide have been reported, as well as one to a related trisaccharide.[164–170] All of these traditional carbohydrate routes drew the tetrasaccharide stereochemistry

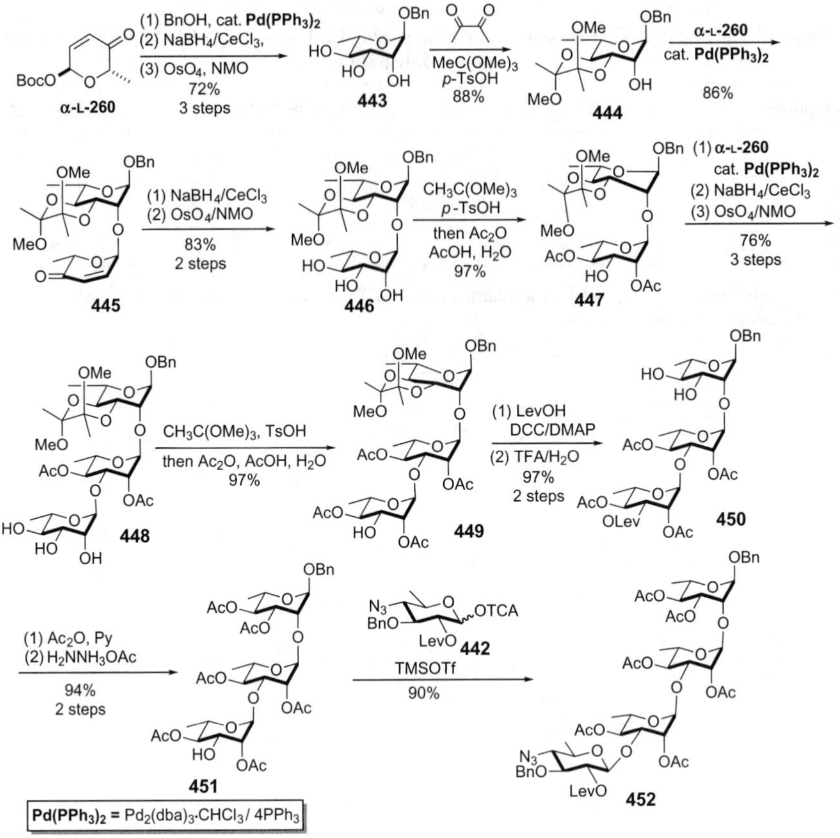

SCHEME 61. Assembly of the anthrax tetrasaccharide.

from L-rhamnose and 6-deoxy-D-galactose (D-fucose, the less common enantiomer). In contrast, this *de novo* approach to the tetrasaccharide **455** uses asymmetric catalysis to install the stereochemistry.

The synthesis of the anthrose precursor **442** is outlined in Scheme 60. The synthesis involves the use of two Pd–π-allylation reactions, one Pd-glycosylation, and one installation of azide at C-4. The route begins with the synthesis of PMB-protected pyranone **433** from **260**, followed by a Luche reduction and formation of the methyl carbonate (**433** to **434**). Once again, Pd-catalyzed C-4 allylic azide chemistry (**434** to **435**) is used to convert pyranone **433** into the allylic azide **435**, via **434**. Dihydroxylation of **435** gives the D-mannose derivative **436**. Next, the

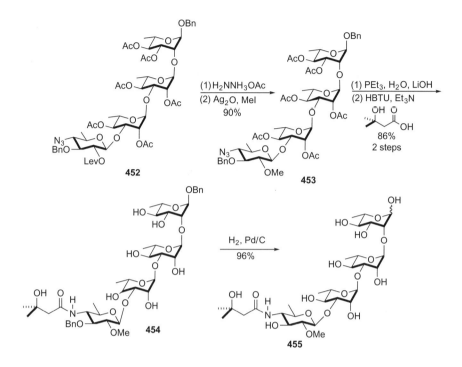

SCHEME 62. Final synthesis of the anthrax tetrasaccharide.

6-deoxy-D-*gluco*-functionality is installed by a protection step (**436** to **437**), followed by inversion strategy at C-2 (**437** to **439**, via **438**) to give the anthrose precursor **439**. Finally a Lev-type spiroketal protection, PMB-deprotection strategy, and trichloroacetimidate formation are used to convert compound **439** into the glycosyl donor **442**.[171]

A *de novo* Achmatowicz approach to the trirhamnosyl portion of the anthrax tetrasaccharide is outlined in Scheme 61. The route begins with the synthesis of disaccharide **447** from the enantiomeric α-L-pyranone **260** and benzyl alcohol. After glycosylation and postglycosylation transformations, the benzyl L-rhamnopyranoside **443** is produced. The 1,2-*trans*-diol of **443** is then protected via a Ley-type spiroketal to provide the monosaccharide derivative **444**. After a similar three-step glycosylation/postglycosylation sequence (**444** to **446**, via **445**), a suitably protected disaccharide **447** is produced, via a one-pot orthoester protocol.

Repeating the same three-step glycosylation and postglycosylation sequence converts disaccharide **447** into trisaccharide **448**. Similarly, a one-pot orthoester formation/acylation/hydrolysis sequence gave the trisaccharide derivative **449**. The instability of

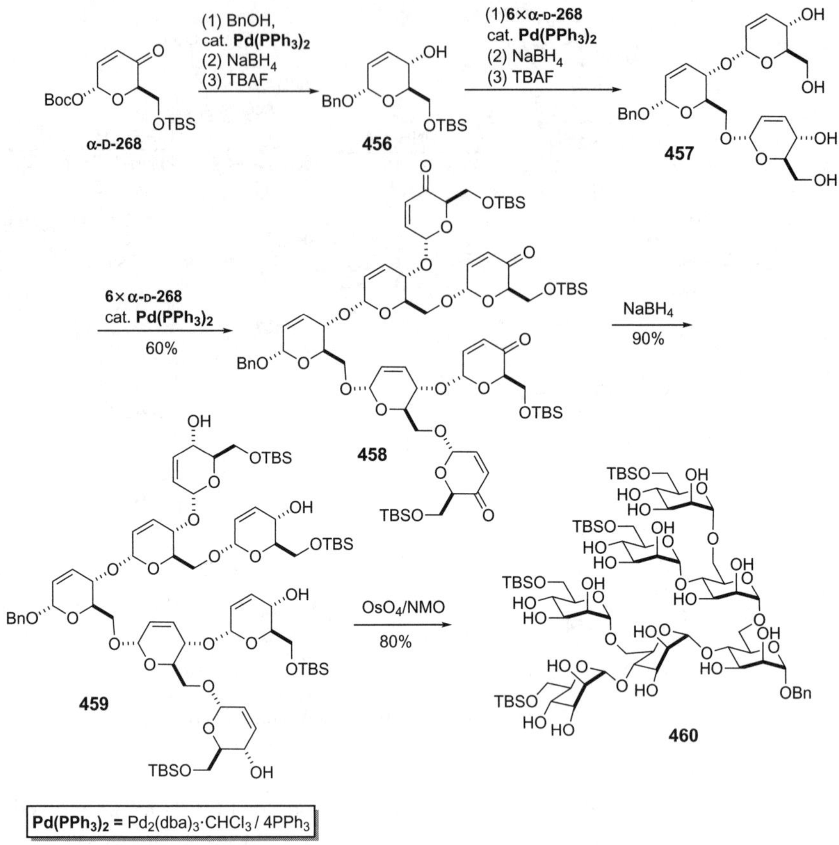

SCHEME 63. Synthesis of highly branched (1→4),(1→6)-heptasaccharides.

the Ley-spiroketal to the Lewis acidic nature of the traditional glycosylation conditions necessitated further protecting-group manipulations. To skirt this problem, the C-3 hydroxyl group is protected as a levulinate ester, and the Ley-spiroketal protecting group is removed to form trisaccharide derivative **450**. Next, an acetylation step and levulinate deprotection afforded compound **451**, which is set up for glycosylation by the anthrose sugar donor. Thus, exposure of **451** to **442** in the presence of TMSOTf affords the tetrasaccharide derivative **452** in excellent overall yield.

The deprotection of tetrasaccharide **452** into the final anthrax tetrasaccharide **455** was accomplished in a five-step sequence (Scheme 62). Removal of the levulinate protecting-group followed by Purdie methylation (MeI/Ag$_2$O) delivers the methyl ether group giving **453**. Azide reduction and amide formation afford the amide **454**. Finally, hydrogenolytic removal of the benzyl groups provided the natural anthrax tetrasaccharide **455**.

While these *de novo* approaches clearly provide complex structures by routes that are complementary to traditional carbohydrate approaches, this synthetic methodology offers its greatest efficiencies when applied bidirectionally. For the synthesis of rare and unnatural monosaccharides, the bidirectional application of this *de novo* Achmatowicz approach is the most powerful when applied to unnatural oligosaccharide motifs. An example of this is seen in the synthesis to the all-α-D-mannopyranose heptasaccharide **460** (Scheme 63). This synthesis of **460** is accomplished from the α-D-donor **268** in nine steps by simply applying the Pd-glycosylation and postglycosylation strategy three times in a tandem $(1 \to 4)$ and $(1 \to 6)$ bidirectional synthesis. Thus, in only six steps, compound **268** is converted in a highly stereoselective fashion into a trisaccharide derivative **457**. Similarly, repeating the glycosylation reaction (**458** to **459**) followed by dihydroxylation produces the heptasaccharide **460** with 35 stereocenters (14 stereocenters in one step, **459** to **460**), in excellent overall yield.[172]

Acknowledgments

We thank the National Institute of General Medical Sciences, NIH (GM090259), and the National Science Foundation (CHE-1213596) for generous financial support of our work.

References

1. E. Fischer and J. Tafel, Synthetische Versuche in der Zuckergruppe. III, *Ber.*, 22 (1889) 97–101.
2. E. Fischer, Synthese der Mannose und Lävulose, *Ber.*, 23 (1890) 370–394.
3. R. Lespieau, Synthesis of hexitols and pentitols from unsaturated polyhydric alcohols, *Adv. Carbohydr. Chem.*, 2 (1946) 107–118.
4. T. Mizuno and A. H. Weiss, Synthesis and utilization of formose sugars, *Adv. Carbohydr. Chem. Biochem.*, 29 (1974) 173–227.
5. M. Černý and J. Staněk, Jr., 1,6-Anhydro derivatives of aldohexoses, *Adv. Carbohydr. Chem. Biochem.*, 34 (1977) 23–177.
6. A. Zamojski, A. Banaszek, and G. Grynkiewicz, The synthesis of sugars from non-carbohydrate substrates, *Adv. Carbohydr. Chem. Biochem.*, 40 (1982) 1–129.
7. H. J. M. Gijsen, L. Qiao, W. Fitz, and C.-H. Wong, Recent advances in the chemoenzymatic synthesis of carbohydrates and carbohydrate mimetics, *Chem. Rev.*, 96 (1996) 443–474.
8. T. Hudlicky, D. A. Entwistle, K. K. Pitzer, and A. J. Thorpe, Modern methods of monosaccharide synthesis from non-carbohydrate sources, *Chem. Rev.*, 96 (1996) 1195–1220.
9. X. Yu and G. A. O'Doherty, *De novo* Synthesis in carbohydrate chemistry: From furans to monosaccharides and oligosaccharides, *ACS Symp. Ser.*, 990 (2008) 3–28.

10. R. H. Crabtree, No protection required, *Science*, 318 (2007) 756–757.
11. M. S. Chen and M. C. White, A predictably selective aliphatic C–H oxidation reaction for complex molecule synthesis, *Science*, 318 (2007) 783–787.
12. S. Y. Ko, A. W. M. Lee, S. Masamune, L. A. Reed, III., K. B. Sharpless, and F. J. Walker, Total synthesis of the L-hexoses, *Science*, 220 (1983) 949–951.
13. S. Ko, A. W. M. Lee, and S. Masamune, Total synthesis of the hexoses, *Tetrahedron*, 46 (1990) 245–264.
14. S. J. Danishefsky, Cycloaddition and cyclocondensation reactions of highly functionalized dienes: Applications to organic synthesis, *Chemtracts Org. Chem.*, 2 (1989) 273–297.
15. For improved catalysis see: S. E. Schaus, J. Branalt, and E. N. Jacobsen, Asymmetric hetero-Diels–Alder reactions catalyzed by chiral (salen)chromium(III) complexes, *J. Org. Chem.*, 63 (1998) 403–405.
16. S. J. Danishefsky and M. P. DeNinno, Totally synthetic routes to the higher monosaccharides, *Angew. Chem.*, 26 (1987) 15–23.
17. S. J. Danishefsky, W. H. Pearson, and B. E. Segmuller, Total synthesis of (±)-3-deoxy-D-manno-2-octulopyranosate (KDO), *J. Am. Chem. Soc.*, 107 (1985) 1280–1285.
18. S. J. Danishefsky, G. Phillips, and M. Ciufolini, A fully synthetic route to the papulacandins: Stereospecific spiroacetalization of a C-1-arylated methyl glycoside, *Carbohydr. Res.*, 171 (1987) 317–327.
19. G. D. Joly and E. N. Jacobsen, Catalyst-controlled diastereoselective hetero-Diels–Alder reactions, *Org. Lett.*, 4 (2002) 1795–1798.
20. A. B. Northrup and D. W. C. Macmillan, Two-step synthesis of carbohydrates by selective aldol reactions, *Science*, 305 (2004) 1752–1755.
21. G. N. Varseev and M. E. Maier, Total synthesis of (+/-)-symbioimine, *Angew. Chem. Int. Ed. Engl.*, 45 (2006) 4767–4771.
22. J. Maresh, J. Zhang, Y. L. Tzeng, N. A. Goodman, and D. G. Lynn, Rational design of inhibitors of VIrA-VirG two-component signal transduction, *Bioorg. Med. Chem. Lett.*, 17 (2007) 3281–3286.
23. W. B. Choi, L. J. Wilson, S. Yeola, D. C. Liotta, and R. F. Schinazi, In situ complexation directs the stereochemistry of N-glycosylation in the synthesis of thialanyl and dioxolanyl nucleoside analogs, *J. Am. Chem. Soc.*, 113 (1991) 9377–9379.
24. A. B. Northhrup, I. K. Mangion, F. Hettche, and D. W. C. MacMillan, Enantioselective organocatalytic direct aldol reactions of α-oxyaldehydes: Step one in a two-step synthesis of carbohydrates, *Angew. Chem. Int. Ed. Engl.*, 43 (2004) 2152–2154.
25. T. Hudlicky, K. K. Pitzer, M. R. Stabile, A. J. Thorpe, and G. M. Whited, Biocatalytic syntheses of protected D-mannose-d_5, D-mannose-d_7, D-mannitol-2,3,4,5,6-d_5, and D-mannitol-1,1,2,3,4,5,6,6-d_8, *J. Org. Chem.*, 61 (1996) 4151–4153.
26. D. T. Gibson, J. R. Koch, and R. E. Kallio, Oxidative degradation of aromatic hydrocarbons by microorganisms. I. Enzymatic formation of catechol from benzene, *Biochemistry*, 7 (1968) 2653–2662.
27. D. Gibson, M. Hemsley, H. Yoshioka, and T. Mabry, Formation of (+)-cis-2,3-dihydroxy-l-methylcyclohexa-4,6-diene from toluene by *Pseudomonas putida*, *Biochemistry*, 9 (1970) 1626–1630.
28. G. J. Zylstra and D. T. Gibson, Toluene degradation by *Pseudomonas putida* F1, *J. Biol. Chem.*, 264 (1989) 14940–14946.
29. C. R. Johnson, A. Golebiowski, D. H. Steensma, and M. A. Scialdone, Enantio- and diastereoselective transformations of cycloheptatriene to sugars and related products, *J. Org. Chem.*, 58 (1993) 7185–7194.
30. J. E. Baeckvall, S. E. Bystroem, and R. E. Nordberg, Stereo- and regioselective palladium-catalyzed 1,4-diacetoxylation of 1,3-dienes, *J. Org. Chem.*, 49 (1984) 4619–4631.
31. I. Henderson, K. B. Sharpless, and C. Wong, Synthesis of carbohydrates via tandem use of the osmium-catalyzed asymmetric dihydroxylation and enzyme-catalyzed aldol addition reactions, *J. Am. Chem. Soc.*, 116 (1994) 558–561.

32. M. S. M. Timmer, A. Adibekian, and P. H. Seeberger, Short *de novo* synthesis of fully functionalized uronic acid monosaccharides, *Angew. Chem. Int. Ed. Engl.*, 44 (2005) 7605–7607.
33. A. Adibekian, P. Bindschädler, M. S. M. Timmer, C. Noti, N. Schützenmeister, and P. H. Seeberger, De novo synthesis of uronic acid building blocks for assembly of heparin oligosaccharides, *Chem. Eur. J.*, 13 (2007) 4510–4527.
34. P. Stallforth, A. Adibekian, and P. H. Seeberger, De novo synthesis of a D-galacturonic acid thioglycoside as key to the total synthesis of a glycosphingolipid from *Sphingomonas yanoikuyae*, *Org. Lett.*, 10 (2008) 1573–1576.
35. L. Bouché and H.-U. Reißig, Synthesis of novel carbohydrate mimetics via 1,2-oxazines, *Pure Appl. Chem.*, 84 (2012) 23–36.
36. F. Pfrengle and H.-U. Reißig, Amino sugars and their mimetics via 1,2-oxazines, *Chem. Soc. Rev.*, 39 (2010) 549–557.
37. A. Dondoni and D. Perrone, Thiazole-based routes to amino hydroxyl aldehydes, and their use of the synthesis of biologically active compounds, *Aldrichim. Acta*, 30 (1997) 35–46.
38. A. Dondoni, G. Fantin, M. Fogagnolo, and A. Medici, Synthesis of long-chain sugars by iterative, diastereoselective homologation of 2,3-O-isopropylidene-D-glyceraldehyde with 2-trimethylsilylthiazole, *Angew. Chem. Int. Ed. Engl.*, 25 (1986) 835–837.
39. A. Dondoni, S. Franco, F. Junquera, F. L. Merchan, P. Merino, T. Tejero, and V. Bertolasi, Stereoselective homologation-amination of aldehydes by addition of their nitrones to C-2 metalated thiazoles. A general entry to a-amino aldehydes and amino sugars, *Chem. Eur. J.*, 1 (1995) 505–520.
40. D. Leonori and P. H. Seeberger, De novo synthesis of D- and L-fucosamine containing disaccharides, *J. Org. Chem.*, 9 (2013) 332–341.
41. D. Leonori and P. H. Seeberger, De novo synthesis of the bacterial 2-amino-2,6-dideoxy sugar building blocks D-fucosamine, D-bacillosamine, and D-xylo-6-deoxy-4-ketohexosamine, *Org. Lett.*, 14 (2012) 4954–4957.
42. A. Adibekian, M. M. Timmer, and P. Stallforth, Stereocontrolled synthesis of fully functionalized D-glucosamine monosaccharides via a domino nitro-Michael/Henry reaction, *Chem. Commun.* (2008) 3549–3551.
43. R. R. Schmidt and P. Zimmermann, Synthesis of D-erythro-sphingosines, *Tetrahedron Lett.*, 27 (1986) 481–484.
44. S. R. Baker, D. W. Clissold, and A. McKillop, Synthesis of leukotriene A4 methyl ester from D-glucose, *Tetrahedron Lett.*, 29 (1988) 991–994.
45. O. Calin, R. Pragani, and P. H. Seeberger, De novo synthesis of L-colitose and L-rhodinose building blocks, *J. Org. Chem.*, 77 (2012) 870–877.
46. P. Stallforth, A. Adibekian, and P. H. Seeberger, De novo synthesis of a D-galacturonic acid thioglycoside as key to the total synthesis of a glycosphingolipid from *Sphingomonas yanoikuyae*, *Org. Lett.*, 10 (2008) 1573–1576.
47. K. Mori and H. Kikuchi, Synthesis of (-)-biopterin, *Liebigs Ann. Chem.*, 10 (1989) 963–967.
48. M. Brasholz and H.-U. Reissig, Oxidative cleavage of 3-alkoxy-2,5-dihydrofurans and its application to the de novo synthesis of rare monosaccharides as exemplified by L-cymarose, *Angew. Chem. Int. Ed. Engl.*, 46 (2007) 1634–1637.
49. J. M. Harris, M. D. Keranen, H. Nguyen, V. G. Young, and G. A. O'Doherty, Syntheses of four D- and L-hexoses via diastereoselective and enantioselective dihydroxylation reactions, *Carbohydr. Res.*, 328 (2000) 17–36.
50. J. M. Harris, M. D. Keranen, and G. A. O'Doherty, Syntheses of D- and L-mannose, gulose, and talose via diastereoselective and enantioselective dihydroxylation reactions, *J. Org. Chem.*, 64 (1999) 2982–2983.
51. Md. M. Ahmed and G. A. O'Doherty, De novo synthesis of a galacto-papulacandin moiety via an iterative dihydroxylation strategy, *Tetrahedron Lett.*, 46 (2005) 4151–4155.

52. D. Gao and G. A. O'Doherty, Enantioselective synthesis of 10-epi-anamarine via an iterative dihydroxylation sequence, *Org. Lett.*, 7 (2005) 1069–1072.
53. Y. Zhang and G. A. O'Doherty, Remote steric effect on the regioselectivity of Sharpless asymmetric dihydroxylation, *Tetrahedron*, 61 (2005) 6337–6351.
54. Md. M. Ahmed, B. P. Berry, T. J. Hunter, D. J. Tomcik, and G. A. O'Doherty, De novo enantioselective syntheses of galacto-sugars and deoxy sugars via the iterative dihydroxylation of dienoate, *Org. Lett.*, 7 (2005) 745–748.
55. M. L. Bushey, M. H. Haukaas, and G. A. O'Doherty, Asymmetric aminohydroxylation of vinylfuran, *J. Org. Chem.*, 64 (1999) 2984–2985.
56. M. H. Haukaas, M. Li, A. M. Starosotnikov, and G. A. O'Doherty, De novo asymmetric approaches to 2-amino-N-(benzyloxycarbonyl)-1-(2'-furyl)-ethanol and 2-amino-N-(t-butoxycarbonyl)-1-(2'-furyl)-ethanol, *Heterocycles*, 76 (2008) 1549–1559.
57. M. Li and G. A. O'Doherty, An enantioselective synthesis of phomopsolide D, *Tetrahedron Lett.*, 45 (2004) 6407–6411.
58. M. Li, J. G. Scott, and G. A. O'Doherty, Synthesis of 7-oxa-phomopsolide E and its C-4 epimer, *Tetrahedron Lett.*, 45 (2004) 1005–1009.
59. M. H. Haukaas and G. A. O'Doherty, Enantioselective synthesis of N-Cbz-protected 6-amino-6-deoxy-mannose, gulose and talose, *Org. Lett.*, 3 (2001) 3899–3992.
60. J. M. Harris, M. D. Keranen, and G. A. O'Doherty, Syntheses of D- and L-mannose, gulose, and talose via diastereoselective and enantioselective dihydroxylation reactions, *J. Org. Chem.*, 64 (1999) 2982–2983.
61. J. M. Harris, M. D. Keranen, H. Nguyen, V. G. Young, and G. A. O'Doherty, Syntheses of various D- and L-hexoses via diastereoselective and enantioselective dihydroxylation reactions, *Carbohydr. Res.*, 328 (2000) 17–36.
62. P. Traxler, J. Gruner, and A. L. Auden, Papulacandins, a new family of antibiotics with antifungal activity. I. Fermentation, isolation, chemical and biological characterization of papulacandins A, B, C, D and E, *J. Antibiot.*, 30 (1977) 289–296.
63. D. Balachari and G. A. O'Doherty, A rapid enantioselective assembly of a manno-papulacandin ring system via the Sharpless catalytic asymmetric dihydroxylation, *Org. Lett.*, 2 (2000) 863–866.
64. D. Balachari and G. A. O'Doherty, Enantioselective synthesis of the papulacandin ring system: Conversion of the mannose diastereoisomer into a glucose stereoisomer, *Org. Lett.*, 2 (2000) 4033–4036.
65. D. Balachari, L. Quinn, and G. A. O'Doherty, Efficient synthesis of 5-aryl-2-vinylfurans by palladium catalyzed cross-coupling strategies, *Tetrahedron Lett.*, 40 (1999) 4769–4773.
66. V. VanRheenen, R. C. Kelly, and D. Y. Cha, An improved catalytic OsO_4 oxidation of olefins to cis-1,2-glycols using tertiary amine oxides as the oxidant, *Tetrahedron Lett.*, 17 (1976) 1973–1976.
67. M. M. Ahmed, B. P. Berry, T. J. Hunter, D. J. Tomcik, and G. A. O'Doherty, De novo enantioselective syntheses of galacto-sugars and deoxy-sugars via the iterative dihydroxylation of dienoate, *Org. Lett.*, 7 (2005) 745–748.
68. M. M. Ahmed and G. A. O'Doherty, De novo asymmetric syntheses of C-4-substituted sugars via an iterative dihydroxylation strategy, *Carbohydr. Res.*, 341 (2006) 1505–1521.
69. Y. Zhang and G. A. O'Doherty, Remote steric effect on the regioselectivity of the Sharpless asymmetric dihydroxylation, *Tetrahedron Lett.*, 61 (2005) 6337–6351.
70. Md. M. Ahmed and G. A. O'Doherty, De novo synthesis of galacto-sugar δ-lactones via a catalytic osmium/palladium/osmium reaction sequence, *Tetrahedron Lett.*, 46 (2005) 3015–3019.
71. D. Gao and G. A. O'Doherty, Enantioselective synthesis of 10-epi-anammarine via an iterative dihydroxylation sequence, *Org. Lett.*, 7 (2005) 1069–1072.
72. D. Gao and G. A. O'Doherty, De novo asymmetric synthesis of anamarine and analogs, *J. Org. Chem.*, 70 (2005) 9932–9939.

73. M. M. Ahmed and G. A. O'Doherty, De novo asymmetric syntheses of C-4-substituted sugars via an iterative dihydroxylation strategy, *Carbohydr. Res.*, 341 (2006) 1505–1521.
74. M. M. Ahmed and G. A. O'Doherty, De novo asymmetric syntheses of D- and L-talose via an iterative dihydroxylation of dienoates, *J. Org. Chem.*, 67 (2005) 10576–10578.
75. M. M. Ahmed and G. A. O'Doherty, De novo asymmetric synthesis of a galacto-papulacandin moiety via an iterative dihydroxylation strategy, *Tetrahedron Lett.*, 46 (2005) 4151–4155.
76. M. H. Haukaas and G. A. O'Doherty, Synthesis of D- and L-deoxymannojirimycin via an asymmetric aminohydroxylation of vinylfuran, *Org. Lett.*, 3 (2001) 401–404.
77. R. S. Babu and G. A. O'Doherty, A palladium-catalyzed glycosylation reaction: The de novo synthesis of natural and unnatural glycosides, *J. Am. Chem. Soc.*, 125 (2003) 12406–12407.
78. Concurrent with these studies was the similar discovery by Feringa, Lee:A. C. Comely, R. Eelkema, A. J. Minnaard, and B. L. Feringa, De novo asymmetric bio- and chemocatalytic synthesis of saccharides—Stereoselective formal O-glycoside bond formation using palladium catalysis, *J. Am. Chem. Soc.*, 125 (2003) 8714–8715.
79. H. Kim, H. Men, and C. Lee, Stereoselective palladium-catalyzed O-glycosylation using glycals, *J. Am. Chem. Soc.*, 126 (2004) 1336–1337.
80. The poor reactivity in Pd-catalyzed allylation reaction of alcohols as well as a nice solution to this problem was reported, see: H. Kim and C. Lee, A mild and efficient method for the stereoselective formation of C-O bonds: Palladium-catalyzed allylic etherification using zinc(II) alkoxides, *Org. Lett.*, 4 (2002) 4369–4437.
81. For a related Rh system, see: P. A. Evans and L. J. Kennedy, Enantiospecific and regioselective rhodium-catalyzed allylic alkylation: Diastereoselective approach to quaternary carbon stereogenic centers, *Org. Lett.*, 2 (2000) 2213–2215.
82. R. S. Babu and G. A. O'Doherty, A palladium-catalyzed glycosylation reaction: The de novo synthesis of natural and unnatural glycosides, *J. Am. Chem. Soc.*, 125 (2003) 12406–12407.
83. H. Guo and G. A. O'Doherty, *De novo* asymmetric synthesis of daumone via a palladium catalyzed glycosylation, *Org. Lett.*, 7 (2005) 3921–3924.
84. T. J. Baiga, H. Guo, Y. Xing, G. A. O'Doherty, A. Parrish, A. Dillin, M. B. Austin, J. P. Noel, and J. La Clair, Metabolite induction of *Caenorhabditis elegans* dauer larvae arises via transport in the pharynx, *ACS Chem. Biol.*, 3 (2008) 294–304.
85. S. R. Guppi, M. Zhou, and G. A. O'Doherty, *De novo* asymmetric synthesis of homo-adenosine via a palladium catalyzed n-glycosylation, *Org. Lett.*, 8 (2006) 293–296.
86. M. Hino, O. Nakayama, Y. Tsurumi, K. Adachi, T. Shibata, H. Terano, M. Kohsaka, H. Aoki, and H. Imanaka, Studies of an immunomodulator, swainsonine. I. enhancement of immune response by swainsonine in vitro, *J. Antibiot.*, 38 (1985) 926–935.
87. M. Patrick, M. W. Adlard, and T. Keshavarz, Production of an indolizidine alkaloid, swainsonine by the filamentous fungus, *Metarhizium anisopliae*, *Biotechnol. Lett.*, 15 (1993) 997–1000.
88. S. M. Colegate, P. R. Dorling, and C. R. Huxtable, A spectroscopic investigation of swainsonine: An α-mannosidase inhibitor isolated from *Swainsona canescens*, *Aust. J. Chem.*, 32 (1979) 2257–2264.
89. R. J. Molyneux and L. F. James, Loco intoxication: Indolizidine alkaloids of spotted locoweed (*Astragalus lentiginosus*), *Science*, 216 (1982) 190–191.
90. D. Davis, P. Schwarz, T. Hernandez, M. Mitchell, B. Warnock, and A. D. Elbein, Isolation and characterization of swainsonine from Texas locoweed (*Astragalus emoryanus*), *Plant Physiol.*, 76 (1984) 972–975.
91. Y. F. Liao, A. Lal, and K. W. Moremen, Cloning, expression, purification, and characterization of the human broad specificity lysosomal acid α-mannosidase, *J. Biol. Chem.*, 271 (1996) 28348–28358.
92. A. D. Elbein, R. Solf, P. R. Dorling, and K. Vosbeck, Swainsonine: An inhibitor of glycoprotein processing, *Proc. Natl. Acad. Sci. U.S.A.*, 78 (1981) 7393–7397.

93. P. C. Das, J. D. Robert, S. L. White, and K. Olden, Activation of resident tissue-specific macrophages by swainsonine, *Oncol. Res.*, 7 (1995) 425–433.
94. P. E. Goss, C. L. Reid, D. Bailey, and J. W. Dennis, Phase IB clinical trial of the oligosaccharide processing inhibitor swainsonine in patients with advanced malignancies, *Clin. Cancer Res.*, 3 (1997) 1077–1086.
95. B. Davis, A. A. Bell, R. J. Nash, A. A. Watson, R. C. Griffiths, M. G. Jones, C. Smith, and G. W. J. Fleet, L-(+)-swainsonine and other pyrrolidine inhibitors of naringinase: Through an enzymic looking glass from D-mannosidase to L-rhamnosidase, *Tetrahedron Lett.*, 37 (1996) 8565–8568.
96. For the first syntheses, see: H. A. Mezher, L. Hough, and A. C. Richardson, A chiral synthesis of swainsonine from D-glucose, *J. Chem. Soc., Chem. Commun.* (1984) 447–448.
97. G. W. J. Fleet, M. J. Gough, and P. W. Smith, Enantiospecific synthesis of swainsonine, (1*S*, 2*R*, 8*R*, 8a*R*)-1,2,8-trihydroxyoctahydroindolizine, from D-mannose, *Tetrahedron Lett.*, 25 (1984) 1853–1856.
98. For a review of swainsonine syntheses, see: A. El Nemr, Synthetic methods for the stereoisomers of swainsonine and its analogues, *Tetrahedron*, 56 (2000) 8579–8629.
99. R. Martin, C. Murruzzu, M. A. Pericas, and A. Riera, General approach to glycosidase inhibitors. Enantioselective synthesis of deoxymannojirimycin and swainsonine, *J. Org. Chem.*, 70 (2005) 2325–2328.
100. L. Song, E. N. Duesler, and P. S. Mariano, Stereoselective synthesis of polyhydroxylated indolizidines based on pyridinium salt photochemistry and ring rearrangement metathesis, *J. Org. Chem.*, 69 (2004) 7284–7293.
101. K. B. Lindsay and S. G. Pyne, Asymmetric synthesis of (−)-swainsonine, *Aust. J. Chem.*, 57 (2004) 669–672.
102. W. H. Pearson, Y. Ren, and J. D. Powers, A synthesis of (1*S*,2*R*,8*R*,8a*R*)-8-hydroxy-1,2-(isopropylidenedioxy)indolizidin-5-one from D-ribose: Improved access to (-)-swainsonine and its analogs, *Heterocycles*, 58 (2002) 421–430.
103. K. B. Lindsay and S. G. Pyne, Asymmetric synthesis of (−)-swainsonine, (+)-1,2-di-epi-swainsonine, and (+)-1,2,8-tri-epi-swainsonine, *J. Org. Chem.*, 67 (2002) 7774–7780.
104. N. Buschmann, A. Rueckert, and S. Blechert, A new approach to (−)-swainsonine by rutheniumcatalyzed ring rearrangement, *J. Org. Chem.*, 67 (2002) 4325–4329.
105. H. Zhao, S. Hans, X. Cheng, and D. R. Mootoo, Allylated monosaccharides as precursors in triple reductive amination strategies: Synthesis of castanospermine and swainsonine, *J. Org. Chem.*, 66 (2001) 1761–1767.
106. J. Ceccon, A. E. Greene, and J. F. Poisson, Asymmetric [2 + 2] cycloaddition: Total synthesis of (−)-swainsonine and (+)-6-epicastanospermine, *Org. Lett.*, 8 (2006) 4739–4742.
107. C. W. G. Au and S. G. Pyne, Asymmetric synthesis of anti-1,2-amino alcohols via the Borono–Mannich reaction: A formal synthesis of (−)-swainsonine, *J. Org. Chem.*, 71 (2006) 7097–7099.
108. I. Dechamps, D. Pardo, and J. Cossy, Enantioselective ring expansion of prolinols and ring-closing metathesis: Formal synthesis of (-)-swainsonine, *Arkivoc*, 5 (2007) 38–45.
109. H. Guo and G. A. O'Doherty, *De novo* asymmetric synthesis of D- and L-swainsonine, *Org. Lett.*, 8 (2006) 1609–1612.
110. J. A. Coral, H. Guo, M. Shan, and G. A. O'Doherty, *De novo* asymmetric approach to 8a-epi-swainsonine, *Heterocycles*, 79 (2009) 521–529.
111. J. N. Abrams, R. S. Babu, H. Guo, D. Le, J. Le, J. M. Osbourn, and G. A. O'Doherty, *De novo* asymmetric synthesis of 8a-epi-swainsonine, *J. Org. Chem.*, 73 (2008) 1935–1940.
112. H. Guo and G. A. O'Doherty, *De novo* asymmetric syntheses of D-, L- and 8-epi-swainsonine, *Tetrahedron Lett.*, 64 (2008) 304–313.

113. J. N. Abrams, R. S. Babu, H. Guo, D. Le, J. Le, J. M. Osbourn, and G. A. O'Doherty, *De novo* asymmetric synthesis of 8a-epi-swainsonine, *J. Org. Chem.*, 73 (2008) 1935–1940.
114. R. S. Babu and G. A. O'Doherty, Palladium catalyzed glycosylation reaction: *De novo* synthesis of trehalose analogues, *J. Carbohydr. Chem.*, 24 (2005) 169–177.
115. M. Shan and G. A. O'Doherty, *De novo* asymmetric synthesis of SL0101 and its analogues via a palladium-catalyzed glycosylation, *Org. Lett.*, 8 (2006) 5149–5152.
116. R. M. Mrozowski, R. Vemula, B. Wu, Q. Zhang, B. R. Schroederd, M. K. Hilinski, D. E. Clarke, S. M. Hecht, G. A. O'Doherty, and D. A. Lannigan, and D. A. Lannigan, Improving the affinity of SL0101 for RSK using structure-based design, *ACS Med. Chem. Lett.*, 3 (2012) 1086–1090.
117. M. Shan and G. A. O'Doherty, Synthesis of cyclitol sugars via Pd-catalyzed cyclopropanol ring opening, *Synthesis*, 19 (2008) 3171–3179.
118. M. Shan and G. A. O'Doherty, Synthesis of carbasugar C-1 phosphates via Pd-catalyzed cyclopropanol ring opening, *Org. Lett.*, 10 (2008) 3381–3384.
119. M. Shan and G. A. O'Doherty, Synthesis of SL0101 carbasugar analogues: Carbasugars via Pd-catalyzed cyclitolization and post cyclitolization transformations, *Org. Lett.*, 12 (2010) 2986–2989.
120. H. He, R. T. Williamson, B. Shen, E. I. Grazaini, H. Y. Yang, S. M. Sakya, P. J. Petersen, and G. T. Carter, Mannopeptimycins, novel antibacterial glycopeptides from *Streptomyces hygroscopicus*, LL-AC98, *J. Am. Chem. Soc.*, 124 (2002) 9729–9736.
121. S. R. Guppi and G. A. O'Doherty, Synthesis of aza-analogues of the glycosylated tyrosine portion of mannopeptimycin-E, *J. Org. Chem.*, 72 (2007) 4966–4969.
122. R. S. Babu, S. R. Guppi, and G. A. O'Doherty, Synthetic studies towards mannopeptimycin-E: Synthesis of a O-linked tyrosine 1,4-α, α-manno, manno-pyanosyl-pyranoside, *Org. Lett.*, 8 (2006) 1605–1608.
123. S. A. Borisova, S. R. Guppi, H. J. Kim, B. Wu, H.-W. Liu, and G. A. O'Doherty, *De novo* synthesis of glycosylated methymycin analogues, *Org. Lett.*, 12 (2010) 5150–5153.
124. J. Haux, O. Klepp, O. Spigset, and S. Tretli, Digitoxin medication and cancer; case control and internal dose-response studies, *BMC Cancer*, 1 (2001) 11.
125. S. K. Manna, N. K. Sah, R. A. Newman, A. Cisneros, and B. B. Aggarwal, Oleandrin suppresses activation of nuclear transcription factor-κB, activator protein-1, and c-jun NH$_2$-terminal kinase, *Cancer Res.*, 60 (2000) 3838–3847.
126. J. Haux, Digitoxin is a potential anticancer agent for several types of cancer, *Med. Hypotheses*, 53 (1999) 543–548.
127. M. López-Lázaro, N. Pastor, S. S. Azrak, M. J. Ayuso, C. A. Austin, and F. Cortes, Digitoxin inhibits the growth of cancer cell lines at concentrations commonly found in cardiac patients, *J. Nat. Prod.*, 68 (2005) 1642–1645.
128. J. M. Langenhan, N. R. Peters, I. A. Guzei, F. M. Hoffman, and J. S. Thorson, Enhancing the anticancer properties of cardiac glycosides by neoglycorandomization, *Proc. Natl. Acad. Sci. U.S.A.*, 102 (2005) 12305–12310.
129. For neoglycoside formation see: F. Peri, P. Dumy, and M. Mutter, Chemo- and stereoselective glycosylation of hydroxylamino derivatives: A versatile approach to glycoconjugates, *Tetrahedron*, 54 (1998) 12269–12278.
130. For neoglycoside formation see: F. Peri, J. Jimenez-Barbero, V. Garcia-Aparico, I. Tvaroška, and F. Nicotra, Synthesis and conformational analysis of novel N(OCH$_3$)-linked disaccharide analogues, *Chem. Eur. J.*, 10 (2004) 1433–1444.
131. M. Zhou and G. A. O'Doherty, *De novo* approach to 2-deoxy-*O*-glycosides: Asymmetric syntheses of digioxose and digitoxin, *J. Org. Chem.*, 72 (2007) 2485–2493.
132. M. Zhou and G. A. O'Doherty, *De novo* asymmetric synthesis of digitoxin via a palladium catalyzed glycosylation reaction, *Org. Lett.*, 8 (2006) 4339–4342.

133. For reduction of pyranones, the CeCl$_3$ is necessary to avoid 1,4-reduction products. For Luche reduction, see: J. L. Luche, Selective 1,2 reductions of conjugated ketones, *J. Am. Chem. Soc.*, 100 (1978) 2226–2227.
134. J. F. King and A. D. Allbutt, Remarkable stereoselectivity in the hydrolysis of dioxolenium ions and orthoesters fused to anchored six-membered rings, *Can. J. Chem.*, 48 (1970) 1754–1769.
135. T. L. Lowary and O. Hindsgaul, Recognition of synthetic O-methyl, epimeric, and amino analogues of the acceptor α-L-Fuc*p*-(1→2)-β-D-Galp-or glycosyltransferases, *Carbohydr. Res.*, 251 (1994) 33–67.
136. A. Iyer, M. Zhou, N. Azad, H. Elbaz, L. Wang, D. K. Rogalsky, Y. Rojanasakul, G. A. O'Doherty, and J. M. Langenhan, A direct comparison of the anticancer activities of digitoxin MeON-neoglycosides and O-glycosides: Oligosaccharide chain length-dependant induction of caspase-9-mediated apoptosis, *ACS Med. Chem. Lett.*, 1 (2010) 326–330.
137. M. Zhou and G. A. O'Doherty, The *de novo* synthesis of oligosaccharides: Application to the medicinal chemical study of digitoxin, *Curr. Top. Med. Chem.*, 8 (2008) 114–125.
138. J. W. Hinds, S. B. McKenna, E. U. Sharif, H.-Y. L. Wang, N. G. Akhmedov, and G. A. O'Doherty, C3'/C4'-stereochemical effects of digitoxigenin -L-/-D-glycoside in cancer cytotoxicity, *ChemMedChem*, 8 (2013) 63–69.
139. H.-Y. L. Wang, W. Xin, M. Zhou, T. A. Stueckle, Y. Rojanasakul, and G. A. O'Doherty, Stereochemical survey of digitoxin monosaccharides, *ACS Med. Chem. Lett.*, 2 (2011) 73–78.
140. M. Shan, Y. Xing, and G. A. O'Doherty, *De novo* asymmetric synthesis of an a-6-deoxyaltropyranoside as well as its 2-/3-deoxy and 2,3-dideoxy congeners, *J. Org. Chem.*, 74 (2009) 5961–5966.
141. H.-Y. L. Wang and G. A. O'Doherty, *De novo* synthesis of deoxy sugar via a Wharton rearrangement, *Chem. Commun.*, 47 (2011) 10251–10253.
142. M. F. Cuccarese, H.-Y. L. Wang, and G. A. O'Doherty, De novo synthesis of ido-pyranoside and 3-deoxy sugar congeners via Wharton rearrangement, *Eur. J. Org. Chem.* (2013) 3067–3075.
143. M. H. Haukaas and G. A. O'Doherty, Enantioselective synthesis of 2-deoxy and 2,3-dideoxyhexoses, *Org. Lett.*, 4 (2002) 1771–1774.
144. M. Zhou and G. A. O'Doherty, A stereoselective synthesis of digitoxin and digitoxigen mono- and bisdigitoxoside from digitoxigenin via a palladium-catalyzed glycosylation, *Org. Lett.*, 8 (2006) 4339–4342.
145. H.-Yu L. Wang, B. Wu, Q. Zhang, S.-W. Kang, Y. Rojanasakul, and G. A. O'Doherty, C5'-alkyl substitution effects on digitoxigenin-L-glycoside epithelial human lung cancer cells cytotoxicity, *ACS Med. Chem. Lett.*, 2 (2011) 259–263.
146. H.-Y. L. Wang, Y. Rojanasakul, and G. A. O'Doherty, Synthesis and evaluation of the α-D-/α-L-rhamnosyl and amicetosyl digitoxigenin oligomers as anti-tumor agents, *ACS Med. Chem. Lett.*, 2 (2011) 264–269.
147. X. Yu and G. A. O'Doherty, *De novo* asymmetric synthesis and biological evaluation of the trisaccharide portion of PI-080 and vineomycin B2, *Org. Lett.*, 10 (2008) 4529–4532.
148. Q. Chen, Y. Zhong, and G. A. O'Doherty, Convergent *de novo* synthesis of vineomycinone B2 methyl ester, *Chem. Commun.*, 49 (2013) 6806–6808.
149. M. S. Abdelfattah, M. K. Kharel, J. A. Hitron, I. Baig, and J. Rohr, Moromycins A and B, isolation and structure elucidation of C-glycosylangucycline-type antibiotics from streptomyces sp. KY002, *J. Nat. Prod.*, 9 (2008) 1569–1573.
150. M. Zhou and G. A. O'Doherty, *De novo* Asymmetric synthesis of the trisaccharide subunit of landomycin A and E, *Org. Lett.*, 10 (2008) 2283–2286.
151. N. Tibrewal, T. E. Downey, S. G. Van Lanen, E. U. Sharif, G. A. O'Doherty, and J. Rohr, Roles of the synergistic reductive *o*-methyltransferase gilm and of *o*-methyltransferase gilmt in the gilvocarcin biosynthetic pathway, *J. Am. Chem. Soc.*, 134 (2012) 12402–12405.

152. E. U. Sharif and G. A. O'Doherty, Biosynthesis and total synthesis studies on the jadomycin family of natural products, *Eur. J. Org. Chem.*, 11 (2012) 2095–2108.
153. M. Shan, E. U. Sharif, and G. A. O'Doherty, Total synthesis of jadomycin A and carbasugar analogue of jadomycin B, *Angew. Chem. Int. Ed. Engl.*, 49 (2010) 9492–9495.
154. P. Tané, J. P. Ayafor, B. L. Sondengam, C. Lavaud, G. Massiot, J. D. Connolly, D. S. Rycroft, and N. Woods, Partially-acetylated dodecanyl tri- and tetra-rhamnoside derivatives from *Cleistopholis glauca* (Annonaceae), *Tetrahedron Lett.*, 29 (1988) 1837–1840.
155. V. Seidel, F. Baileul, and P. G. Waterman, Partially acetylated tri- and tetrarhamnoside dodecanyl ether derivatives from *Cleistopholis patens*, *J. Phytochem.*, 52 (1999) 465–472.
156. J.-F. Hu, E. Garo, G. W. Hough, M. G. Goering, M. O'Neil-Johnson, and G. R. Eldridge, Antibacterial, partially acetylated oligorhamnosides from *Cleistopholis patens*, *J. Nat. Prod.*, 69 (2006) 585–590.
157. P. Shi, M. Silva, B. Wu, H.-Y. L. Wang, N. G. Akhmedov, M. Li, P. Beuning, and G. A. O'Doherty, Structure activity relationship study of the cleistrioside/cleistetroside natural products for antibacterial/anticancer activity, *ACS Med. Chem. Lett.*, 3 (2012) 1086–1090.
158. B. Wu, M. Li, and G. A. O'Doherty, Synthesis of several cleistrioside and cleistetroside natural products via a divergent *de novo* asymmetric approach, *Org. Lett.*, 12 (2010) 5466–5469.
159. H. Guo and G. A. O'Doherty, De novo asymmetric synthesis of the anthrax tetrasaccharide by a palladium-catalyzed glycosylation reaction, *Angew. Chem. Int. Ed. Engl.*, 46 (2007) 5206–5208.
160. H. Guo and G. A. O'Doherty, De novo asymmetric synthesis of anthrax tetrasaccharide and related tetrasaccharide, *J. Org. Chem.*, 73 (2008) 5211–5220.
161. H.-Y. L. Wang, H. Guo, and G. A. O'Doherty, De novo asymmetric synthesis of rhamno di- and trisaccharides related to the anthrax tetrasaccharide, *Tetrahedron*, 69 (2013) 3432–3436.
162. M. Mock and A. Fouet, Anthrax, *Annu. Rev. Microbiol.*, 55 (2001) 647–671.
163. P. Sylvestre, E. Couture-Tosi, and M. Mock, A collagen-like surface glycoprotein is a structural component of the *Bacillus anthracis* exosporium, *Mol. Microbiol.*, 45 (2002) 169–178.
164. D. B. Werz and P. H. Seeberger, Total synthesis of antigen *Bacillus anthracis* tetrasaccharide—Creation of an anthrax vaccine candidate, *Angew. Chem. Int. Ed. Engl.*, 44 (2005) 6315–6318.
165. R. Adamo, R. Saksena, and P. Kovac, Synthesis of the anomer of the spacer-equipped tetrasaccharide side chain of the major glycoprotein of the *Bacillus anthracis* exosporium, *Carbohydr. Res.*, 340 (2005) 2579–2582.
166. R. Saksena, R. Adamo, and P. Kovac, Synthesis of the tetrasaccharide side chain of the major glycoprotein of the *Bacillus anthracis* exosporium, *Med. Chem. Lett.*, 16 (2006) 615–617.
167. R. Adamo, R. Saksena, and P. Kovac, Studies towards a conjugate vaccine for anthrax: Synthesis of the tetrasaccharide side chain of the bacillus anthracis exosporium, *Helv. Chim. Acta*, 89 (2006) 1075–1089.
168. A. S. Mehta, E. Saile, W. Zhong, T. Buskas, R. Carlson, E. Kannenberg, Y. Reed, C. P. Quinn, and G. J. Boons, Synthesis and antigenic analysis of the BclA glycoprotein oligosaccharide from the *Bacillus anthracis* exosporium, *Chem. Eur. J.*, 12 (2006) 9136–9149.
169. D. Crich and O. Vinogradova, Synthesis of the antigenic tetrasaccharide side chain from the major glycoprotein of *Bacillus anthracis* exosporium, *J. Org. Chem.*, 72 (2007) 6513–6520.
170. D. B. Werz, A. Adibekian, and P. H. Seeberger, Synthesis of a spore surface pentasaccharide of *Bacillus anthracis*, *Eur. J. Org. Chem.*, 12 (2007) 1976–1982.
171. H.-Y. L. Wang, H. Guo, and G. A. O'Doherty, *De novo* asymmetric synthesis of rhamno di- and trisaccharides related to the anthrax tetrasaccharide, *Tetrahedron*, 69 (2013) 3432–3436.
172. R. S. Babu, Q. Chen, S.-W. Kang, M. Zhou, and G. A. O'Doherty, *De novo* synthesis of oligosaccharides using green chemistry principles, *J. Am. Chem. Soc.*, 134 (2012) 11952–11955.

RECENT ADVANCES TOWARD THE DEVELOPMENT OF INHIBITORS TO ATTENUATE TUMOR METASTASIS VIA THE INTERRUPTION OF LECTIN–LIGAND INTERACTIONS

Rachel Hevey and Chang-Chun Ling

Alberta Glycomics Centre, Department of Chemistry, University of Calgary, Calgary, Alberta, Canada

I. Introduction	126
1. Siglecs	127
2. Galectins	130
3. Selectins	134
II. Roles of Carbohydrates in Tumor Development	137
1. Abnormal Glycosylation in Tumors	137
2. Abnormal Glycosylation and Cancer Metastasis	140
III. Roles of Lectins in Tumor Development	141
1. Siglecs	141
2. Galectins	142
3. Selectins	144
IV. Developing Anticancer Approaches by Targeting Tumor-Associated Carbohydrate–Lectin Systems	145
1. Efforts Toward Galectin Inhibitors	146
2. Efforts Toward Selectin Inhibitors	171
V. Conclusions	187
References	187

Abbreviations

4-F-GlcNAc, 2-acetamido-2,4-dideoxy-4-fluoro-D-glucose; cDNA, complementary DNA; CP, citrus pectin; CRD, carbohydrate-recognition domain; EGF, epidermal growth factor; ELISA, enzyme-linked immunosorbent assay; FUT1, α-1,2-fucosyltransferase I; HUVEC, human umbilical-vein endothelial cell; Ig, immunoglobulin; Lac-L-Leu, *N*-(1-deoxy-lactulos-1-yl)-L-leucine; mAb, monoclonal antibody; MCP, modified citrus pectin; Neu5Ac, *N*-acetylneuraminic acid; Neu5Gc, *N*-glycolylneuraminic acid; PEG, polyethylene glycol; PP, pectic polysaccharide; PSGL-1, P-selectin glycoprotein ligand-1; RT-PCR, reverse-transcription polymerase chain reaction; SCLC, small cell lung cancer; Sd^a, β-D-GalNAc-(1→4)-[α-Neu5Ac-(2→3)]-β-D-Gal-(1→4)-D-GlcNAc; siRNA, small interfering RNA; sLe^a, sialyl Lewis A; sLe^x, sialyl Lewis X; STD, saturation transfer difference; TDG, di-β-D-galactopyranosyl sulfide, "thiodigalactoside"; trNOE, transferred nuclear Overhauser effect

I. INTRODUCTION

Carbohydrate structures coat the surfaces of cells and often act as a first point of interaction with other cells, commonly functioning as receptor ligands; this provides them with an important role in many biological processes in both healthy and diseased cells, including the transmission of signals, adhesion of other molecules (toxins, lectins, antibodies, and the like), regulation of cell–cell interactions, and cell differentiation and proliferation.[1] As compared to peptides, carbohydrates exhibit much more structural variation and thus can convey more structural information through different chain lengths or sequences in the glycan as well as through the regiochemistry of glycosylation and the anomeric configurations. Glycans can comprise from 1 to greater than 20 monosaccharide units and are often present as glycoconjugates, either glycolipids, typically linked to a ceramide chain, or glycoproteins, which can be either N-linked (via Asn) or O-linked (via Ser or Thr).[2]

Lectins are a group of carbohydrate-binding proteins that mediate a wide range of cellular processes upon ligand binding. Unlike enzymes, lectins do not chemically modify their substrates. The word "lectin" comes from the Latin word "legere" which means "to select or choose." Although lectins were first discovered in plants, they were later found to exist in all types of organisms.[3] In general, plant and animal lectins display no structural homology; however, they can bind to similar carbohydrate structures.[4] This indicates that they arose via convergent evolution. Because of the substrate similarity, plant lectins (such as concanavalin A, phytohemagglutinin, and wheat germ agglutinin) have been widely used to characterize glycan expression on mammalian cells. Healthy cells often succumb to the cytotoxicity induced by these lectins; on the other hand, malignant cells are often found to be resistant—this correlates well to a phenomenon termed "aberrant glycosylation" at the surface of malignant cells.[4,5]

As compared to protein–protein binding interactions (often having K_d values in the nanomolar range), carbohydrates bind to proteins with significantly weaker affinities (usually with micromolar K_d values).[6] Carbohydrate–lectin binding occurs through hydrogen bonding, along with electrostatic and van der Waals interactions, and these are similar to those observed for protein–protein interactions. Although both types of interaction exhibit a similar enthalpic free-energy component of binding,[7] carbohydrate–protein binding involves an unfavorable entropic contribution; this has been attributed to both a loss of conformational flexibility upon binding and changes in solvation.[8] Because of this weak interaction, the binding of carbohydrates to lectins (as on leukocytes) often results in a "rolling" process along the endothelial surface; eventually a stronger protein–protein interaction (such as leukocyte–integrin binding to endothelial receptors) takes place, which causes the cell to arrest. Upon strong binding, cytokines are released and this promotes extravasation and transmigration, enabling cells ultimately to reach the site of inflammation or create metastatic lesions.

To generate potent biological signals, carbohydrate–lectin interactions rely on multivalency for enhanced binding. Lectins typically display multiple carbohydrate-recognition domains (CRDs) and can effectively cross-link cell-surface carbohydrates; this triggers a cytoskeletal rearrangement that can induce tyrosine phosphorylation on proteins.[9] Phosphorylation triggers neutrophils to release destructive enzymes and reactive oxygen species that are cytotoxic to the cell, resulting in cell death.[10] Endogenous lectins play an important role as a first defense against pathogens or aberrant somatic cells and are important in inflammatory and immune responses. In mammals, endogenous lectins have been implicated in a number of processes, including cell adhesion, inflammation, immunity, cell proliferation, apoptosis, and tumor progression.[4] Many cell types expressing lectins also display carbohydrate structures at their cell surface, allowing for lectin–carbohydrate binding to afford either homo- (same cell type) or heterotypic (different cell type) interactions.

Three major classes of lectins exist: siglecs, galectins, and selectins. Siglecs and selectins are C-type lectins, which means that they require Ca(II) for ligand binding.

1. Siglecs

Siglecs (*s*ialic acid *Ig*-like *lec*tins) are transmembrane, sialic acid (Neu5Ac)-binding lectins that are expressed in a variety of cell types, including immune cells, nerve tissue, and others, but are mainly found on cells of the hematopoietic system.[11] All siglecs have 2–17 extracellular Ig-like domains, with the most distal N-terminal domain being the one responsible for the binding of sialoside ligands; upon binding, the cytoplasmic tail is able to initiate receptor-mediated signal transduction.[12,13]

Sialic acids are mainly found at the terminal positions of oligosaccharides, giving them an important role in cell–cell interactions and communication. Sialosides with α-(2→3)-, α-(2→6)-, and α-(2→8)-linkages are found at cellular surfaces, and siglecs are effective at differentiating them by exhibiting binding specificity: for example, siglec-2 preferentially binds α-(2→6)-linked sialosides,[14–16] siglec-4 preferentially binds to α-(2→3)-linked sialosides,[14] and siglec-7 preferentially binds to α-(2→8)-linked disialosides (Fig. 1).[17] Similarly, differences between the species can also be observed: human siglec-2 binds both Neu5Ac and N-glycolylneuraminic acid (Neu5Gc), murine siglec-2 favors Neu5Gc, while human siglec-1 recognizes only Neu5Ac.[12]

The binding of different sialosides is typically mediated by interactions of the lectin with the neuraminic acid residue (Fig. 2). In the X-ray structure of sialyl-(2→3)-lactose or 3′-sialyl-lactose-bound SnD1 (a truncated siglec containing only the CRD region), it was observed that three sialoadhesin residues (Trp2, Arg97, and Trp106) were responsible for Neu5Ac binding; these three amino acids are found to be conserved throughout the siglecs.[18] To confirm their roles, mutagenesis studies at

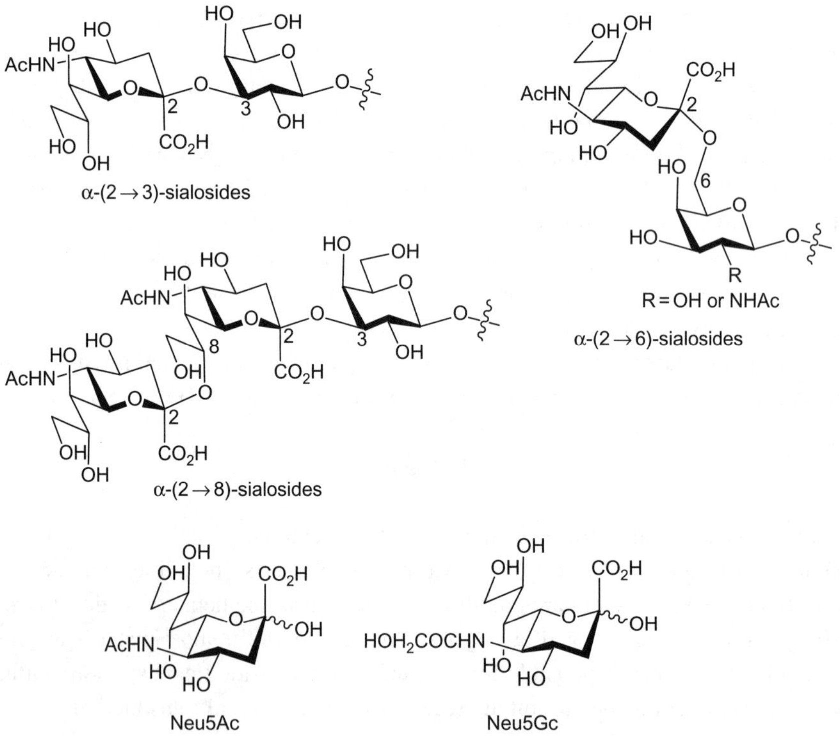

FIG. 1. Examples of different sialoside linkages, and the N-acetylneuraminic acid (Neu5Ac) and N-glycolylneuraminic acid (Neu5Gc) structures.

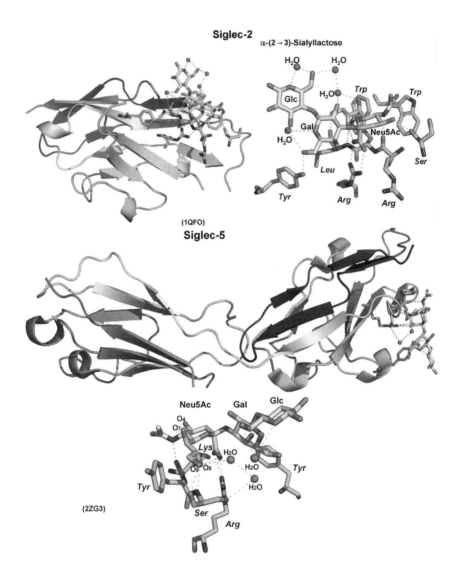

FIG. 2. The X-ray crystal structures of siglec-2 and siglec-5, which illustrate the key interactions between the binding site and Neu5Ac. (For the color version of this figure, see Color Insert and the online version of this chapter.)

any of these three residues significantly decreased productive binding to sialic acids.[19] Furthermore, acetylation of the 9-OH group or cleavage of the sialic acid glycol chain (via periodate cleavage) attenuated the binding in a number of siglecs, indicating that these sites play a key role in binding interactions.[12,16,17]

2. Galectins

Galectins are Ca(II)-independent lectins that contain a conserved amino acid-sequence motif, which binds β-galactosides (or LacNAc) with high affinity.[20] They are unique in that they do not contain a transmembrane domain, but instead are typically found in a soluble form, present in the cytoplasm or nucleus; they can also be secreted extracellularly or are found at the cellular surface in dimerized form. There are approximately 16 types of galectins, found throughout mammals and other vertebrates, invertebrates, plants, and protists.[21,22] Galectins play a key role in inflammatory function, as well as cancer growth and metastasis, by mediating a number of processes, including cell adhesion, angiogenesis, and apoptosis[23–25]; they regulate cell growth and differentiation, and play important roles in embryogenesis and immune function. In addition, they are required factors in pre-RNA splicing.[26,27]

The galectins can be divided into three groups: prototype, tandem repeat type, and chimera type. Prototype galectins (such as galectins-1, -2, -5, -7, -10, -11, -13, and -14) contain a single CRD and are present as either monomers or noncovalent homodimers. Tandem repeat-type galectins (such as galectins-4, -6, -8, -9, and -12) contain two CRDs in a single polypeptide chain, which are connected via a peptide linker (up to 70 amino acids).[28] Chimera-type galectins (such as galectin-3) have a single CRD at the C-terminus (approximately 140 amino acids) that is linked to an N-terminus collagen-like stalk (8–129 amino acid residues, rich in Pro and Gly), which promotes the intermolecular association of multiple galectins at high concentrations of galectin.[29] Galectins-1, -3, -8, and -9 are much more widely distributed as compared to others, such as galectins-2, -4, and -7,[30] with galectins-1 and -3 being the most thoroughly studied. Galectin-1 is present as a homodimer that folds into a globular structure, and this can bind a number of native substrates: both O- and N-linked glycans expressed on endothelial cells, T lymphocytes, or components of the extracellular matrix.[31] Galectin-3 (31–42 kDa)[32] plays a role in cell growth, adhesion, invasion, apoptosis, and metastasis; it is expressed on both healthy and malignant cells, can be localized at either the cell surface or intracellularly, and can also bind single-stranded nucleotide sequences at a non-CRD site.[33]

X-ray crystal structures have been reported for galectins-1, -2, -3, -7, -8, -9, and -10, and studying the molecular interactions at their CRDs have greatly aided in understanding the binding affinities and substrate specificities (Fig. 3).[34] Each CRD (approximately 130 amino acid residues) of galectins contains a β-sheet sandwich with a conserved amino acid sequence of approximately 7 residues.[35,36] In the X-ray crystal structures of substrate-bound galectins, several homologies have been observed including (i) a conserved Lac/LacNAc binding site and (ii) a groove of

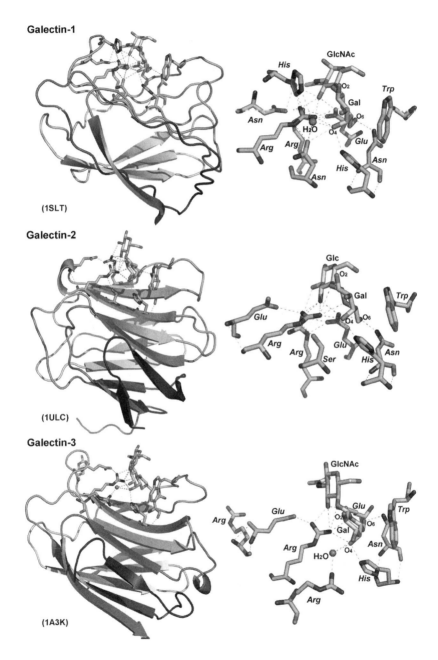

FIG. 3. The X-ray crystal structures of substrate-bound galectins-1, -2, -3, -7, -8, and -9. (For the color version of this figure, see Color Insert and the online version of this chapter.)

(Continued)

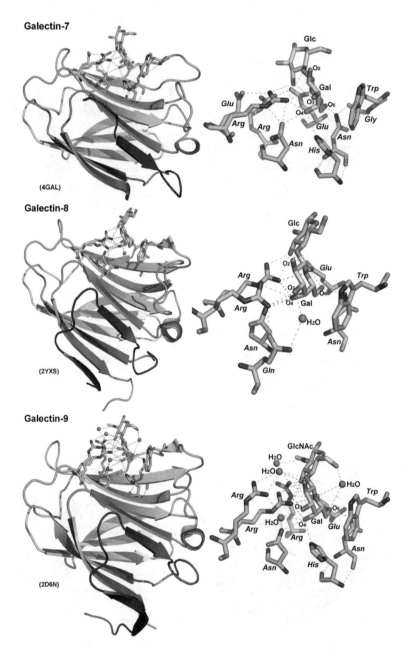

FIGURE 3—Cont'd

variable structure around the region of Gal 3-OH.[37] In the binding site, the 4-OH and 6-OH groups form integral hydrogen bonds with arginine, asparagine, glutamic acid, and/or histidine residues (as with His168, Asp160, and Arg162 in galectin-3), while the 2-OH and 3-OH groups are not directly involved. In addition, face-to-face stacking of the hydrophobic galactose surface (through multiple C—H bonds on one face) and a nearby Trp residue (such as Trp181 in galectin-3) provides additional favorable interactions.[38] The GlcNAc residue in LacNAc can also contribute to a minor extent, through a favorable H-bonding interaction between the galectin and 3-OH on the GlcNAc residue.[39] A number of basic (Arg and His) residues surround the binding site and lead to the particular ligand specificity for the different galectins.[40]

Although there is homology in the binding site of the different galectins, some fine specificity exists that allows ligands to bind preferentially one galectin family over another. Extensive studies have been performed in order to better understand which substrates bind most strongly to which galectins. Both galectins-1 and -3 bind type I [β-D-Gal-(1→3)-D-GlcNAc] and type II [β-D-Gal-(1→4)-D-GlcNAc] galactosides, although an extended binding site in galectin-3 provides stronger binding of galectin-3 to such longer oligosaccharides as α-Neu5Ac-(2→3)-β-D-Gal-(1→4)-D-GlcNAc and α-D-GalNAc-(1→3)-[α-L-Fuc-(1→2)]-β-D-Gal-(1→4)-D-GlcNAc (human blood group A) (Fig. 4)[41,42]; also it has been observed that a greater expression of more highly branched N-glycans enhanced the binding of galectin-3.[43] In general, galectin-1 binds the nonreducing end of oligosaccharides, whereas galectin-3 can bind both internal and terminal regions.[44,45]

In order to better understand the binding specificity of galectins, affinity chromatography has been used to analyze human serum to identify possible native substrates.[30] Galectins-2, -4, and -7 were found to bind only trace levels of the natural ligands (over 500 times less binding than galectin-3), whereas galectins-1, -3, -8, and -9 could bind a range of serum proteins. Galectin-3 displayed the requirement for a LacNAc substrate and was observed to bind a number of abundant glycoproteins. Isothermal titration microcalorimetry and hemagglutination-inhibition assays were also used to study the ligand preference of a number of galectins (bovine heart galectin-1, recombinant murine galectin-3, and recombinant human galectin-7).[44] Galectin-7 appeared to bind Lac and LacNAc-II with equal affinities, although the LacNAc-II-containing saccharides displayed significantly lower affinities in comparison to galectins-1 and -3 (6- and 11-fold decreases, respectively). For a series of sialylated oligosaccharides, sialyl-(2→6)-LacNAc-II did not bind any of the three galectins, but sialyl-(2→3)-LacNAc-II bound all three.[44] Sialyl-(2→6)-diLacNAc and sialyl-(2→6)-lacto-N-neotetraose both bound all three galectins studied, but displayed a significantly weaker binding to galectin-1. Therefore, it was concluded

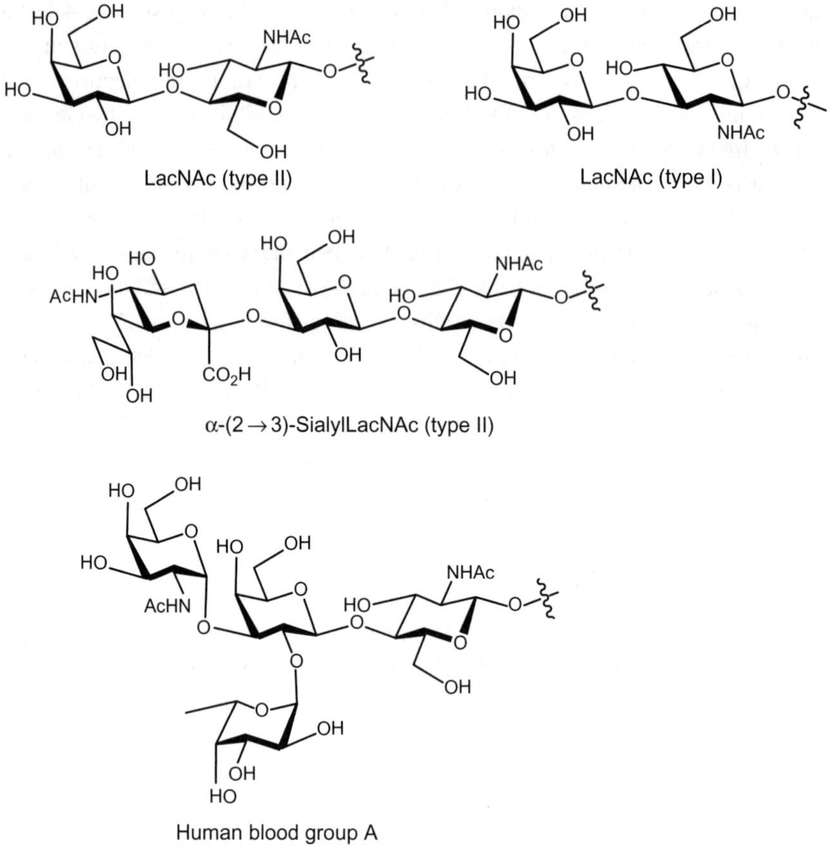

LacNAc (type II)

LacNAc (type I)

α-(2→3)-SialylLacNAc (type II)

Human blood group A

FIG. 4. Examples of substrates recognized by galectins-1 and -3.

that galectin-1 preferentially binds terminal LacNAc-II sequences, while galectins-3 and -7 bind both terminal and internal sequences.[44] Multivalency may also play a role in binding affinities, as galectin-1 was found to bind monovalent N-glycans more strongly than galectin-3.[46]

3. Selectins

Selectins are C-type transmembrane glycoproteins, comprising an N-terminal lectin domain, a transmembrane domain, and a cytoplasmic tail, which bind surface glycans containing Lewis blood group family-related structures such as sialyl Lewis

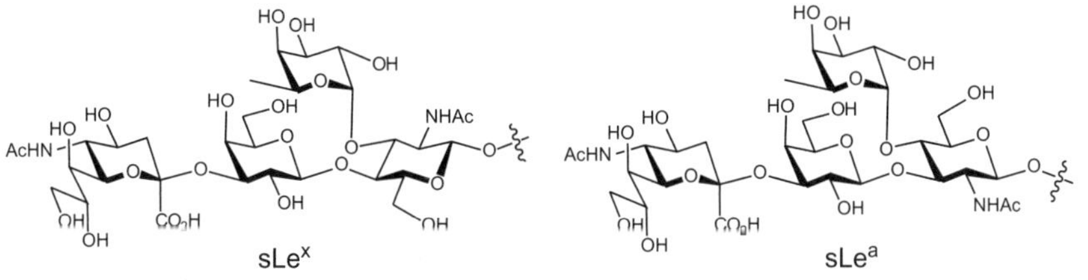

FIG. 5. Examples of substrates recognized by selectins.

X (sLe^x) or sialyl Lewis A (sLe^a) (Fig. 5). Selectins are expressed on the surface of leukocytes, endothelial cells, and platelets; they assist in the migration of naive lymphocytes to lymphoid organs and have been found to play an important role in inflammation and the migration and metastatic spread of tumor cells.[47] For example, in the case of injury, selectins can mediate leukocyte rolling on endothelium, followed by integrin binding and extravasation.[47]

Selectins are grouped into three families: E-selectins (CD62E) present on endothelial cell surfaces, L-selectins (CD62L) present on leukocyte cell surfaces, and P-selectins (CD62P) present on platelets and endothelial cells. The expression of E- and P-selectin is typically induced upon activation of endothelial cells, while L-selectin is constitutively expressed on neutrophils, monocytes, and certain groups of T, B, and NK cells. The common substrate for most selectin ligands is fucosylated and sialylated type I [(1→3) linkage] or type II [(1→4) linkage] lactosamine derivatives (for example, sLe^x, sLe^a).[48] These types of oligosaccharides are typically located at the nonreducing termini of glycosphingolipids or N- and O-linked glycans.[4] E-Selectin has been linked to neoangiogenesis, which can be initiated by sLe^x interactions *in vitro*,[49] and it was demonstrated that the treatment of a rat cornea with soluble E-selectin induced neovascularization.[50] E-Selectin is also involved in leukocyte rolling and adhesion processes associated with inflammation.[51] P-selectin is normally stored within platelets and endothelial cells, but upon activation, it is translocated to the cell surface.[52] In contrast, L-selectin is constitutively expressed on leukocytes and has been linked to the leukocyte–tumor-cell interactions that occur during metastatic processes. In addition, L- and P-selectins can also bind to a number of sulfated polysaccharides (such as heparin, fucoidan, dextran sulfate) or inositol polyanions.[53]

PSGL-1 is a well-studied ligand of P-selectin, containing sialylated and fucosylated O-linked glycans as well as sulfated Tyr residues in the peptide chain.[54–56] The X-ray crystal structure of bound PSGL-1 indicates the necessity of sulfated Tyr residues as well as a particular sLe^x arrangement displayed on a core 2 O-glycan (Fig. 6).[54,57]

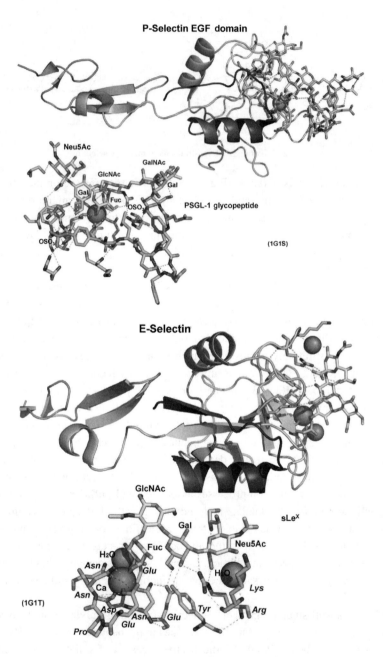

FIG. 6. The crystal structures of substrate-bound E- and P-selectins. (For the color version of this figure, see Color Insert and the online version of this chapter.)

PSGL-1 also binds L- and E-selectins, although in the latter, sulfated Tyr residues are not involved and a 50-fold decrease in the affinity is observed as compared to L-selectin.[55,56] In regard to the fine specificity of binding observed, L-selectin displays enhanced binding with heavily glycosylated mucin-like structures, such as Gly-CAM-1, CD34, and MadCAM-1, heparan sulfate glycosaminoglycans, or sulfated Le^x structures, while E-selectin binds well to sialylated and fucosylated N-linked glycans, such as those found on human neutrophils or in 3′-sulfated Le^a and Le^x.[58]

II. Roles of Carbohydrates in Tumor Development

1. Abnormal Glycosylation in Tumors

Aberrant glycosylation is known to occur in tumors, due to the transcriptional up- or down-regulation of particular glycosyltransferases and glycosidases, which can lead to the overexpression, deletion, or truncation of certain glycan structures that can be tumor associated or tumor specific. Gangliosides and blood group-related antigens are common complex carbohydrates associated with cancer biology. Another class of antigens are the mucins, which are glycopeptide regions attached to glycoproteins that protrude far out from the mammalian cell membrane to form a glycocalyx and are also implicated in cancer.[59] Gangliosides (for example, GM_2, GD_2, GD_3, fucosyl GM_1, polysialic acid) are neuraminic acid-containing glycosphingolipids that have a ceramide tail embedded into the cell membrane (Fig. 7); they are often found to be overexpressed on tumors of neuroectodermal origin.[60] The other blood group-related antigens, such as Tn, sTn, TF, sLe^a, and sLe^x, are commonly expressed on a variety of epithelial cancers (breast, prostate, lung, colon, pancreas, and ovary) and mucins (Fig. 8).[59,60]

These antigens result from incomplete glycosylation on tumor cells that leads to truncated carbohydrate sequences (Tn, sTn, or TF)[61,62] due to a downregulation of β-1,6-*N*-acetylglucosaminyltransferase (TF) or β-1,3-galactosyltransferase (Tn and sTn) and an upregulation of GalNAc α-2,6-sialyltransferase (sTn, increased 8- to 10-fold in cancer cells compared to normal cells).[63,64] On the other hand, the expression of sLe^a and sLe^x arises from the downregulation of GlcNAc-α-2,6-sialyltransferase (to produce a disialylated derivative) and 6-sulfotransferase, as a result of methylation in the promoter region of these enzymes.[64] Increased glycosyltransferase expression has been associated with the β-1,6-*N*-acetylglucosaminyltransferase enzyme, which results in increased size and branching of N-linked glycans on cancer

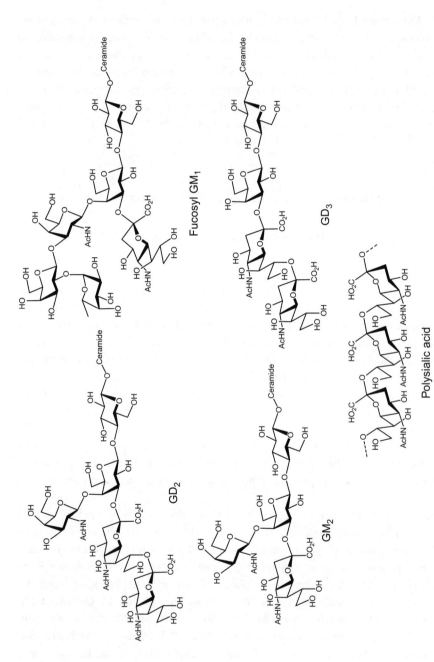

FIG. 7. Examples of gangliosides commonly overexpressed on cancer cells.

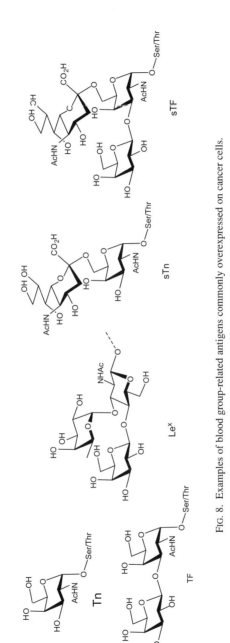

FIG. 8. Examples of blood group-related antigens commonly overexpressed on cancer cells.

cells.[65,66] Since sialyltransferases are typically overexpressed, this results in more potential sites for sialylation resulting in an increase in the number of terminal sialic acid residues on tumor-cell glycans[67,68]; ultimately, the excess aberrant sialylation results in significant changes in charge and sterics of the protein, eventually leading to conformational changes which can inhibit the normal activity of the biomolecule.

Since aberrant glycosylation results in truncated carbohydrate structures on mucins, this results in an increased exposure of the peptide backbone (normally buried in healthy cells) that establish as tumor-specific peptide epitopes.[63] The overexpression of different mucins is correlated to different tumor types: for example, MUC1 in cancers of the breast, lung, ovaries, and endometrium; MUC2 in colon and prostate cancers; MUC4 in colon and pancreatic cancers; and MUC5AC in breast and gastric cancers.[69] The truncated carbohydrates expressed on mucins are displayed as molecular clusters on the surface of tumor cells, indicating that multivalency may aid in carbohydrate–protein binding processes.[70] Just as different cells of the same cancer type are known to express different quantities and types of tumor-associated glycans, the clustering of carbohydrates may also vary between different cell types, creating a considerable challenge in targeting therapeutics to these structures.[71,72]

2. Abnormal Glycosylation and Cancer Metastasis

Several studies have suggested a positive correlation between sialylation on cell-surface saccharides and the ability of a primary tumor to metastasize, although in other studies no correlation has been observed[73]; these mixed results may arise based on how the sialic acid is incorporated into the saccharide structure. Instead of quantifying the total sialic acid content, it is probable that distributions of some particular sialosides are more biologically significant than others: for example, an increase in α-$(2 \rightarrow 3)$-sialylation afforded more metastatic cells in the comparison of two different cell lines,[74] and the overexpression of ST3Gal-I sialyltransferase has been shown to promote mammary tumorigenesis.[75] An improved ability to metastasize has also been correlated to an increase in the β-$(1 \rightarrow 6)$-branching of N-linked glycans, but this may instead be the result of an increased number of α-$(2 \rightarrow 3)$-sialic acid residues.[76] sTn expression in epithelial cancers has also been correlated to a more-aggressive phenotype and subsequently linked to a poorer overall prognosis.[61,67,77]

The tumor-associated TF antigen has also been associated with the adhesion of breast and prostate cancer cells onto endothelium via interactions involving galectin-3.[78] The overexpression of TF antigen, as determined via immunohistochemical staining with an anti-TF mAb (49H.8), is well correlated with a more metastatic cell line.[79] TF antigen is an example of substrate that has also been associated with the primary interaction of circulating emboli with endothelial cells.[41]

Endothelial selectins that recognize either sLe^x or sLe^a structures enable the metastasis of carcinoma cells.[80] Since both of these carbohydrates can be either O- or N-linked, a study attempted to determine whether the particular linkage dictated its ability for metastasis; histochemical analysis of primary versus metastatic lesions indicated very different carbohydrate profiles, suggesting that a change in glycosylation may promote the ability of a cell to metastasize.[66,81–83] An increase in the expression of sLe^x has been shown to enable improved binding of tumor cells to endothelial cells, resulting in an increased occurrence of liver metastasis,[84] while another study had the opposite approach where isolated colon carcinoma cells, which were more prone to liver metastasis, were identified as having increased sLe^x expression at the cell surface as compared to less metastatic cell lines.[85] Increases in sLe^x expression have also been correlated to an increased chance of bone metastasis in breast cancers, which occur via binding of the carbohydrate to E-selectin.[86] Likewise, an increased sLe^a expression was also found to correlate to enhanced metastatic ability and adhesion processes.[85,87] sLe^a is expressed on a number of other cancers, including breast,[88] colon,[81] pancreatic,[89] and small cell lung cancer (SCLC),[71] and its elevated expression has been correlated to more-aggressive phenotypes, presumably a result of its ability to act as a ligand for E-selectin, enhancing its ability to adhere to endothelial cells.[81,87,88,90]

III. Roles of Lectins in Tumor Development

Along with integrins, cadherins, and other immunoglobulin-like proteins, many lectin molecules have been found to be upregulated in cancer cells and are attributed to being involved in the cell adhesion to other biomolecules.[24,91] The most well studied are galectins and selectins.

1. Siglecs

There have been only a few reports on the involvement of siglecs in tumor development, even though increased sialylation has clearly been linked to improved invasion and metastasis of tumor cells.[92] For example, siglec-15, which is expressed on a subset of macrophages, was recently reported to preferentially recognize tumor-associated sTn antigen.[93] Upon sTn binding, TGF-β secretion is upregulated, which induces a change in macrophage function that has been correlated to an enhanced likelihood of metastasis.

Furthermore, the expression of siglec-4a (MAG) was observed to be upregulated in pancreatic adenocarcinoma; its binding to MUC1 has been associated with improved chance of perineural invasion, which causes severe lower-back pain in affected patients and may also play a role in metastasis.[94] Other siglecs have also been correlated to certain cancer types, such as siglec-9, which can play a role in evading immunosurveillance, and siglec-12, which is overexpressed in breast and prostate carcinomas.[95]

2. Galectins

A number of galectins have been linked to cancer progression, especially galectins-1, -3, -7, and -9. For example, in cancers of the head and neck, thyroid, central nervous system, pancreas, and gastrointestinal tract, increased expression of galectin has been linked to a more-aggressive cancer.[32,96] Since galectins can present multiple CRDs simultaneously, either via poly-CRD-containing galectins (for example, tandem repeat type) or upon oligomerization (for example, prototype, chimera type), this enables them to cross-link with other cancer cells, serum glycoproteins, platelets, or immune cells. As a first step of metastasis, cancer cells detach from the primary tumor and enter the lymphatic or blood circulatory system. Galectin-mediated binding of these cells to surrounding serum components can result in the formation of aggregates which continue to circulate throughout the bloodstream; it is believed that formation of these emboli can actually protect circulating tumor cells by preventing the cytotoxic activity of NK cells, which provides them with enhanced opportunity for metastasis.[91,97] These embolisms can eventually become lodged in the pulmonary vasculature or adhere to vascular endothelial tissue, resulting in the formation of secondary lesions.[4]

To date, galectins-1 and -3 have been by far the most comprehensively studied. Galectin-3 levels in a number of cell lines (HEK293, A431, B16F10, MCF-7, HeLa, MDA-MB-231) were assayed and compared to normal cell lines (NIH-3T3, VERO); 5- to 15-fold greater levels of galectin-3 expression (at the cell surface and in the media) were observed for the metastatic MDA-MB-231 and HeLa cells as compared to normal cell lines, as determined by anti-galectin-3 mAb binding in ELISA.[22] Alternatively, in certain carcinomas, such as those of the ovary or breast, both the up- and downregulation of galectins have been observed, depending on the particular cell line studied.[24,98] This may relate to the location of expression of the particular galectin, as galectin-3 overexpression in the cytoplasm has been correlated to a more-aggressive cancer and worse prognosis, whereas no effect has been correlated to galectin-3 expression in the nucleus.[99] In another study, a colon cancer cell line (LS174T) with a low chance of metastasis was transfected with the cDNA for galectin-3, and upon inoculation into the spleen or cecum of nude mice, an increased rate of metastatic lesion formation was observed.[100] Alternatively, transfecting highly metastatic colon cancer cell lines (LSLiM6 and HM7) with the antisense galectin-3 cDNA resulted in a downregulation of galectin-3 expression and reduced the occurrence of metastasis in an experimental model.[100] In a study of galectin-3 involving five different human breast carcinoma lines, three lines expressed galectin-3 and were able to establish lesions upon injection into nude mice, while the two cell lines not expressing galectin-3 were unable to form lesions[101]; when the latter two lines were transfected with galectin-3 cDNA and challenged in nude mice, the development of experimental metastatic lesions was observed. Another study showed that exogenous galectin-3 enabled transmigration of human breast carcinoma cells through extracellular matrix- and basement membrane-like media.[102]

In another mechanism that promotes cancer survival and more-rapid tumor growth, galectins play a regulatory role in the thymocyte population surrounding cancer cells. Galectin-1 in humans and galectin-9 in mice have both been observed to induce cytotoxicity in surrounding T lymphocyte populations,[31,103] while galectin-3 can also have an antiapoptotic effect.[104] This latter effect seems to arise from the C-terminal NWGR motif present in galectin-3, while the other galectins contain an XWGXEER domain. Bcl-2 also contains an NWGR domain and has also been shown to elicit an antiapoptotic effect; a Gly to Ala point mutation in this sequence in galectin-3 was shown to eliminate antiapoptotic function.[104] T lymphocytes are commonly heavily glycosylated, and galectin-1 overexpressed and secreted by cancer cells can bind to CD45 or CD7 and induce apoptosis in activated T cells.[31] Poly-N-acetyllactosamine ligands expressed on T cells can bind to galectin-1,[105] which can then stimulate a number of processes that block and/or destroy T cell activity[37,106]; by blocking these galectin interactions, an enhanced immune response can be generated, which also aids in tumor destruction.

Changes in the microenvironment surrounding a primary tumor can also cause regulatory changes in the expression of galectins. For example, hypoxia in dense, solid tumors results from an abnormal and poorly organized arrangement of the vasculature within the tumor, which not only reduces immune-cell access but also causes poor blood flow and supply in certain areas of the tumor. This results in regions of hypoxia, which promotes the production of reactive oxygen species and initiates a pathway that can cause the upregulation of galectin-1,[107] and increased secretion of this lectin from the affected cells, which promotes emboli formation.[108]

Upon adherence of emboli to vascular endothelial cells, the cells become stimulated, destabilization of the surrounding extracellular environment and basal membrane occurs, and the cells and neovasculature are allowed to penetrate into the tissue to form a permanent site of metastasis.[109]

A galectin-binding protein, LGalS3BP, has been recognized to play an essential role in tumor metastasis, and higher levels have been correlated to cell lines that are more metastatic.[110] LGalS3BP has been isolated from the sera of breast cancer patients and is known to bind galectins-1, -3, and -7, aiding in homotypic cell aggregation.[24,111] Other ligands, such as CD44 and CD326, that bind galectin-1 expressed on endothelial cells are also involved in the attachment of metastatic cancer cells to endothelial tissue.[112]

Interestingly, galectins have also been found to serve different functions at different stages of tumor progression. For example, galectin-9 has been shown to enhance antitumor immunity at early stages, while at later stages, it was found to enhance tumor development and survival.[113] This seems to suggest that galectins could be playing multiple roles in the mechanisms of tumor growth and metastasis. Indeed, they have been implicated in a number of prometastatic processes, including tumor-cell adhesion,[114] apoptosis,[99,115] homotypic cell aggregation,[111] heterotypic cell aggregation,[116] angiogenesis,[27] more-rapid cell growth,[117] and immune system evasion.[91,118]

3. Selectins

Primary tumors are known to have local regions of hypoxia, and cancer cells that are cultured in a hypoxic atmosphere have been shown to upregulate the expression of selectin ligands.[2] More-advanced tumors often overexpress selectin ligands and can bind to other selectin-expressing cells, such as platelets. Selectins are known to play an important role in the recruitment of leukocytes to sites of inflammation; since glycan binding to vascular selectins at the cell surface has been shown to increase the ability of tumor cells to metastasize, and overexpression of selectin ligands has been associated with more-aggressive cancer phenotypes, it is likely that the selectins are also involved with extravasation during the formation of metastatic lesions.

Normal cells do not express significant levels of E- or P-selectins; however, cancer metastasis has been correlated to increased levels of both.[119] In fact, both P- and L-selectins have been linked to carcinoma metastasis, based on the results of studies in mouse models.[120,121] P-selectin-mediated binding has also been observed in a number of cancers, including lung, breast, gastric, colon, melanoma, and neuroblastoma cell lines.[122,123] It has been observed that platelets are able to form a protective layer that surrounds tumor cells, which enhances the chance for metastasis[120]; indeed, upon cleavage of mucins from the tumor-cell surface, significantly reduced platelet protection and greater macrophage association with tumor cells were observed, which aid in the elimination of tumor cells from circulation.[120,124] Therefore, selectins are involved in tumor progression by aiding in the formation of cell–platelet complexes in the circulatory system.[125] Binding of P-selectin to leukocytes has been associated with activating the leukocytes, and this results in the release of procoagulant molecules, which can further aid in formation of emboli.[126]

In addition, selectins can also be involved in the preliminary rolling of emboli at the endothelial cell surface, which occurs in the first stages of metastatic lesion formation.[127]

sLe^a and sLe^x are both ligands of vascular selectins and have been associated with promoting the adhesion of tumor cells to endothelial tissue[64,128,129]; the overexpression of these ligands has been described in leukemia and a number of human carcinomas (colon, bladder, breast, and lungs) and, in addition, has been linked to other metastatic malignancies.[81,83,130,131] The expression of these ligands is higher in metastatic lesions as compared to primary tumors, suggesting that the overexpression of particular tumor-associated carbohydrates occurs in advanced stages of the disease.[132]

In an effort to increase the amount of cell-surface sLe^x, B16-F1 cells were transfected with α-1,3-fucosyltransferase III, which induces the tumor cells to express sLe^x[133]; transfection of these tumor cells was associated with the formation of a greater number of lung-tumor lesions. In addition, in cells that had been activated with thrombin (which induces expression of P-selectin on platelets), it was observed that human colon carcinoma cells bound platelets having elevated selectin expression significantly better.[120] In the opposite approach, when metastatic carcinoma cells were treated with the enzyme O-sialoglycoprotease, which partially cleaves sialylated fucosylated mucins, the binding of platelets to tumor cells was significantly decreased as a result of cleavage of the terminal portion of Le blood group-related ligands.[120,121]

IV. Developing Anticancer Approaches by Targeting Tumor-Associated Carbohydrate–Lectin Systems

Since glycosylation varies between cancerous and healthy cells, and the abnormal carbohydrate structures present on tumor cells have been closely correlated to increased risks of cancer metastasis, lectins and tumor-associated carbohydrate ligands become important targets in developing novel cancer therapies. This variation can be exploited by a number of potential therapeutic approaches: (i) conjugate vaccines to target the immune system against cancer cells, (ii) inhibiting lectin binding to attenuate the risk of metastasis, (iii) using antibodies against aberrantly glycosylated glycoproteins for targeted drug delivery to tumor cells, or (iv) inhibiting glycosyltransferases and glycosidases responsible for the aberrant glycosylation in order to revert tumor cells to their normal phenotype. Since carbohydrate epitopes are extremely difficult to isolate from natural sources and are likely not viably obtained on a scale sufficient for clinical use, chemical or chemoenzymatic syntheses need to be established for the glycan, or a functional mimetic of the glycan, in order to proceed with biological testing.

Antibody therapy is an ideal treatment option for the elimination of circulating tumor cells and micrometastases; using either passive or active administration of antibodies, a number of preclinical studies have been demonstrated to eliminate circulating tumor cells effectively.[134] However, there have been considerable challenges in raising high titers of carbohydrate-specific antibodies against tumor-associated carbohydrate antigens, as they are often self-antigens. To best prevent the risk of forming metastatic lesions, antibody therapy could be coadministered with an inhibitor therapy that aims to block lectin binding to tumor antigens. Clinical trials have revealed the benefits of using sLe^x mimetics, which can potentially inhibit selectin binding and subsequent tumor metastasis; likewise, the blocking of sLe^a with an anti-sLe^a antibody was successful at inhibiting metastasis of human pancreatic adenocarcinoma cells in nude mice.[135]

Unfortunately, carbohydrates often exhibit poor pharmacokinetics, and so there exists a need to develop inhibitors that are more resistant to hydrolysis and metabolic breakdown. The ideal inhibitor needs to be hydrolytically and metabolically stable and nontoxic, and still display the correct chemical-group arrangement and valency necessary for high-affinity lectin binding.[136]

Developing a suitable assaying method that allows accurate comparisons of different inhibitors is very important. When measuring binding affinities, it has been observed that different assay methods afforded slightly different inhibitory concentrations, indicating that instead of comparing absolute IC_{50} or K_d values between studies, the comparison of relative inhibitory activity based on a standard compound would be more accurate.

1. Efforts Toward Galectin Inhibitors

Generating inhibitors having high affinity and specificity for galectin ligands would allow for both the detection of galectins in tissues and the development of therapeutic agents that may serve to attenuate cancer growth and metastasis.[29]

Because of the involvement in inflammatory processes, inhibitors for galectin-3 have been well studied in an effort to develop treatments for such immune system-mediated diseases as asthma.[137] Because of their role on macrophages, it is important to monitor closely whether or not the inhibition of galectins to decrease the incidence of metastasis would also have adverse effects on immune function. Although this is a possibility, treatment of a host with exogenous galactose showed no changes in immune functions (IgM production, monocyte and complement activation, and so on) when compared to control groups.[136]

Galectins (such as galectins-1 and -3) are able to inhibit T-cell immune function by blocking proinflammatory cytokine secretion and by initiating growth arrest and apoptosis of activated T cells, impairing T-cell function.[138] In addition, they have been correlated with neoangiogenesis of tumor cells and are capable of stimulating capillary-tube formation.[4,27]

a. Inhibition of Galectin Expression.—A direct approach to inhibiting galectin-mediated metastasis is to attenuate the expression of galectin on the cell surface. For example, blocking the expression of galectin-1 by treatment with small-hairpin RNA resulted in decreased tumor-cell migration and invasion, and decreased tumor growth and metastasis *in vivo*.[139] Similarly, siRNA silencing was used in a human colorectal cancer cell line, where the downregulation of galectin-3 induced apoptosis of tumor cells.[140] In glioma and glioblastoma cells, both antisense interfering RNA and antisense oligodeoxynucleotides resulted in anti-galectin effects.[25] In B16 murine melanoma cells, an antisense sequence was used to knock down galectin-1 gene expression, which resulted in an improved T-cell-mediated response.[91] Unfortunately, antisense galectin-1 is difficult to target selectively to cancer cells, and this may limit its potential as a therapeutic.[141]

Instead of using an antisense sequence, B16 melanoma knockdown transfectant cells were injected into a mouse model, and this resulted in improved tumor rejection and also successfully generated a T-cell-dependent tumor-specific immune response.[91]

Alternatively, curcumin [(1*E*,6*E*)-1,7-bis(4-hydroxy-3-methoxyphenyl)-1,6-hepta-diene-3,5-dione], a natural product isolated from rhizomes of *Curcuma* species, was found to inhibit the expression of galectin-3 (Fig. 9).[142] Expression levels were quantified by using a mAb (M3-38) and Western-blot analysis, and it was observed that after 5 h of incubation with curcumin a decrease in the level of galectin-3 could be detected.

b. Anti-Galectin mAbs or Soluble Recombinant Galectins.—As an alternative to blocking gene expression, binding can be blocked by using an anti-galectin mAb to shield the CRD of the galectin. It was shown that anti-galectin Abs could disrupt

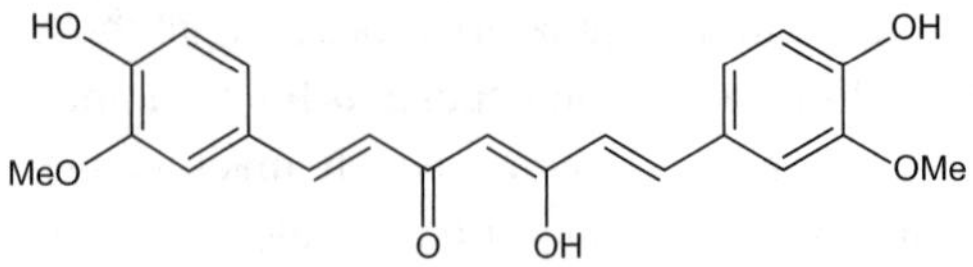

FIG. 9. Curcumin, a natural product that inhibits galectin-3 expression.

tumor-cell aggregation and adhesion processes, resulting in reduced experimental metastases. For example, treatment with a mAb against galectin-3 inhibited the metastasis of melanoma (B16) and fibrosarcoma (UV-2237) cells.[143] Alternatively, since TF antigen is associated with the adhesion of breast and prostate cancer cells onto endothelium via interactions with galectin-3, upon treatment of tumor cells with an anti-galectin-3 antibody, a 90% reduction in the occurrence of lung metastasis was observed[78,143]; an anti-TF mAb (A78-G/A7) was also shown to inhibit the proliferation of BeWo cells in a concentration-dependent manner by blocking TF interactions with galectin-1.[144]

Soluble recombinant galectins can also be generated to inhibit binding, since the recombinant galectin will bind the carbohydrate substrate thereby making it less available to bind any cell-surface galectins. In a human breast cancer murine model, an N-truncated galectin-3 (the C-terminal contains the CRD) was used to compete with the endogenous galectin-3, but since the N-terminus is required for oligomerization, any endogenous ligands that bound the truncated construct would be incapable of assembly and subsequent signaling pathways.[145] In human T-leukemia cells, treatment with exogenously added recombinant galectin-3 induced apoptosis in the T-cell population,[146] even though intracellular galectin-3 is known to reduce apoptosis in tumor cells.[147]

c. Inhibition of Biosynthesis of Tumor-Associated Carbohydrate Antigens.—Ligands of galectin-1 include N- or O-linked LacNAc (either a type I or type II), and glycoproteins such as CD4, CD7, CD43, and CD45.[148] Because effector T cells express a significant number of galectin-1 ligands, they are able to bind tumor cells, which results in immune-cell apoptosis as a defense mechanism. In an effort to overcome this, melanoma- or lymphoma-bearing mice were treated with 2-acetamido-2,4-dideoxy-4-fluoro-D-glucose (4-F-GlcNAc), which has been found to inhibit the biosynthesis of cell-surface LacNAc glycans.[149] Resulting from the reduced expression of LacNAc, a decrease in the apoptosis of immune cells (both T cells and NK cells) was observed, together with a reduction in the proliferation of B16 melanomas and EL-4 lymphomas. However, 4-F-GlcNAc was unsuccessful in slowing tumor growth in immunodeficient mice, confirming that its activity is due to changes in immune-cell populations and not its involvement in angiogenesis; similar observations were also reported in a lung-carcinoma mouse model.[150]

Alternatively, nonanomeric aldoxime ethers can also attenuate cell-surface carbohydrate synthesis by inhibiting mucin-type O-glycosylation pathways.[151] The aldoxime ethers are typically stable under physiological conditions and resistant to enzymatic hydrolysis, and their increased hydrophobicity may also aid in enhanced cell permeability, resulting in improved pharmacokinetic properties.[152]

d. Modified Citrus Pectin (MCP) and Other Dietary Polysaccharides.—Citrus pectin (CP) is an acidic polysaccharide that has been isolated from citrus fruits, and it comprises "smooth" and "hairy" regions. The smooth regions contain partially esterified galacturonic acids, whereas the hairy regions comprise galacturonic acid with randomly inserted rhamnose residues, further functionalized with a number of neutral sugar residues (such as arabinose, galactose, glucose, mannose, and xylose).[153] Because of its large size, CP is water insoluble and displays poor pharmacokinetic properties. In order to improve its characteristics, modified citrus pectin (MCP; GCS-100) has been developed; its production involves first treating the polymer at high pH (pH 10) under mild heating (50–60 °C), which induces β-elimination at positions along the galacturonic acid chain, breaking up the polymer into smaller fragments.[153,154] The fragments are then further degraded at low pH (pH 3), which partially cleaves at the neutral residues to afford simpler oligosaccharides of lower molecular weight (approximately 10 kDa), with a similar sugar composition, but with improved pharmacokinetic properties. After these modifications, these oligosaccharides become orally absorbable and constitute a better ligand for galectin-3.[153,155–157] In comparing the carbohydrate content of CP and MCP, it was observed that the latter contains a greater fraction of galactose, rhamnose, and xylose residues.[153]

MCP was shown to attenuate angiogenesis of human umbilical-vein endothelial cells (HUVECs) *in vitro*, while a significant decrease in the density of tumor-associated blood vessels was observed *in vivo*.[153] In addition, MCP inhibited binding of galectin-3-expressing breast cancer cells (MDA-MB-435) to endothelial cells.[153] MCP has also successfully inhibited both lung metastasis[156] and hetero- and homotypic interactions of B16-F1 melanoma cells,[155] and even the oral administration of MCP in rats successfully attenuated lung metastasis of prostate cancer cells.[157] In nude mice, MCP inhibited the tumor growth, angiogenesis, and metastasis of human breast and colon carcinoma cells,[153] induced apoptosis in human prostate cancer cells,[158] and attenuated proliferation and enhanced dexamethasone sensitivity in myeloma cells.[159]

Because it was unclear what the active region of MCP was, the activity of a number of CPs and MCPs was screened in assays; it was reported that β-(1 → 4)-galactan (present in potato) could inhibit galectin-3 function,[160] and it was suggested that the β-(1→4)-galactan in the rhamnogalacturonan I domain branching chains may play an important role (Fig. 10).[161] In another study to establish ligand specificity, MCP was first fractionated on a column of diethylaminoethyl cellulose, and then individual fractions were tested in a hemagglutination assay with galectin-3.[154] The results of this study concluded that the binding region may not reside in the internal region of the oligosaccharide, but instead may lie in the terminal residues.

MCP has also been used in cotherapies, since the inhibition of galectins has been shown also to mitigate galectin-mediated antiapoptotic functions, resulting in an enhancement of the sensitivity of tumor cells to conventional cytotoxic agents. Galectin-3 was previously shown to decrease tumor-cell sensitivity to a number of cytotoxic agents: cisplatin, staurosporine, etoposide, bortezomib, dexamethasone, and doxorubicin.[159,162,163] MCP inhibition of galectin-3 successfully attenuated the antiapoptotic mechanism, causing multiple myeloma cells to exhibit enhanced sensitivity to bortezomib and dexamethasone, breast carcinoma cells to taxol,[164] or hemangiosarcoma cells to doxorubicin.[163] Although late-stage prostate cancer is resistant to cisplatin treatment, the inhibition of galectin-3 (via either MCP or siRNA treatment) reestablished the sensitivity of PC3 cells to cisplatin-induced apoptosis.[147,165]

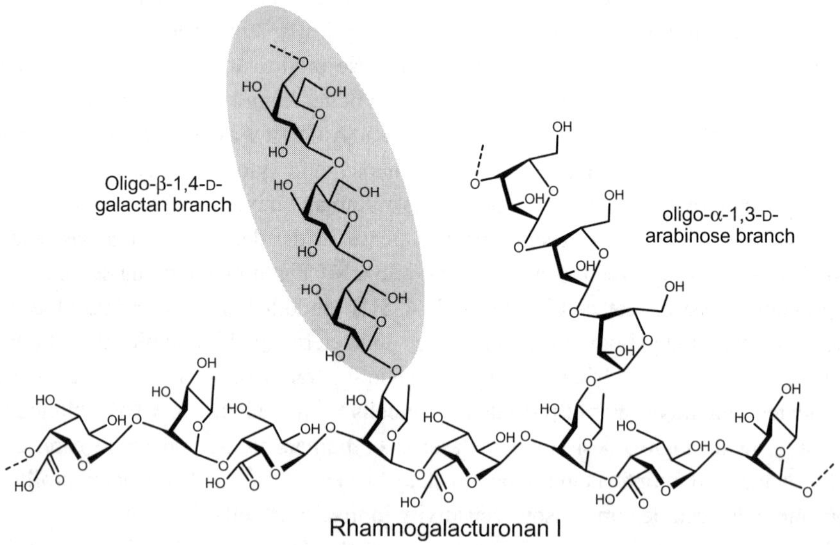

FIG. 10. The β-(1→4)-galactan region in rhamnogalacturonan I may play an important role in the inhibition of galectins.[161]

In a number of other studies, the oral administration of MCP has resulted in reduced metastasis and inhibition of angiogenesis.[157,166] In studying the MCP-mediated inhibition of galectin-3, mice were fed MCP in their drinking water and then injected with either human breast carcinoma cells (MDA-MB-435) or human colon carcinoma cells (LSLiM6); it was observed that angiogenesis, metastasis, and tumor growth were significantly reduced in MCP-treated mice as compared to the control, and in addition, MCP treatment *in vitro* inhibited capillary-tube formation in a dose-dependent manner in HUVECs.[153] In another study, mice treated with MCP that were subsequently injected with melanoma B16-F1 cells developed significantly reduced levels of lung colonization when compared to the control group.[156] Further studies showed that the oral administration of MCP to rats that were inoculated with the prostate cancer cell line MAT-LyLu displayed reduced lung-tumor formation in a dose-dependent manner, and tumor growth was decreased in mice upon oral administration of MCP.[157,167] These studies indicate that MCP is able to bind galectin-3 in a dose-dependent manner.

Other dietary compounds have been suggested as galectin-3 inhibitors; for example, swallow-root pectic polysaccharide (PP) displayed a 14-fold greater inhibition as compared to other pectins.[168] A number of dietary pectic polysaccharides have been examined other than swallow-root PP, including *Hemidesmus indicus* PP, black cumin PP, *Andrographis serpylligolia* PP, ginger PP, and citrus PP.[22] The polysaccharides were studied for their ability to inhibit galectin-3 binding, and it was determined that a greater arabinogalactan content was better for inhibition, such as was the case for swallow-root PP and citrus PP.

No inhibition was observed for ginger PP (which lacks galactose content), indicating that galactose (likely in a particular arrangement) may play an important role in galectin inhibition. On the other hand, swallow-root PP was observed to inhibit metastatic invasion of breast carcinoma cells (MDA-MB-231) by preventing the heterotypic adhesion of tumor cells to the extracellular matrix, and its antiapoptotic characteristics were also abolished[22]; since it is nontoxic and inexpensive, it therefore may be promising as a potential therapeutic agent.

e. Small Carbohydrate-Based Inhibitors for Galectins.—It was illustrated that the repeated intraperitoneal injection of D-galactose, methyl α-D-galactopyranoside, or arabinogalactan inhibited the adhesion of metastatic sarcoma (L-1), melanoma (B16), or lymphoma (ESb) cells into the liver of mice.[65,136,169] In addition, the use of allyl β-Lac and methyl β-LacNAc further demonstrated that small molecules could be used as inhibitors of galectins-1 and -3 and that by preventing interaction with galectins, both homotypic aggregation and antiapoptotic processes could be severely diminished.[170] Unfortunately, as expected, these small molecules based on native galectin substrates (such as Gal, Lac, LacNAc) display relatively low binding affinities (0.1–1 mM IC_{50}): LacNAc binds slightly better than Lac, while Gal binds 50-fold worse than these disaccharides.[44,45,171] Other problems exist with developing natural saccharides as inhibitors, since their glycosidic linkages are sensitive to both chemical and enzymatic hydrolysis, and their high polarity severely limits their ability to permeate cell membranes.

Although a multivalent presentation probably exists in native ligands, the development of high-affinity monovalent inhibitors for galectins is still possible because of the deep binding pocket, which has been well characterized.[172,173] As already mentioned, there are a number of ionic and polar amino acid residues (such as Arg, His, Lys) that surround the binding site; although they are not involved in the key β-galactoside-binding interactions,[172] they may still interact with external regions of extended oligosaccharide substrates, and thus synthetic modifications to potential substrates to maximize interactions with these residues could be performed in order to generate inhibitors having significantly improved binding affinities. Because of the absence of a glycosidic linkage, modified monosaccharide-based inhibitors will likely display improved pharmacokinetics when administered *in vivo*; thus they could afford an improved half-life and membrane permeability.[174] In addition, the monosaccharide (Glc/GlcNAc) moiety could be further replaced with a simpler, less-polar component in order to further improve the pharmacokinetics and synthetic efficiency.

Since X-ray crystallography revealed that only the 4-OH and 6-OH groups of galactopyranosides form key H-bonds with galectins in the binding site, and 3-OH and 2-OH are not directly involved in binding, potential chemical modifications at C-3, C-2, or C-1 center are suggested for developing high-affinity carbohydrate-based small-molecule inhibitors.

(i) 3-O-Aromatization. Several C-3 derivatives have been prepared, such as amide, sulfonamide, and urea LacNAc derivatives to probe as galectin inhibitors.[175] By replacing the LacNAc 3′-OH group by an amine, a variety of modifications can be made to create a library of derivatives; since this position can interact with the binding site through an H-bond in the natural substrate, substitution of the hydroxyl group by an amine or amide function should not disrupt this interaction. Amide groups at this position can also improve the *in vivo* stability of inhibitors and also show slightly improved binding affinities as compared to the ester analogues: the X-ray crystal structure shows additional H-bonding to the amide function through solvent water molecules, which may provide the enhanced binding affinity.[171]

3′-Benzamido-LacNAc derivatives were the first small-molecule inhibitors developed that had a submicromolar affinity for galectins (Fig. 11).[175,176] Based on NMR

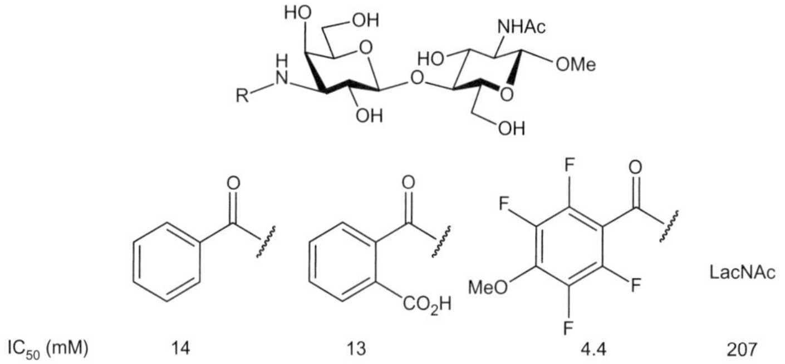

FIG. 11. One of the earliest small-molecule analogues found to display sub-micromolar affinities for galectins were the 3′-benzamido-LacNAc derivatives.[175]

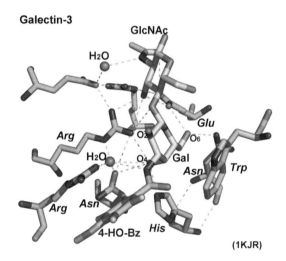

FIG. 12. X-ray structure of galectin-3 with a 3′-p-hydroxybenzamido-LacNAc derivative. The aromatic ring is closely facing the guanidinium group of an arginine residue to engage in a cation–π interaction. (For the color version of this figure, see Color Insert and the online version of this chapter.)

studies and analysis of the crystal structure (Fig. 12), this enhanced affinity arises from a cation–π interaction between the guanidinium group of an Arg residue (for instance, Arg144 in galectin-3) and the aromatic group on the sugar substituent.[177] A library of potential inhibitors containing N-acylated or N-sulfonylated derivatives was prepared and screened as inhibitors of galectin-3 in comparison to a LacNAc standard.[175] The binding was assayed via competitive ELISA studies using

Gal-α-(1 → 3)-LacNAc–horseradish-peroxidase conjugates and galectin-3-coated microwells, where a relative decrease in the binding of the horseradish-peroxidase conjugate could be measured for each compound. The best inhibitors had an aromatic C-3′ amide function on a Gal residue, with the best inhibitor showing a 50-fold affinity improvement over LacNAc.

Due to the success of the amides, a number of other functional groups were explored at this position, including 1,2,3-triazoles, obtainable through high-yielding conjugation reactions that can be performed under mild reaction conditions. Because of its foreign nature, the triazole function can decrease susceptibility to enzymatic hydrolysis, and in addition, it adds hydrophobic nature to the molecule, which should result in improved bioavailability. Upon synthesis of a library of potential inhibitors from alkynes and 3-azido-3-deoxy-β-D-galactopyranosides, binding studies indicated that a 3-deoxy-3-(triazol-1-yl)-galactopyranoside was unable to inhibit galectin-3, but upon triazole substitution at the C-4′ position, binding was restored and even showed enhancement as compared to the reference galactoside (Fig. 13)[174]; up to 40-fold enhancement as compared to methyl β-D-galactopyranoside was observed with a phenyl or amide group on the triazole ring, suggesting that although the triazole ring itself decreases binding affinity, it is probably able to enhance binding potency by acting as a spacer, placing the triazole C-4′ substituent in an orientation that enables favorable interactions with the galectin surface. In general, these inhibitors were demonstrated to successfully inhibit galectin-3, showing binding affinities almost comparable to those of the C-3 aromatic amides.[174,178]

Alternatively, a C-3 modification with triazolylmethyl ethers (such as triazol-1-yl-methyl ether 1-thiogalactopyranoside) afforded inhibitors that bound galectin-1 40 times better than Gal, and an analogous modification at C-3 with a thiourea functional group generated a 10-fold better inhibitor for both galectins-7 and -9N as compared to the LacNAc standard.[179,180] Additionally, 3-O-alkynylbenzyl galactosides were prepared, using a solid-phase approach via the 3,4-O-stannylene acetal, and their binding affinities to galectins-1, -3, -7, -8N, and -9N were studied in a

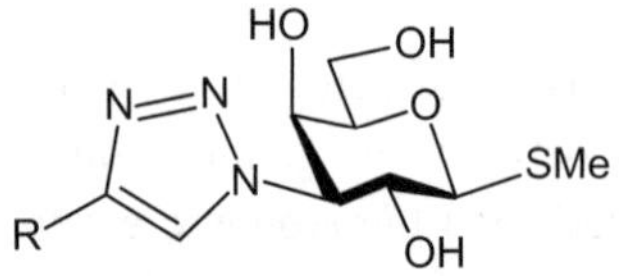

FIG. 13. Methyl 3-deoxy-(1,2,3-triazol-1-yl)-1-thio-β-D-galactopyranosides were developed as galectin inhibitors; the substituent R is involved in favorable interactions with the binding site.

competitive fluorescence-polarization assay. The 3-*O*-alkynylbenzyl substituent was shown to improve binding affinity to galectin-7 by an order of magnitude.[35]

In addition to the triazoles, the isoxazole function (via reaction with a nitrile oxide) has also been explored. Isoxazoles are easy to synthesize and also stable under physiological conditions, which may lead to improved pharmacokinetic properties.[179] A library of triazole and isoxazole lactosides and galactosides were screened as galectin-1, -3, and -4 inhibitors, via testing in a hemagglutination assay in red blood cells, using Lac and Gal as reference compounds.[179] None of the inhibitors tested were observed to bind galectin-4, and different binding profiles were observed for galectins-1 and -3. Binding to galectin-3 was most effective with triazoles that had been prepared from a 3-*O*-propynyl glycoside, and the *C*-galactosides (with enhanced *in vivo* stability) bound galectin-1 effectively but were unable to inhibit galectin-3. Several isoxazole analogs also showed enhanced affinities as compared to the control compounds, with approximately 40-fold improved binding.

(ii) 2-O-Aromatization. Similar modifications at C-2 were also carried out. For example, the incorporation of a benzoyl ester group at C-2 enhances binding affinities (as compared to LacNAc) by generating a favorable cation–π interaction with Arg74 (galectin-1), Arg186 (galectin-3), Arg75 (galectin-7), or Arg87 (galectin-9N) residues (Fig. 14).[181] Lacto-*N*-biose disaccharide derivatives containing modifications at the NH and 2-OH groups were synthesized via solid-phase methods, and their inhibition

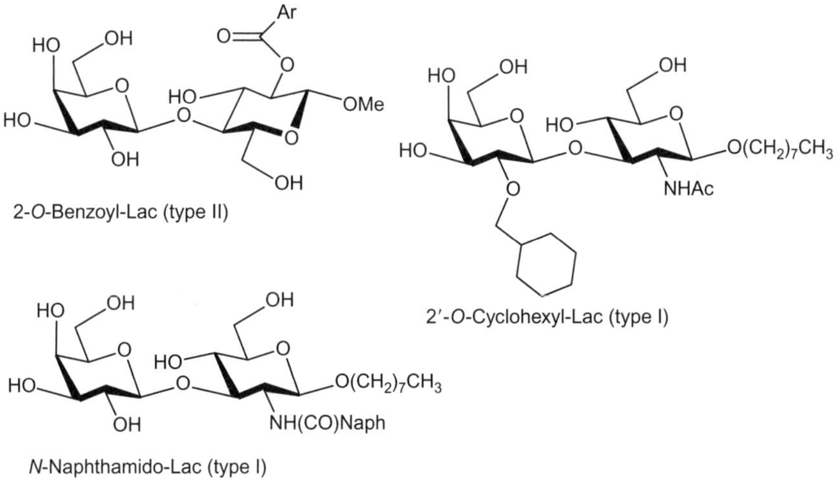

FIG. 14. Some lactose-based inhibitors of galectins.

against galectin-3 was assayed.[39] It was observed that hydrophobic substitution at NH resulted in inhibitors having the highest binding affinities; the best inhibitor identified was an N-naphthoyl derivative, with an affinity sevenfold better than the N-acetyl standard used for the study (Fig. 14).[39] This further confirmed that the enhanced binding was due to a favorable cation–π interaction between the naphthoyl substituent and a neighboring guanidine-containing Arg residue.[173]

In support of the idea that noncritical sugar residues can be replaced by aglycone units in order to improve synthesis and stability (see Section IV.2.d), a 2′-O-cyclohexylmethyl-modified LacNAc derivative appeared to act as a fucose mimic, presumably retaining the important interaction with a nearby aromatic residue, such as Trp181 in the galectin-3 model (Fig. 14).[39,173] Alternatively, substitution by a more-polar group also improved inhibitor binding to galectin-3, although to a lesser extent, which suggests the possibility of invoking H-bonding interactions to nearby amino acids.[39] In human galectin-7, a 2-O-benzylphosphate-galactoside was shown to be more than 60-fold more potent than its unsubstituted analogue; the crystal structure indicated that conventional interactions were maintained, while additional interactions with the phosphate moiety resulted in improved affinity.[182] As an alternative to targeting favorable cation–π interactions, 2-naphthamido substitution was also performed, since upon binding, this aryl moiety should project toward the Trp181 residue, likely generating a favorable π–π T-stacking interaction.[176]

(iii) C-1-Aromatization. Aryl substitution at the anomeric position has also been studied as a possible approach to enhance binding affinities and generate inhibitor selectivity for particular galectins. Aryl 1-thio-β-D-galacto- and lactopyranosides were synthesized using solid-phase synthesis, with various aromatic substituents installed for comparison.[183] It was observed that *p*-nitrophenyl 1-thiogalactopyranoside, *o*-nitrophenyl 1-thiolactoside, and naphthylsulfonyl lactoside were the best for binding galectin-1, with a 20-fold affinity enhancement observed as compared to the native ligand. The improved potency was attributed to the electronic inductive effect generated by the aglycone component, which concomitantly enhanced H-bonding interactions involving the sugar ring. Phenyl 1-thio-β-D-galactopyranoside derivatives were also found to be relatively good inhibitors of galectin-7 activity.[184]

(iv) C-2-Epimerization. Since the X-ray crystal structure indicated that, in the galactoside ligand, the 2-OH substituent points away from the protein surface and into the bulk solvent, it was hypothesized that using the C-2 epimers (talosides) may create better inhibitors by generating favorable interactions with the nearby polar Arg and His residues that interact with the β face of the galactose ring.[40] Since the Arg and

His residues are electron poor, substituents at an inverted C-2 position should complement them by being electron rich and/or aromatic in order to maximize potential binding by creating ionic or cation–π interactions. In addition to improved binding affinities, talopyranosides should also exhibit improved pharmacokinetics, as they will have longer half-lives *in vivo*: since they are not naturally present in mammals, endogenous talose-processing enzymes are not abundant in serum.[40]

In order to probe this hypothesis, a series of 2-*O*-functionalized methyl 3-*O*-(4-toluoyl)-β-D-talopyranosides were synthesized as inhibitors and screened against galectins-1, -2, -3, -4N, -4C, -8N, and -9N, comparing them to the D-galactopyranoside as a standard (Fig. 15).[40] Although the majority of the galectins displayed very weak binding, it was observed that both galectins-4C and -8N bound the talopyranoside more strongly than the galactopyranoside, and it was possible to generate selective inhibitors based on the correct functionalization at O-2 and/or O-3. Since the disaccharides Lac and LacNAc showed improved binding affinities as compared to Gal, it is likely that the replacement of galactose by talose in disaccharides will lead to significantly better inhibitors.[40]

In the previous study, the binding of galectin-1 to talopyranosides was found sensitive to the nature of the O-2 substituent, while for galectin-3, binding to 3-*O*-substituted talosides was negligible and affinities were measured to be two to three times lower than the corresponding galactoside.[40] Galectin-4N was not observed to bind any talopyranosides, although galectin-4C binding was evident and was further enhanced by 3-*O*-toluoyl substitution, while 2-*O*-substitution had very little effect on binding affinity. Binding to talopyranosides was enhanced in galectin-8N, when compared to the corresponding galactopyranoside; the interaction was significantly enhanced upon 3-*O*-toluoylation, while 2-*O*-substitution appeared to have a detrimental effect on binding. The improved binding to galectins-4 and -8 indicates that targeting talopyranosides may be useful for these particular lectins, since inhibiting only a single domain on the galectin was previously observed to inhibit activity and

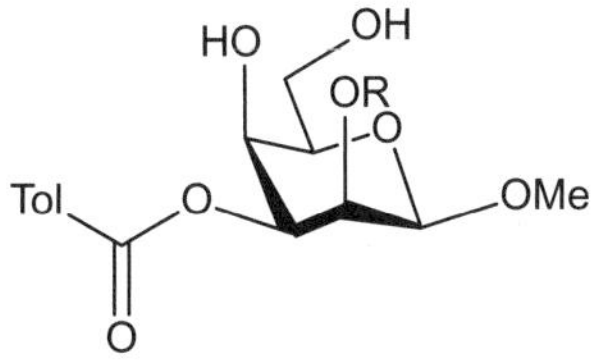

FIG. 15. 3-*O*-(4-Toluoyl)-taloside derivatives with various 2-*O*-substitution (ester, sulfate, phosphonate) have been studied as galectin inhibitors.

that binding to both the N- and C-terminal domains is necessary for function.[185] Galectin-9N did not bind talopyranosides well, and binding was further diminished upon 2- and 3-O-substitutions.[40]

In other talopyranoside analogues, selective binding has been observed to occur for galectin-1, while binding to galectin-3 remains significantly weaker. Based on analysis of their X-ray crystal structures, this selectivity has been attributed to differences in the loop portion of the binding site that is in the region of O-2 binding.[20] In galectin-1, the loop contains a His residue (His52), is 7 amino acids long, and is in closer proximity to the substrate. In contrast, galectin-3 is slightly truncated (only 6 amino acid residues) and is more conformationally restricted, being held in place by a salt bridge between Glu165 in the loop region and nearby Arg186. In addition, a slight tilting of the taloside ring (as compared to galactoside substrates) in galectin-1 results in enhanced affinities for talosides with an aromatic substituent at O-2 due to the formation of a favorable cation–π interaction with His52, which is not observed in galectin-3.[40,186] Sequence-alignment studies indicate that only galectin-12 (which has a very limited expression profile) contains a His residue in this same region, which suggests that this residue could be exploited in order to generate galectin-1-specific inhibitors.[20,187] Analysis of the remainder of the binding-site structure showed that both galectins-1 and -3 still retained the critical cation–π interaction with Trp (Trp181 in galectin-3), and the O-4 and O-6 H-bonding interactions, and both structures illustrated that the axial O-2 substituent was directed toward the protein face.[20]

(v) Anomeric Oximes. Galactosyl oximes have also been used as inhibitors, since the aglycone moiety can act as a mimic for Glc and GlcNAc residues. Since these compounds are more hydrophobic in comparison to their native analogues, they may exhibit higher bioactivity because of their increased ability to permeate the cell membrane and improved stability to enzymatic hydrolysis.[152] The galactosyl oximes could be synthesized from the corresponding β-D-galactopyranosylhydroxylamine and upon reaction with aldehydes generated a series of potential inhibitors that were screened against galectin-3 in a competitive fluorescence-polarization assay, with K_d values obtained that were up to 24-fold enhanced as compared to methyl β-D-galactopyranoside.[152]

In order to further increase inhibition, the galactosyl oxime function could be used in combination with C-3-triazolyl substitution.[188] The identity of the aglycone substituent had a significant effect on the ability of the inhibitors to bind galectin-3: when compared to the standard methyl β-D-galactopyranoside, some inhibitors were found to bind better while others were observed to have inferior binding.[152] Aminooxy and oxime ether derivatives were also tested, but showed no enhancement in galectin binding, thereby supporting the conclusion that this linking functional group has only a very minor influence on the binding affinity. The best inhibitors were found to contain polycyclic aromatic functions (such as naphthalene, quinolone, and indole), indicating that a favorable cation–π interaction is probably generated at this site, as suggested for the aromatic glycosides described previously.

(vi) C-2 and C-4 Double Epimerizations. As galactose mimics, triazolyl β-D-mannopyranosides have been developed as another class of galectin inhibitors. The *cis*-O1,O2 substituents in mannose are able to mimic the *cis*-O3,O4 substituents in galactose, and the C-1 of mannose can be modified with a triazolyl moiety as in the technique used for C-3 in galactose (Fig. 16).[189] This feature presents a major advantage, as the C-1 position of mannose is more easily modified, as compared to the synthesis of 3-azido-3-deoxy-β-D-galactose, which is comparatively lengthy. Studies in galectin-10 have indicated that binding to mannosides may even be preferred in comparison over galactosides.[190]

A library of mannosides bearing substituted triazoles at the anomeric position was synthesized, and binding to the various galectins subsequently studied.[189] In general, it was observed that mannoside binding to galectins was weak or nonexistent for galectins-1, -7, and -8N, while some reasonable affinities were observed for galectins-3 and -9N. Triazole-substituted mannosides containing a methyl-, ethyl-, or butylamide linker displayed poor binding, while the affinity for propylamide was enhanced, implying that the propyl alkyl chain is the correct size to fit into the extended binding groove of galectin-3. Unfortunately, in comparison to the related galactoside analogues, binding affinity to the mannosides was inferior. However, binding to galectin-9N was significantly improved: 4-benzylaminocarbonyl-1*H*-[1,2,3]-triazol-1-yl β-D-mannopyranoside bound galectin-9N with an affinity comparable to that of methyl LacNAc, and sixfold improved as compared to methyl β-D-galactopyranoside.

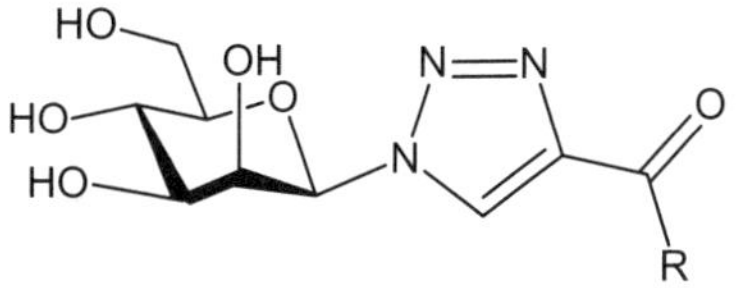

FIG. 16. Triazolyl β-D-mannopyranosides have been developed as galectin inhibitors, with affinities comparable to those for LacNAc.

(vii) Anionic Inhibitors. Since the interactions of basic residues (Arg and His) with galactose at the binding site of galectins have been well established, as an alternative to aromatic modifications that create favorable cation–π interactions, the appendage of anionic functional groups could also generate inhibitors, having enhanced binding affinities. It was illustrated that a sulfate modification at 2-OH successfully produced a strong inhibitor of galectin-3, and molecular modeling suggested it was due to interaction with a nearby Arg144 residue.[191]

An X-ray crystal structure of a 2-*O*-benzylphosphate galactoside bound to human galectin-7 indicated that the phosphate moiety at O-2 mediates a number of other favorable interactions with the galectin binding surface, including several direct and water-mediated H-bonds (with Arg31 and Asn51), and a nearby Arg53 residue provides the possibility for additional enhancements to the binding affinity (Fig. 17). Upon further analysis of the crystal structure, it was determined that the benzyl group projects away from the CRD and appears not to be involved in galectin binding. The benzamide group is situated among a cluster of polar residues near the CRD, indicating that possibly changing the aromatic moiety to strengthen cation–π binding (as by using 2,3,5,6-tetrafluoro-4-methoxybenzamido)[192] may further enhance the potency of these inhibitors.

(viii) C-Galactosyl Compounds. *C*-Glycosyl analogues have also been prepared as inhibitors, as they are resistant to hydrolysis *in vivo* as compared to the 1-thioglycosides, using various transition metal-catalyzed cross-coupling reactions (Sonogashira, Heck, and Suzuki).[193] The monovalent inhibitor methyl (*E*)-2-phenyl-4-(β-D-galactopyranosyl)but-2-enoate **A** (Fig. 18) was effective against galectin-1

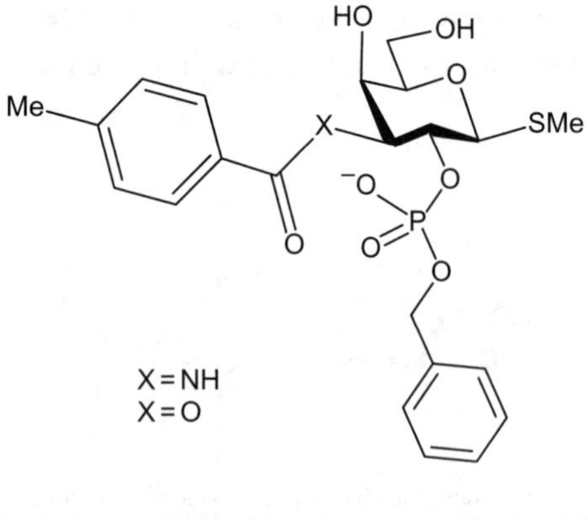

Fig. 17. Synthesized 2-*O*-benzylphosphate galactosides to inhibit galectin-3.

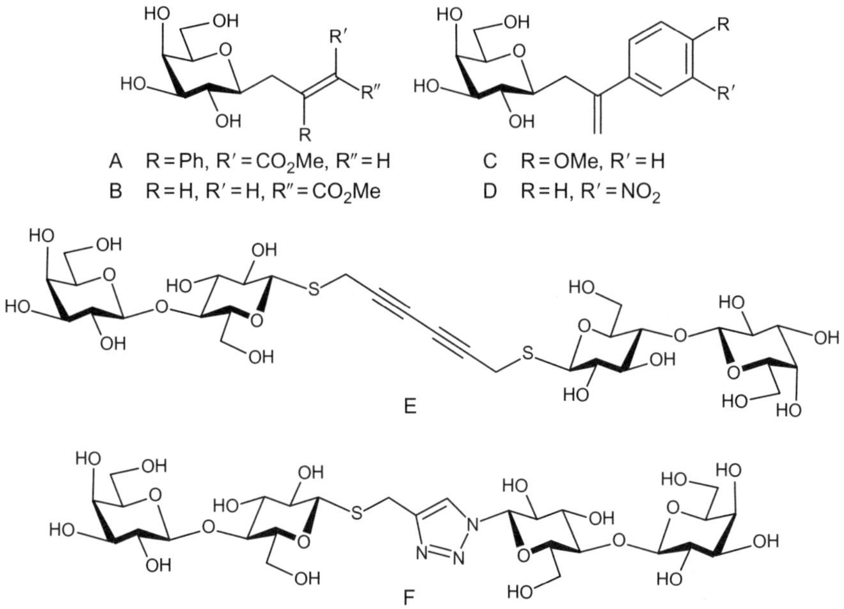

FIG. 18. *C*-Glycosyl inhibitors of galectin-1 and -3.

(160-fold improvement over galactose) and, together with enhanced binding for such other compounds as structures **B**, **C**, and **D**, suggested that possibly an sp^2-hybrized carbon atom next to C-1 may enhance galectin binding. Further improvement was observed upon dimerization: bis-lactosides **E** and **F** showed enhanced binding to galectin-3,[193] with affinities 2.5 times improved over the galactose standard (calculated per Lac residue), while no improvement in binding was observed for Gal–Lac heterodimers.

(ix) 1-Thiogalactosides. Phase-transfer catalysis has been used to synthesize 1-thiogalactoside inhibitors, with a significant focus on aryl aromatic sulfides.[194] In addition to their enhanced hydrolytic stability *in vivo*, aryl groups also increase the lipophilic character of the inhibitors, which should simultaneously result in enhanced cell-membrane permeability. In a set of potential thiogalactoside inhibitors, it was observed that none were able to bind galectin-4, but binding to both galectins-1 and -3 was observed (Fig. 19).[183,194] The sulfide substrates were also oxidized to sulfones to elucidate whether this modification could further enhance binding through the generation of more interactions with the lectin binding site.[183] For galectins-1 and -3, the

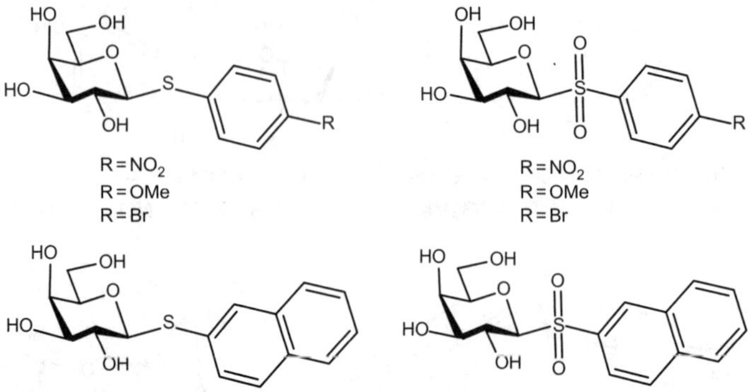

FIG. 19. Synthesized aryl 1-thiogalactopyranosides and their sulfone derivatives to inhibit galectins.

thiogalactosides displayed increased affinities, especially the aromatic glycans, with the best inhibitor of galectin-1 being a *p*-nitrophenyl 1-thiogalactopyranoside and the best inhibitor of galectin-3 a *p*-methoxyphenyl 1-thiogalactopyranoside (both showed 20-fold enhanced binding as compared to the galactose control) (Fig. 19). In addition, an anomeric sulfone also resulted in 20-fold enhanced binding to galectin-1 as compared to lactose; this increase was attributed to electron-withdrawing effects of the sulfone, which by decreasing the electron density on the 3-OH group of lactose effectively strengthened the H-bond to a nearby Glu residue.[183] The location of the aryl nitro-substituent was observed to have an effect on binding, but in order to elucidate whether this was a steric or an electronic effect, the X-ray crystal structure of O-linked nitrophenyl lactoside was studied.[183] It was observed that the aryl substituent projects out of the CRD pocket, making enhancement due to steric effects less likely. Instead, the electron-withdrawing effects of the nitro group more probably enhances the H-bonding ability of hydroxyl groups projecting from the sugar ring; this was supported by electron-density calculations, which demonstrated that electron-withdrawing groups at C-1 decreased electron density at O-3, resulting in a stronger interaction with the Glu71 binding-site residue. Other inhibitors containing different functional groups attached via a non-*O*-glycosidic linkage have also been studied because of their enhanced *in vivo* stability, such as lactulose amines and lactosyl steroids.[164,195–197]

f. Multivalent Inhibitors.—Since lectin binding has been established to occur via multivalent interactions, the generation of multivalent ligand constructs may exhibit higher binding affinities in comparison to monovalent ligands.

The 1-thiogalactosides can be dimerized via the sulfide linkage to generate the thiodigalactosides (TDGs), which constitute another class of potential inhibitors; with their thioglycosidic linkage, they have been shown to display an improved metabolic stability. In TDGs, one galactose residue binds normally, while the second mimics Glc/GlcNAc binding.[176] Because of the C_2-symmetry in these inhibitors, the free energy of binding is increased entropically; upon aromatic substitution at both C-3 positions, favorable enthalpic cation–π interactions between both Arg144 and Arg186 are established, resulting in significantly improved binding affinities.[198] TDGs containing the 3,3′-substituted aryl amides, previously observed in the monovalent constructs to enhance binding, resulted in significantly enhanced nanomolar binding affinities (33–61 nM) (Fig. 20).[199]

A number of studies have examined the effect of TDG treatment on tumor growth, for example, a reduction in the occurrence of lung metastasis was observed to occur in murine breast (4T1) and colon (CT26) cancer cells via inhibition of galectin-1.[200] Intratumoral injection with TDG in galectin-1-expressing B16F10 melanoma and 4T1 breast cancer cells afforded elevated levels of $CD8^+$ lymphocyte cells[201]; this inhibition of the antiapoptotic effect of galectin-1 resulted in reduced tumor growth. Upon induced apoptotic stress (H_2O_2 treatment), galectin-1 was found to have a protective effect on the tumor cells, while after treatment with TDG, the cells became sensitive to oxidative stress-induced apoptosis. Since galectin-1 contains six Cys residues and is secreted extracellularly in its reduced form, it is possible that, in addition to antiapoptotic effects, it may also play a role directly in protection against oxidative stressors.[202] In addition, TDG treatment in Balb/c nude mice (which are defective in T-cell immunity) also reduced both the rate of tumor growth (by a third) and the rate of angiogenesis, indicating that galectin-1 inhibition is able to block multiple metastasis-related pathways simultaneously. In a galectin-1 knockout line, tumor growth was significantly reduced, and treatment with TDG had effects that were extremely less pronounced, confirming that indeed TDG acts via galectin-1

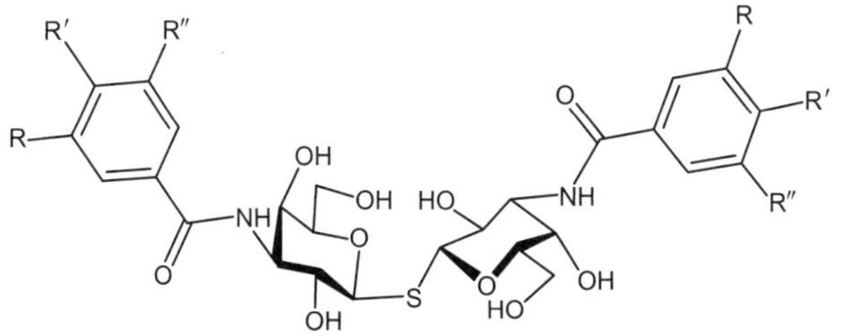

FIG. 20. A series of thiodigalactoside derivatives have attained nanomolar binding affinities.

inhibition. 3,3′-Ditriazolyl-thiodigalactosides have also exhibited high (nanomolar K_d values) binding affinities to galectin-3, but in comparison to 3,3′-benzamido derivatives, they displayed different selectivity profiles for the different galectins.[176] In a cotherapeutic approach, where treatment with both TDG and a whole-cell breast-cancer vaccine was administered in combination, greater reduction in tumor growth was observed, probably because galectin inhibition was able to block the immunosuppressive properties of galectins, and this allowed for an enhanced immune response to the vaccine.[141]

Aromatic 3,3′-diester-TDGs were also shown to reduce tumor-cell motility[203] and afforded enhanced sensitivity to proapoptotic chemotherapeutic agents in human thyroid cancer cells.[204] It was observed that the two ester moieties interacted favorably with Arg144 and Arg186 of galectin-3, resulting in a relatively high affinity and specificity for this particular galectin.[203] Inhibition of galectin-3 using the inhibitor was capable of enhancing apoptosis, radiosensitivity, and doxorubicin sensitivity, most probably due to inhibition of the galectin-3-mediated antiapoptotic mechanism.[204] Inhibition of these pathways is promising as a cotherapeutic regimen, since it enhances the sensitivity of tumor cells to conventional treatments.[204]

In addition to the TDGs, oligovalent lactulose amines have also been synthesized as potential inhibitors; *N*-lactulos-1-yl-octamethylenediamine, *N*,*N*′-dilactulos-1-yl-octamethylenediamine, and *N*,*N*′-dilactulos-1-yl-dodecamethylenediamine were synthesized and studied. Using solid-phase assays, these compounds were found to exhibit different levels of inhibition of galectins-1 and -3 to 90 K (a densely glycosylated glycoprotein). Assays indicated that each inhibitor displayed different regulatory effects on the tumorigenesis process (for example, inhibition of cell aggregation, apoptotic sensitivity, and the like).[195]

Mono-, di-, and trivalent lactoside derivatives have also been generated using regioselective 1,3-dipolar cycloaddition: 2-azidoethyl β-D-galactopyranosyl-(1 → 4)-β-D-glucopyranoside was coupled to acetylene-functionalized phenylalanine, phenethylamine, phenyl-bis(alanine), and phenyl-tris(alanine).[205] Their inhibitory potencies were evaluated against galectins-1, -3, -4N, -4C, -4, -7, -8N, and -9N, and it was observed that the divalent derivative could enhance binding by 30-fold as compared to Lac (calculated per Lac residue) and was also mildly selective for galectin-1, since binding to this galectin was at least an order of magnitude greater than for the other galectins examined. As a general trend in this study, it was concluded that multivalency improved binding to both galectins-1 and -4N. In another study, a trivalent presentation of lactoside was shown to improve galectin-1 binding by 13-fold as compared to the monovalent Lac (calculated per Lac residue), although multivalency did not greatly improve binding to galectin-3.[179]

In the presence of a TF–polyacrylamide conjugate, binding of BeWo chorion carcinoma cells to galectin-1 could also be inhibited.[144] In another example, a pseudo-polyrotaxane consisting of a chain bearing a number of threaded cyclodextrin units, each containing a Lac residue, was able to inhibit galectin-1-mediated agglutination of cultured human T-leukemia cells, although the binding affinity was only one order of magnitude better than monovalent Lac.[206]

To make the design of multivalent inhibitors more difficult, it was observed that lectin specificity arises not from the number of ligands available, but instead results from their particular steric presentation.[207,208] Because different lectins have different arrangements of CRDs, different multivalent presentations may create selectivity for a particular inhibitor. This is easier for the cell-surface lectins, in contrast to soluble galectins, since the latter are less organized, can be dimeric or oligomeric, or remain as monomers in regions of low concentration, and therefore may be more difficult to target.[173,209] In fact, because of the CRD arrangement in most galectins (for example, prototype, tandem repeat type), several studies have suggested that the divalent inhibitors are unlikely to chelate the same oligomeric galectin, but instead probably enhance cross-linking between different galectin complexes, influencing aggregation, rather than directly affecting binding affinities via entropic factors.[210,211] In contrast, a study of mono- and divalent lactoside inhibitors with a galectin-1 mutant, which had been mutated to attenuate its ability to dimerize, indicated that, in the divalent lactosides, the second substrate most likely bound a secondary lactose site on the same CRD, but did not in fact aid in the cross-linking of galectin-1.[212]

In some instances, multivalency can play an extremely important role, as the formation of galectin-3 pentamers can abrogate the apoptotic effect of dimeric galectin-1. Upon exposure to multivalent ligands, the N-terminal domains of galectin-3 associate to form pentavalent constructs, which bind more strongly to the multivalent substrates.[213] However, without the N-terminal domain, galectin-3 is unable to associate and binding to galectin-1 cannot be overcome.[214]

Although divalent LacNAc constructs exhibited minor improvements in binding affinity upon multivalency, increased valency and the introduction of more-rigid backbones have seen much greater enhancements of binding[215,216]; for example, a tetravalent lactoside demonstrated a greater than 1000-fold improvement per Lac residue in binding galectin-3 as compared to Lac, while binding to galectins-1 and -5 showed only minor improvements (Fig. 21).[210] The lactose inhibitor, containing terminal lactose-2-aminothiazoline ligands, was studied in a solid-phase inhibition assay using asialofetuin that had been immobilized on microtiter plate wells. Another series of lactose dendrimers, based on a 3,5-di-(2-aminoethoxy)benzoic acid scaffold, were synthesized to obtain a series of wedge-like dendrimers containing either two,

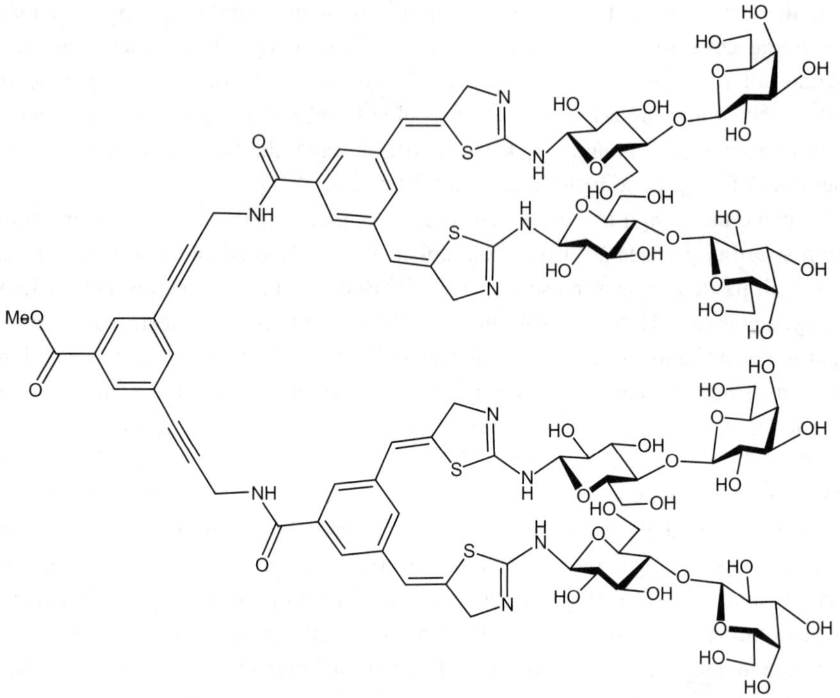

FIG. 21. A tetravalent lactose derivative was able to inhibit galectin-3 binding 1000-fold better than monovalent Lac.[210]

four, or eight sugar moieties.[209] The binding of dendrimers to galectins-1, -3, -7 (AB)2 (a plant toxin), and lactose-binding human IgG was assayed in several solid-phase competition studies and cell assays, the results of which demonstrated that both galectins-1 and -3 showed improved binding to multivalent constructs.

Cyclic platforms are beneficial because they provide a more-rigid backbone; several examples of rigid scaffolds have been used: cyclodextrins, cyclic oligopeptides, cyclophanes, cyclic imide ring systems, and calixarenes (phenol–formaldehyde oligomers).[216–218] Systems such as calixarenes and oligopeptides present interesting scaffolds, since their ring size, flexibility, valency, and symmetry can all be varied to better determine the particular ligand presentation that best binds the galectin of interest. Calixarenes using a thiourea linker have been studied, since the linker region could potentially participate in a favorable H-bonding interaction, and in addition, the conjugation reaction between an amine and isothiocyanate is relatively efficient.[219] Fourteen different calix[n]arenes were synthesized ($n=4$, 6, or 8) with either terminal galactose or lactose moieties; in cell assays, it was observed that galectins-1 and -4 had the highest affinity interactions with lactose-derivatized calix[4]arene (Fig. 22).[218]

Many of the multivalent constructs examined have used neutral saccharide ligands that more closely resemble the natural substrate, but have not yet fully explored the potential affinity enhancements from the chemical modifications known to improve binding in the small-molecule inhibitors.

g. Glycopeptides and Peptidomimetics.—Peptides have also been generated as inhibitors of galectins, as they can either mimic the natural substrate by acting as competitive inhibitors at the CRD, or alternatively, since noncarbohydrate binding domains exist, it is possible that nonmimetic peptides can also act as effective inhibitors.[220] Since solid-phase synthesis can be used to readily prepare peptide and glycopeptide inhibitors, they are more easily obtained and produced on larger scales. Both solid-phase peptide and phage-display libraries have been used to identify potential galectin inhibitors.[221,222] Solid-phase libraries containing β-Gal-*O*-Thr-, β-Gal-*S*-Cys-, β-Gal-*N*-Asn-, and β-Lac-*O*-Thr-based glycopeptides were studied using fluorescently labeled human galectins-1 and -3[223]; the most successful inhibitors identified contained lactose-bound peptides. These glycopeptides bound with stronger affinities than free lactose, indicating the presence of favorable interactions between galectin and the peptide region. This enhancement likely arises via favorable H-bonding and hydrophobic interactions; peptide fragments containing Cha (cyclohexylalanine), Phe, Arg, and Ile appeared to provide the greatest affinities. Differences in the binding profiles were observed between galectins-1 and -3, and a few noncarbohydrate containing peptides were identified as effective inhibitors.

A number of synthetic carbohydrate–amino acid conjugates were screened for their activity against human breast carcinoma (MDA-MB-435) xenograft metastasis.[197,224] The peptides were initially isolated, purified, and identified from human and rodent serum, but were subsequently synthesized chemically in order to obtain them in sufficient yields for biological testing. The most effective metastatic inhibitors were determined to be *N*-(1-deoxy-D-lactulos-1-yl)-L-leucine (Lac-L-Leu) and *N*-(1-deoxy-D-fructos-1-yl)-D-Leu, while the *N*-(1-deoxy-D-fructos-1-yl)-L-Phe and *N*-(1-deoxy-D-fructos-1-yl)-L-Leu were identified as less-active inhibitors (Fig. 23). The activity of the conjugates in nude mice was attributed to either apoptotic induction or the inhibition of aggregate formation.

It has been confirmed that Lac-L-Leu is able to bind galectin-3 in competition with the native TF antigen, and in another study using breast carcinoma cells (MDA-MB-435), which express high levels of both galectin-3 and TF antigen, it was observed that the inhibitor was able to reduce both homo- and heterotypic aggregation.[144]

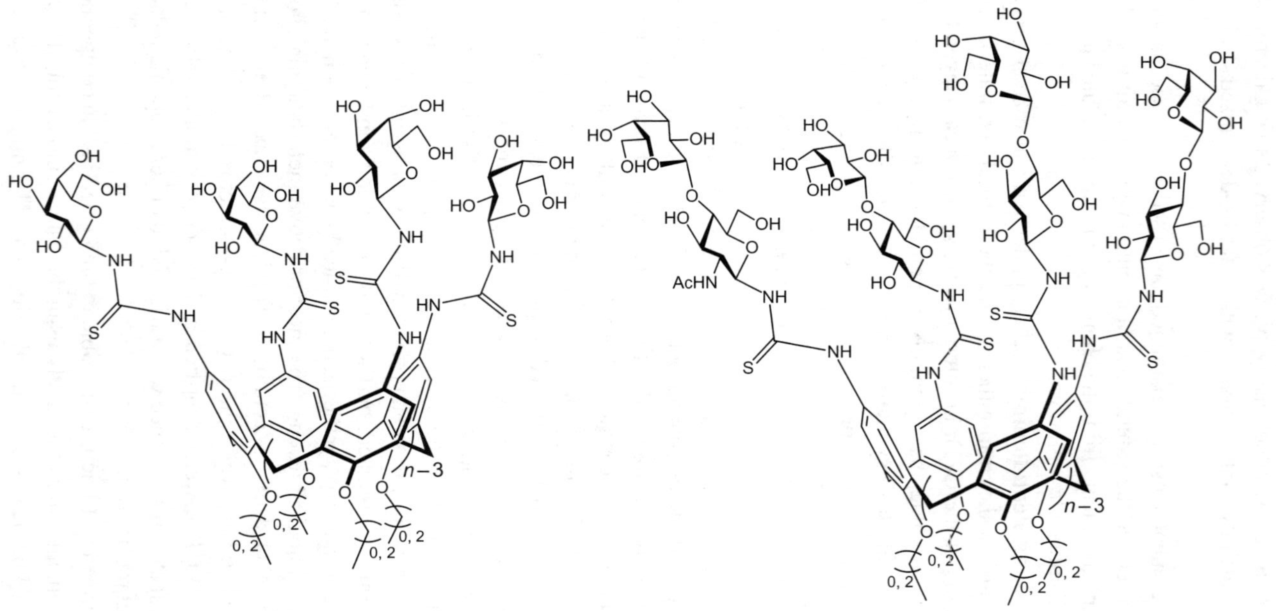

Thiourea linked calix[n]arene-galactose and lactose inhibitors with either a methyl or propyl tail

FIG. 22. Thiourea-linked calix[n]arenes have been studied as galectin inhibitors ($n = 4$, 6, or 8).

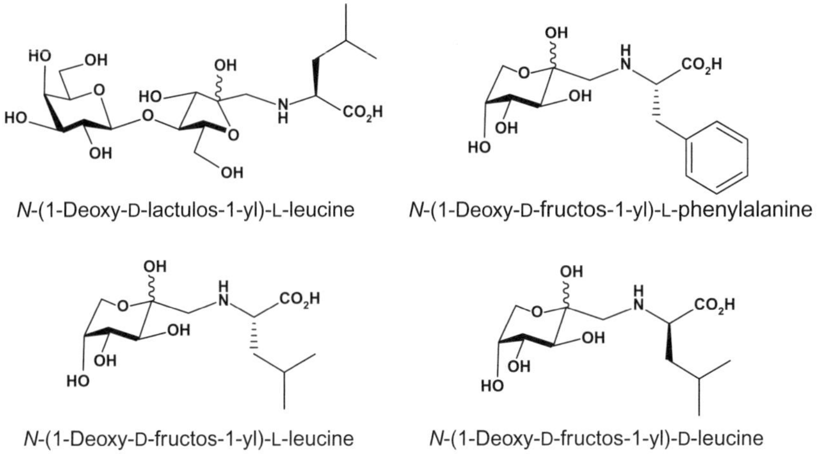

FIG. 23. A number of glycosylamines have been studied as galectin inhibitors.

In addition to breast carcinoma cells, Lac-L-Leu has been shown to reduce the metastatic effects of a number of cancer types, including melanomas, prostate carcinomas, and hemangiosarcomas.[163,224,225] The intraperitoneal administration of Lac-L-Leu resulted in a threefold decrease in tumor metastasis, as compared to the untreated control.[226] In a negative control experiment using lactitol-L-Leu, no galectin-3 binding was observed.[78]

Since galectin-3 displays antiapoptotic effects, inhibiting galectin-3 should make tumor cells more sensitive to cytotoxic agents, which typically act by inducing apoptosis via the mitochondrial apoptosis pathway. When treated separately, Lac-L-Leu resulted in a 5.5-fold decrease in the median number of pulmonary metastases in each mouse, but did not reduce the overall incidence of whether or not metastasis occurred, while treatment with paclitaxel did not affect the median number or the incidence of metastases.[164] Upon cotreatment, a decrease in both the incidence of metastases and the number of metastases formed was observed: overall, a fivefold increase in metastasis-free animals was observed (70% metastasis free, as compared to 14% in the control group).

The peptide motif Tyr-Xxx-Tyr was also reported to act as a glycomimetic inhibitor of galectins; first reported as a ligand of the plant lectin concanavalin A,[227] it was later identified as a galectin substrate upon the screening of a library of pentapeptides.[221,228] Solid-phase assays and surface plasmon resonance studies were performed using galectin-1 and asialofetuin, a 48 kDa protein containing three triantennary N-linked carbohydrates and known to bind galectin-1.[229] Although the

Tyr-Xxx-Tyr motif (e.g., TYDYF-NH$_2$, WYKYW-NH$_2$, TYPYFR-NH$_2$) was capable of disrupting the interaction of galectin-1 with asialofetuin, it was unable to inhibit binding of galectin-1 to leukemic Jurkat T-cell-surface glycans.[183,228,229] Subsequent STD and trNOE NMR experiments in fact revealed that these peptides do not bind to galectin-1 but instead bind to asialofetuin.[229]

h. Allosteric Inhibitors.—Examples of noncarbohydrate synthetic inhibitors have also been reported which are able to act via allosteric binding. Two tetrahydroisoquinoline natural product analogues (DX-52-1 and HUK-921) were shown to act as inhibitors of galectin-3, but appeared to bind outside the CRD region.[230] In cells that overexpressed galectin-3, the cell sensitivity to DX-52-1 and HUK-921 was greatly reduced, although they resulted in cells returning to a more healthy morphology and a change in localization of the galectin-3.

Studies involved with targeting galectins for drug delivery used the galectin-1 ligand anginex (a 33-amino acid peptide) to target cytotoxic 6-hydroxypropyllactylfulvene to human ovarian cancer cells in mice.[231] It was observed that the conjugated drug inhibited tumor growth better than either compound administered alone, although immunofluorescence studies indicated that the conjugate bound to the tumor vasculature to attenuate angiogenesis. To improve pharmacokinetic properties, efforts have been directed toward reducing the size and peptidic character of anginex. Initially, the dibenzofuran-based partial peptide 6DBF7 was developed, which was found to successfully target galectin-1.[232,233] Anginex and 6DBF7 were successful in attenuating angiogenesis and tumor growth, and both constructs displayed amphipathic and polar (for example, ionic) substituents at the proper locations required for galectin binding.[233,234]

Pharmacokinetic properties were further improved upon generation of the inhibitor calixarene 0118, which successfully inhibited angiogenesis and tumor growth in murine tumor models using human ovarian carcinoma (MA148) and murine melanoma (B15) cancer cells (Fig. 24).[235] It was shown to be an effective inhibitor of galectin-1-mediated cell agglutination in a dose-dependent manner, via inhibiting the binding of galectin-1 to cell-surface glycans.[236] Calixarene 0118 is more likely to be a successful therapeutic inhibitor because of its smaller size (937 Da), decreased risk of hydrolysis *in vivo* (lacks carbohydrate and peptide components), and improved metabolic stability; as a result of these promising attributes, the compound has progressed into phase I clinical trials.[236]

Further studies have demonstrated that all three inhibitors (anginex, 6DBF7, and calixarene 0118) function similarly *in vitro* and *in vivo*, by inhibiting angiogenesis and proliferation of tumor cells,[234] although the ligand display in calixarene 0118 appears to differ from those in the other two inhibitors. The latter contain a

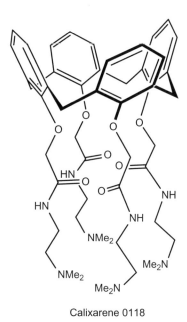

FIG. 24. Calixarene 0118 is a noncarbohydrate-based allosteric inhibitor of galectins.

hydrophilic surface comprising mainly cationic Lys residues and a hydrophobic surface of alkyl and aromatic residues (Leu, Val, Ile, Phe, and Trp).[236] In contrast, the hydrophilic surface of calixarene 0118 comprises mainly tertiary amines, and the hydrophobic surface mostly aryl residues. Based on structure–activity relationships, it appears that all three inhibitors interact with galectin-1 through the hydrophobic face; ^{15}N–^{1}H HSQC NMR spectroscopy has confirmed that calixarene 0118 binds at the back face of the lectin relative to the CRD, therefore acting as an allosteric inhibitor, since it significantly attenuates the binding of galectin-1 to cell-surface glycans.[236]

2. Efforts Toward Selectin Inhibitors

Because of the involvement in metastatic formation, the inhibition of selectins may be most useful at stages of high metastatic risk, such as during surgeries to remove a primary tumor where the number of tumor cells in circulation may become temporarily increased.[237] The selectins can bind a number of substrates, including antigens of the Le blood group, heparan and chondroitin sulfates, and PSGL-1,[238] but unfortunately inhibitors based on these ligands have been hampered by their poor selectivity, weak affinities, limited availability, reduced membrane permeability, and poor *in vivo* stability, all of which have limited their clinical use.[239,240] However, the development of potent inhibitors to interfere with selectin–carbohydrate interactions remains a valid approach. For example, treatment of tumor cells with an E-selectin mAb (BBA2) was found to be successful at disrupting tumor-cell adhesion,[241] and the use of anti-E-selectin mAb was found to effectively inhibit experimental metastases by 97%[119,129]; similarly, the use of anti-L-selectin mAbs was shown to partially diminish the adhesion of neutrophils to HO-8910 tumor cells by 50%.[242] The results from the past studies suggested that there could exist another adhesion mechanism between the two cell types.

a. **Soluble Recombinant Proteins, Inhibition of Selectin Expression.**—By downregulating their expression, selectin-mediated tumor growth and metastasis can be significantly reduced; this was observed for a number of agents that successfully reduced selectin expression in the cell. For example, andrographolide (Fig. 25), a natural product isolated from the Chinese herb *Andrographis paniculata*, was observed to significantly reduce the levels of E-selectin mRNA, which resulted in reduced adhesion to gastric cancer cells and anticancer activity.[243] Likewise, both lovastatin and a low-molecular-weight polysaccharide from *Agaricus blazei* were able to reduce gene expression of E-selectin, resulting in reduced cell-surface expression[244]; treatment with either compound successfully interfered with the adhesion of colon carcinoma cells to HUVECs. Similarly, treatment with C-raf kinase in human colorectal carcinoma (CX-1) cells inhibited experimental metastases by preventing the cytokine-mediated upregulation of E-selectin.[245]

Interestingly, the compound cimetidine (Fig. 25) has also been successful in reducing E-selectin expression (as in gastric cancer cell lines SGC-7901, MGC-803, and BGC-823), but has no effect on E-selectin mRNA levels, indicating that its

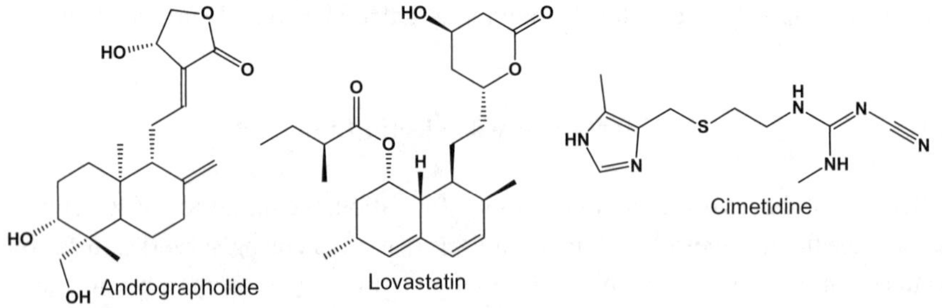

FIG. 25. The chemical structures of andrographolide, lovastatin, and cimetidine.

involvement in the downregulation is occurring posttranscriptionally, during either protein synthesis or cell-surface translocation.[130,246] Cimetidine has been shown to suppress the occurrence of metastasis in a nude-mouse model and has also been used clinically in patients with colorectal cancer, since it can block the adhesion of tumor cells to the endothelial surface.[246,247] Treatment with cytotoxic agents used in chemotherapeutic approaches has been associated with the upregulation of selectin expression, thereby potentially increasing the risk of metastasis. Cytotoxic agents (5-fluorouracil, doxorubicin, and cisplatin) were cultured with HUVECs and, using quantitative RT-PCR and immunohistochemical staining, were quantitatively confirmed to upregulate the selectin levels of both mRNA and protein expression.[248] When these cells were pretreated with cimetidine prior to exposure to the cytotoxic agent, the observed mRNA levels were still elevated, but cell-surface E-selectin expression levels were reduced, suggesting that cotreating patients with cimetidine during conventional chemotherapeutic courses could be beneficial in reducing the risk of metastasis.[248] Similarly, ionizing radiation has been shown to induce E-selectin gene expression, resulting in elevated levels of protein expression, but this upregulation could be inhibited by treatment with either lovastatin or all-*trans* retinoic acid.[249]

Instead of altering selectin expression levels, soluble E-selectins can be used to diminish binding of cancer-associated selectins. For example, treatment with a soluble E-selectin, based on a fusion E-selectin-Ig chimera, successfully inhibited the lung colonization of tumor cells in nude mice upon intravenous administration of human colon carcinoma (HT-29) cells with cytokine-induced, activated endothelial cells.[250] The carcinoma cell line studied was known to express ligands for both E- and L-selectin, although in this particular study, L-selectin did not appear to play a significant role in the metastatic process, as only soluble E-selectin chimera was capable of attenuating lung metastasis.

b. **Glycosyltransferase Transfection, Inhibition of sLe^x Expression, Anti-Le Abs.**—Binding can also be attenuated by inhibiting access to selectin ligands. For example, treatment with anti-sLe^x or anti-sLe^a Abs can interrupt binding of HUVECs and colon carcinoma cells.[87] Alternatively, biosynthesis of the ligands can be interrupted, and studies with 4-F-GlcNAc, a compound also used to attenuate galectin binding via inhibiting sLe^x expression, were carried out.[251] Unfortunately, 4-F-GlcNAc treatment was less successful at inhibiting the formation of endothelial selectin ligands, such as LacNAc expressed on glycolipids, in contrast to the galectin-1-binding LacNAc expressed on glycoproteins.

Alternatively, the cell-surface expression of sLe^x on tumor cells could be altered upon treatment with peracetylated β-GlcNAc-(1 → 3)-β-Gal-*O*-naphthalenemethanol, while the expression of sLe^a was unaffected.[240,252] Inhibition of sLe^x could also be achieved by generating peracetylated analogues of the β-GlcNAc-(1→3)-β-Gal-*O*-naphthalenemethanol modified by C-3′ and C-4′ substitution with -H, -F, -N$_3$, -NH$_2$, or -OMe (Fig. 26). Acetylation serves to increase permeability of the cell membrane, and upon entering the cytosol, the disaccharide is deacetylated by carboxyesterases and can then act as a substrate in the Golgi complex for enzymes involved in the normal biosynthesis of sLe^x, such as human β-1,4-galactosyltransferases.[253] The treatment of monocytic leukemia (U937) cells with peracetylated β-GlcNAc-(1→3)-β-Gal-*O*-naphthalenemethanol successfully attenuated sLe^x biosynthesis.

Early studies indicated that alterations in cell-surface fucosylation were closely correlated to the biosynthesis of tumor-associated antigens, and therefore, alterations in the biosynthetic pathways involving these enzymes have been studied as a therapeutic option.[254] Early reports had indicated that the transfection of α-1,2-fucosyltransferase I (FUT1) was able to upregulate FUT1 expression in CHO cells, which severely limited the cell-surface expression of sLe^x.[255] Likewise, in human hepatic (HepG2), colonic (HT-29), and pancreatic (BxPC3) tumor cell lines, transfection of FUT1 was shown to reduce sLe^x expression in HepG2 and HT-29 cells, resulting in a reduced adherence to activated E-selectin-expressing endothelial cells.[255] An increase in the expression levels of cell-surface Le^y and Le^b was observed, but no effect on sLe^a expression was evident. FUT1 upregulation in GxPC3 cells had no effect on tumor progression, suggesting that this cell line may

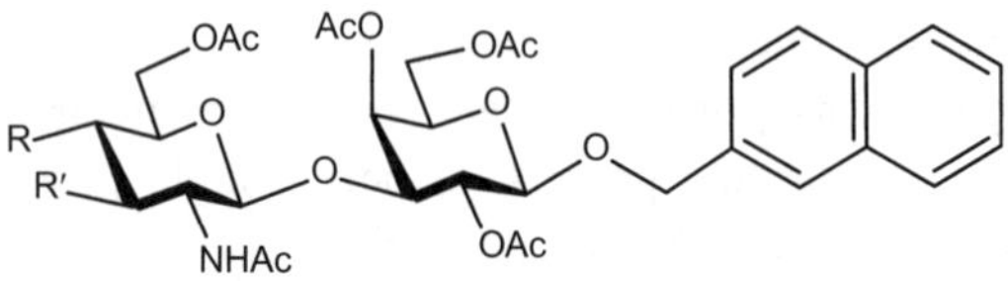

R	R′
OAc	OAc
H	OAc
F	OAc
N$_3$	OAc
NH$_2$	OAc
OMe	OAc
OAc	H
OAc	F
OAc	OMe

FIG. 26. A series of peracetylated 3′- or 4′-substituted β-GlcNAc-(1→3)-β-Gal-*O*-naphthalenemethanol analogues were used to attenuate sLe^x biosynthesis *in vitro*.[253]

follow a different adhesion mechanism independent of E-selectin; it was observed that FUT1 preferentially utilized glycoprotein substrates over glycosphingolipids, in agreement with earlier observations.[255,256] Previously, the overexpression of FUT1 had been shown to reduce sLe^x expression, although the inhibition of binding and tumor-cell adherence was reported to be less successful.[257]

Alternatively, antisense techniques can be used to alter gene expression and truncate cancer-associated glycans. The adherence of adenocarcinoma cells to HUVECs, which is mediated by E-selectin binding, was inhibited by transfecting FUT3 antisense sequences into mice.[254] Although antisense oligomers are typically less immunogenic and display improved pharmacokinetics as compared to Ab-based therapies, a major difficulty lies in delivering these sequences to the nucleus.[258] In addition, other studies have demonstrated that inhibiting glycosylation can have a number of independent, adverse effects as other nonrelated biosynthetic pathways can be simultaneously modulated.[4]

β-GalNAc-(1 → 4)-[α-Neu5Ac-(2 → 3)]-β-Gal-(1 → 4)-GlcNAc (Sd^a) is a blood-group antigen normally expressed in healthy gastrointestinal tissue, but in cancer cells, its expression is significantly downregulated (Fig. 27).[259] Reduced Sd^a expression is a result of the downregulation of β-1,4-N-acetylgalactosaminyltransferase, an enzyme involved in the normal biosynthetic pathway of the antigen. Sd^a, sLe^a, and sLe^x are all known to compete for the same precursor in the biosynthetic pathway[260]; therefore, in an effort to attenuate the expression of tumor-associated glycans, cancer cells (KATOIII and HT29) known to express elevated levels of sLe^x and sLe^a antigens were treated with β-1,4-GalNAc-transferase cDNA. This resulted in a significant

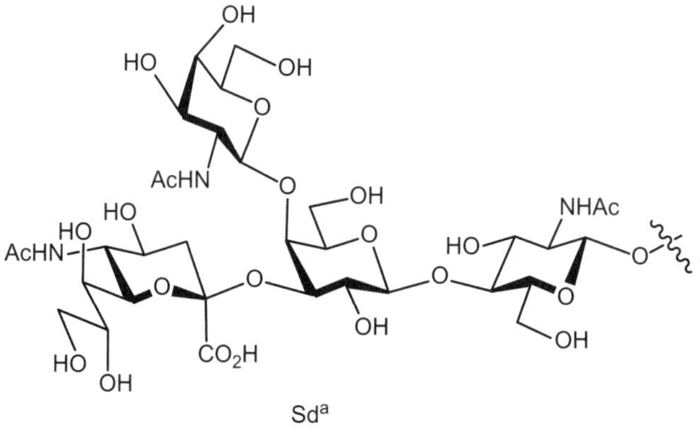

FIG. 27. Structure of Sd^a antigen.

increase in the expression of Sd^a and a concomitant reduction in the levels of sLe^x and sLe^a.[259] This approach has an advantage over FUT targeting, since multiple tumor-associated carbohydrates can be targeted simultaneously with the same approach, which is beneficial against tumor-cell heterogeneity. With cDNA treatment, the adhesion of tumor cells to HUVECs was greatly decreased, but as not much is known about the regulation of Sd^a expression *in vivo*, it is difficult to predict whether its induced overexpression will have any adverse effects by potentially modifying the regulation of other biosynthetic pathways.[259]

    c. Heparin, Chitosan, Dermatan Sulfate Derivatives.—It has been illustrated that the preferred native ligands of selectins are sialylated, fucosylated, or sulfated glycans, commonly expressed as mucins on cancer cells.[124,261] Heparins are a family of densely sulfated glycosaminoglycans, which have also been shown to act as P-selectin ligands; they are linear polysaccharides, comprising a repeating uronic acid-(1 → 4)-D-glucosamine disaccharide unit, and have been used therapeutically as an anticoagulant for over 75 years (Fig. 28).[239] Heparin resembles tumor-associated mucin glycoproteins in that they both express dense regions of anionic charge density, and fortunately chemical modifications to heparin can decrease its anticoagulant activity, and this has generated interest in using them as inhibitors of selectin binding, since they have already been proved safe for clinical use.[124,239]

    E-Selectin can play a role in both angiogenesis and the adhesion of tumor cells to endothelial tissue, and a competitive solid-phase assay performed using endothelial cells fixed onto a poly-L-lysine-coated surface indicated that soluble E-selectin-Ig chimeric structures were effective inhibitors.[262] The soluble E-selectins were able to bind both bovine capillary endothelial cells and human dermal microvascular endothelial cells, but this binding could be inhibited by treatment with either heparin or syndecan-1 ectodomain (which contains both heparan-sulfate and chondroitin-sulfate regions), and to a lesser degree by chondroitin sulfate; sLe^x-containing ligands were unable to inhibit binding. The binding of recombinant soluble selectins to tumor-cell

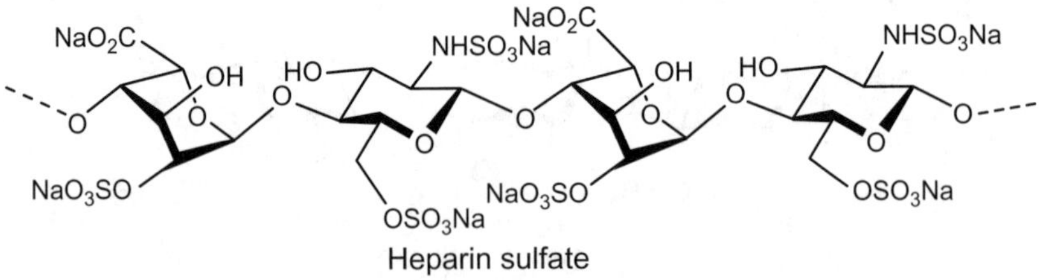

Heparin sulfate

FIG. 28. A tetrasaccharide fragment of heparin sulfate, an anticoagulant and selectin inhibitor.

mucins *in vitro* was inhibited by heparin treatment, and in addition, the treatment was able to attenuate the binding of activated platelets to tumor cells, resulting in reduced formation of metastatic lesions *in vivo*.[120] Both heparan-sulfate-like glycans and mucin-type glycoproteins acting as P-selectin ligands are expressed at the cell surface of different COLO320 cells, indicating a heterotypic expression pattern among this cell line.[239] Changes in expression have been attributed to either different stages of carcinoma progression or the ability of P-selectin ligands to contain both sialylated and heparin sulfate-like regions.[239] This conclusion is in agreement with a previous report which suggested that both sialylated and sulfated heparin ligands could bind near the selectin binding site, suggesting that instead of a distinct, well-defined binding site, a general binding region may instead select substrates based on an appropriate display of polar functional groups, such as hydroxyl, carboxylate, and sulfate.[263]

Many early studies demonstrated that heparin was capable of limiting cancer metastasis in both animal and human models. This was initially attributed to its anticoagulant effects, but when heparin treatment was replaced with vitamin K, no effect on the survival of carcinoma patients was observed, suggesting that heparin may be acting through an alternative pathway.[52] It was later established that heparin was able to block metastasis through inhibition of P- and L-selectin-mediated binding to tumor cells.[264] Unfortunately, therapeutic heparin use can result in a number of severe adverse effects, such as heparin-induced thrombocytopenia, and it has become desirable instead to identify heparin derivatives that are capable of inhibiting selectins, but have decreased anticoagulant activity. This has been achieved through structural modifications to heparins, such as carboxylate-reduced heparins that retain their ability to bind P-selectins, but show diminished anticoagulant activity.[263,265] In a study of a library of heparins, it was even suggested that non-anticoagulant heparins often displayed enhanced binding affinities to selectins, as compared to anticoagulant heparins.[266] Heparin and low-molecular-weight heparin have both been demonstrated to be capable of reducing secondary metastases in experimental tumor models[267]; this observation was directly linked to P-selectin expression, since P-selectin −/− mice had improved outcomes upon tumor challenge, and treatment with heparin at clinically relevant doses showed no further improvement in the inhibition of metastatic lesions.

Heparins of lower molecular weight are often considered beneficial in comparison to unfractionated heparin because of their improved pharmacokinetic properties, reduced risk of heparin-induced side effects, and lower anticoagulant activities.[52] Unfractionated heparin has a poor bioavailability, but sequential periodate oxidation and borohydride reduction affords "RO-heparin," which has attenuated anticoagulant properties, reduced heparin-related side effects, and improved pharmacokinetic properties because of its lower molecular weight (3–9 kDa).[52] Although preliminary studies using two low-molecular-weight heparins indicated that the smaller heparins were less active *in vitro* as compared to unfractionated heparin,[263] their improved bioavailability may make them comparatively more active *in vivo*. A number of examples of cancer patients that had been treated with low-molecular-weight heparins for unrelated reasons were observed to display improved survival as compared to untreated patients.[268]

Heparin and RO-heparin were both capable of attenuating the adhesion of L-selectin-mediated human ovarian carcinoma (HO-8910) cells (known to express heparan-sulfate-like cell-surface structures) to L-selectin-expressing CHO cells (induced via cDNA transfection) and neutrophils[242]; heparinase treatment of the tumor cells was also able to interfere with L-selectin adhesion to the carcinoma cell surface. Interestingly, it was observed that the heparins were able to inhibit carcinoma cell binding more effectively than an anti-L-selectin mAb, suggesting that the heparins are able to block simultaneously multiple pathways of adhesion.

The observation that heparins may be inhibiting various pathways is well supported by other studies. The adhesion of non-small cell lung cancer cells to P-selectin was inhibited *in vitro* upon treatment with both heparin or chemically modified heparin,[269] but the use of anti-heparin sulfate mAb to block heparin sulfate-like glycans on the tumor cell surface resulted in only a partial reduction in cell adhesion. In addition, treatment of the cancer cells with heparinase again resulted in only partial inhibition of binding, further supporting the conclusion that multiple binding mechanisms must be involved. Neuraminidase treatment to remove cell-surface sialic acids had no effect on cell adhesion, indicating that sialic acid-containing ligands are not involved in these additional interactions, but the cleavage of glycoprotein glycans reduced binding yet did not fully abrogate it, and so some other carbohydrate-based interaction is probably involved (for example, glycolipid).[269]

In another study using a melanoma cell line that expressed low levels of sLe^x, which bound only low levels of recombinant P-selectin, it was observed that heparin was still able to inhibit cell metastasis.[52] This observation could suggest the role of alternative adhesion molecules, or alternatively, it is possible that ligand–selectin interactions not directly involving the tumor cells may be aiding in tumor metastasis. These endogenous interactions could be a result of platelet aggregates surrounding tumor cells, which are known to protect circulating emboli from attack by immune cells; it was previously reported that these platelet aggregates were able to form around tumor cells not expressing selectin ligands,[270] and this may be another mechanism whereby heparin inhibits metastasis, suggesting that heparin treatment may not necessarily be appropriate only for selectin ligand-expressing tumor cell lines.[52]

Many studies of heparin inhibition used large doses of heparin, and so in an effort to better examine the role of heparin at clinically relevant doses, the inhibition of five different heparins already approved for clinical use was examined.[52] The inhibition of P- and L-selectin-mediated binding to carcinoma and melanoma cells in a murine model was examined, as well as their effects on subsequent metastasis. It was observed that Fondaparinux (an anticoagulant that binds selectins poorly) had no effect on tumor metastasis, while the other heparins were able to attenuate metastasis, supporting the involvement of selectin binding (Fig. 29). In this series of heparins, their ability to inhibit selectin binding *in vitro* was proportional to their ability to prevent metastasis *in vivo*; it was suggested that these differences in inhibitory potency arose from the different size distributions for each heparin.

The inhibition of P-selectin binding by heparin can attenuate platelet binding to emboli that form in the circulatory system as a preliminary step in the formation of metastatic lesions. Heparin and chemically modified heparins were studied as inhibitors of P-selectin, using a solid-phase assay with P-selectin-expressing CHO cells (transfected with human P-selectin DNA) and activated platelets.[239] Binding to several human colon carcinoma (COLO320, LS174T, and CW-2) cell lines was examined, in an effort to identify inhibitors that displayed high antimetastatic activity but low anticoagulant activity. Four potential candidates were isolated that displayed these properties, RO-heparin, CR-heparin, 2-*O*,3-*O*-desulfated-heparin, and *N*,2-*O*,3-*O*-desulfated-heparin, which suggests that 6-*O*-sulfation is critical for inhibitory activity. All three of the colon carcinoma cell lines were observed to express heparan-sulfate-like glycans at their cell surface, and treating COLO320 cells with heparinase reduced the adhesion of tumor cells to P-selectin. Although *N*-sulfation did

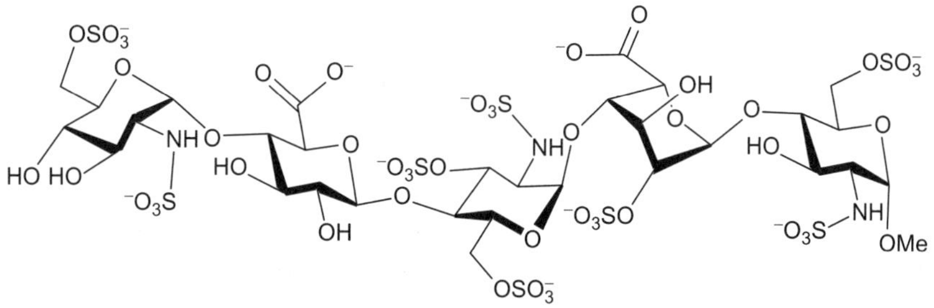

FIG. 29. Fondaparinux, an anticoagulant having poor affinity for selectins, was unable to attenuate tumor metastasis in a mouse model.[52]

not attenuate selectin binding, it was reported that complete N-acetylation of heparin yielded decreased inhibitory activity.[266]

The heparan-sulfate-like glycans expressed on tumor cells are able to undergo structural changes during tumor progression, where the glycans on more-malignant cells have been observed to undergo a decrease in 2-O-sulfated iduronic acid residues and a concomitant increase in 6-O-sulfated glucosamine residues.[123] These same changes were also observed in human colon carcinoma (CaCo-2) cells[271] and provide an explanation as to why heparan-sulfate-like glycans on normal cells do not act as P-selectin ligands.[239] Studying the binding characteristics of these heparins confirmed that a 6-O-sulfate modification on glucosamine residues was necessary for P-selectin-mediated adhesion of tumor cells,[239] but using a mAb (9E1) to block P-selectin only reduced COLO320 cell–platelet binding by 50%, again indicating that an alternative mechanism of adhesion is likely involved; the integrin receptor $\alpha_{IIb}\beta_3$, which is involved in the permanent arrest of cells after initial rolling, may play a role, as it was previously reported to act cooperatively with P-selectin in the adhesion of activated platelets to LS174T cells.[125] Since heparin was a better inhibitor than 9E1, again heparin appears to be blocking multiple-adhesion pathways and acts not only as an inhibitor of selectin binding,[239] a conclusion further supported by a previous study which reported that both *in vitro* and *in vivo* heparin was able to bind $\alpha_{IIb}\beta_3$ expressed on platelets.[272]

Another study of the effects of heparin sulfation on P-selectin inhibition indicated that N,6-O-sulfated and 6-O-sulfated heparins displayed high binding affinities, with 6-O-sulfation being the most critical modification.[273] Analogues 2-O- and 3-O-sulfated did not bind P-selectin well, but 2-O,3-O-desulfated heparins could inhibit the adhesion of platelets to human melanoma (A375) cells and, encouragingly, displayed diminished anticoagulant effects.[273] In accordance with observations that sulfation patterns affect binding specificities, the charge density likely plays an important role in the inhibition by heparins, glucan sulfates, and related oligosaccharides.[274] P-selectin binding to breast cancer (ZR-75-30) cells, which overexpress heparan-sulfate-like glycans at the cell surface, used modified heparins having decreased anticoagulant activity to study selectin inhibition.[275] Interference with the P-selectin-mediated adhesion of tumor cells was observed and appeared to be dependent on the pattern of heparin sulfation; similar to the previous reports, an anti-heparan-sulfate mAb did not completely abrogate cell binding, indicating that an interaction involving another adhesion ligand was probably involved, although the adhesion with sialic acid-terminated ligands was again ruled out.[275]

Fortunately, human P-selectin is even more sensitive than mouse P-selectin, and so inhibition studies have reported IC_{50} values that are acceptable for clinical use.[124]

Heparin treatment may be most beneficial at times when tumor cells are more likely to enter the circulatory system, such as during the removal of a primary tumor. Although concerns may arise that using heparin during surgical intervention could have adverse effects (as on anticoagulation), the use of heparin in patients at clinically relevant doses during surgery has already been widely used with minimal complications reported.[52]

Other oligosaccharides have been examined as potential therapeutics, including the use of dermatan sulfates as P-selectin inhibitors (Fig. 30); it was observed that those substrates containing the same core structure, $[(1 \rightarrow 4)\text{-}\beta\text{-IdoA}(2S)\text{-}(1 \rightarrow 3)\text{-}\beta\text{-GalNAc}]_n$, were the most active inhibitors.[276] The dermatan sulfates were isolated from the ascidians *Styela plicata* (sulfated at 4-*O*-GalNAc) and *Phallusia nigra* (sulfated at 6-*O*-GalNAc), and by inhibiting P-selectin-mediated adhesion, they were both successful at preventing metastasis in a mouse model by cells of both colon carcinoma (MC-38) and melanoma (B16–BL6).

As alternatives, chitosans, linear $\beta\text{-}(1 \rightarrow 4)$-linked polymers of D-glucosamine, have also been studied as selectin inhibitors (Fig. 30). The chitosans are desirable because of their nontoxicity and biodegradability, and upon sulfation, they can act as heparin mimics. Native chitin polysaccharide was first cleaved into smaller fragments via acid-catalyzed hydrolysis and then subsequently *N*-deacetylated to provide water-soluble fragments, which could then be differentially sulfated in order to observe which sulfation pattern resulted in the best binding characteristics.[277] *N*-Sulfated

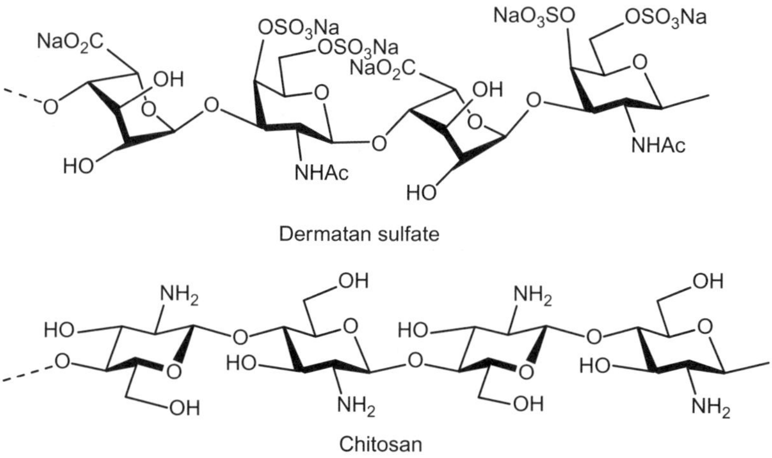

FIG. 30. Other oligosaccharides, such as dermatan sulfates and chitosans, have also been studied as inhibitors of selectins.

chitosan, 6-*O*-sulfated chitosan, 2-*N*,6-*O*-disulfated chitosan, and 3-*O*,6-*O*-disulfated chitosan were all prepared and then analyzed based on their ability to inhibit the adhesion of human malignant melanoma (A375) cells to P-selectin. It was concluded that, similarly to the sulfation required for high-affinity heparin binding, the 6-*O*-sulfation of chitosan was necessary for melanoma cell adhesion, while additional *N*-sulfation or 3-*O*-sulfation both enhanced selectin binding, but to a lesser extent.

d. **Monovalent, Small-Molecule Inhibitors—sLe^x/sLe^a Derivatives, Mannosides, *C*-Glycosyl Derivatives.**—In accordance with the inherently poor binding of monovalent carbohydrates to lectins, the affinity of sLe^x to selectins is also quite low (K_d: 0.1–5.0 mM), and so efforts have been made to generate more potent small-molecule inhibitors.[278] As determined through X-ray crystal structure analysis, trNOE NMR experiments, and structure–activity relationships, the key interactions involved in binding E-selectins involve the fucose OH groups, the 4-OH and 6-OH groups of galactose, and the carboxylate on sialic acid, while the glucosamine residue has been implicated in properly orienting Fuc and Gal residues in the binding site.[57,208,279]

Since sLe^x is difficult to synthesize, displays poor bioavailability, and has a relatively weak binding affinity, there is motivation to generate inhibitors that have improved characteristics. Attempts have been made at replacing the fucose and sialic acid residues by mimics, in order to simplify the synthetic routes necessary to obtain these inhibitors, such as with the generation of α-Gal-(1↔1)-β-Man, which was shown to have a binding affinity similar to that of sLe^x. Replacing L-fucose by D-arabinopyranose or L-galactose can also increase the hydrolytic and enzymatic stability of the inhibitors, since the latter two sugars are not naturally expressed in mammals.[280,281] Efforts have also advanced in the replacement of sialic acid residues with (*S*)-cyclohexyl lactic acid.[282]

The antiviral agent NMSO3, a sulfated sialic acid derivative containing two alkyl chains, was identified as an inhibitor of P-selectin (Fig. 31).[283] Immobilized NMSO3 could bind to a soluble P-selectin-IgG chimera, and treatment with NMSO3 inhibited binding between this chimera and PSGL-1, and also inhibited HL60 from adhering to P-selectin-expressing CHO cells. Sulfated and phosphorylated β-D-galactosides and

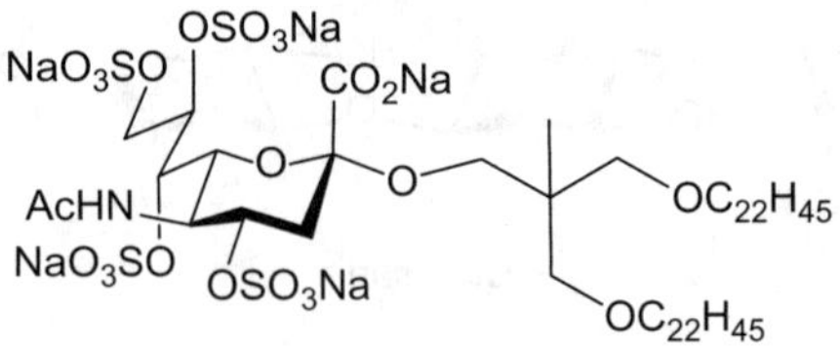

FIG. 31. The chemical structure of NMSO3.

lactoside glycolipids, containing a branched aglycone alkyl chain in place of ceramide, were synthesized and studied as selectin inhibitors. Regioselective modifications were achieved via the protected (for instance, dibutylstannylene acetalated) sugar and SO_3-NMe_3 or dibenzyloxy(diisopropylamino)phosphine.[284] *In vitro* competitive-binding assays involving sLe^x-expressing HL-60 cells and immobilized P-, L-, and E-selectins were performed. A number of the substrates showed enhanced binding as compared to the sLe^x tetrasaccharide; longer aglycone alkyl chains demonstrated improved binding inhibition against P- and L-selectins. In addition, stronger binding to these two lectins was achieved with 3-*O*- and 6-*O*-sulfated galactosides. Alternatively, 3,4-*O*-bisphosphorylation of the Gal residues in both galactosides and lactosides bound P- and L-selectins less well, but bound E-selectin with enhanced affinity.

In efforts to further improve the half-life of inhibitors *in vivo*, *C*-glycosyl analogues have been generated. These have improved chemical and enzymatic stability but are significantly more flexible due to lack of the *exo*-anomeric effect; this flexibility can result in losses of binding affinity because of unfavorable entropic costs upon ligand binding. In an effort to minimize the degree of conformational flexibility in *C*-glycosyl derivatives, fluorination of the methylene linker has been studied, since the electron-withdrawing characteristics of fluorine may be able to induce an *exo*-anomeric-effect type of conformational preference.[285] Upon fluorination, to give the desired disaccharide conformer, the binding affinity was improved by twofold (Fig. 32). Difluorinated *C*-mannopeptides were also synthesized as inhibitors of E- and P-selectins.[286] Both the α and β analogues, as well as their 1-hydroxy derivatives, were prepared, and the inhibition ability was comparable to that of the methylene *C*-glycosyl compounds. Competitive-inhibition studies were performed

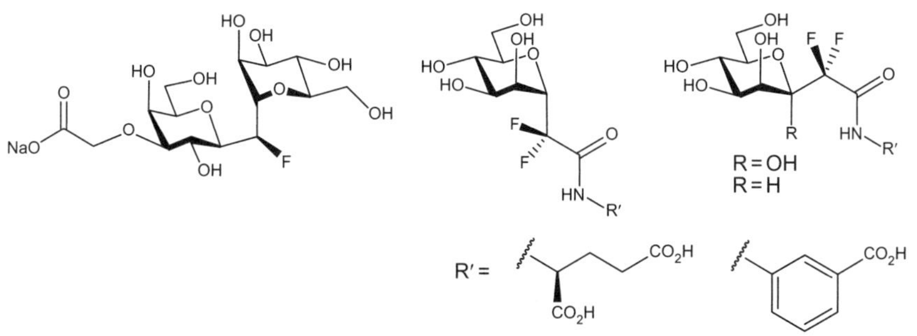

FIG. 32. Fluorinated *C*-glycosyl compounds can be used as inhibitors of selectin; the *C*-glycosyl analogues display improved stability *in vivo*, and fluorination can induce an *exo*-anomeric-like conformational preference.

with sLe^x-expressing HL-60 cells and E- and P-selectins that had been immobilized on microtiter wells. The percentage inhibitions were measured, indicating that α-mannosides were the best inhibitors, but in general, no enhancement of binding was observed for the fluorinated derivatives. The difluorinated C-mannosyl derivatives were both structurally flexible and displayed similar binding properties.

A 16-membered macrolide isolated from *Microsphaeropsis* sp. FO-5050, Macrosphelide B, was shown to inhibit E-selectin-mediated HUVEC binding to sLe^x-expressing tumor cells (HL-60).[287] Macrosphelide B treatment in mouse melanoma (B16/BL6) cells was shown to decrease lung colonization *in vivo*, but upon challenge with mouse lymphoma (L5178Y0ML) cells, no inhibition of spleen or liver metastasis was observed. Flow cytometry indicated that expression of sLe^x was significantly elevated in the melanoma cells but not the lymphoma cells, and since complete inhibition of metastasis was not possible in either model, it is likely that an alternative pathway of adhesion also exists through which metastasis can occur.[287]

e. Multivalent Displays.—The binding of selectin to native ligands is estimated at 10^4–10^5 times greater than for the monomeric sLe^x ligand, which may be a result of the multivalent presentation in natural substrates.[288] In an effort to improve binding affinities, multivalent inhibitor constructs have been designed and studied. Sulfated trimannose C—C-linked dimers were synthesized and were shown to act as inhibitors of both selectin and heparanase (Fig. 33).[289] Heparanase is an endoglycosidase that catabolizes heparan sulfate, and its activity has been linked to increased rates of tumor

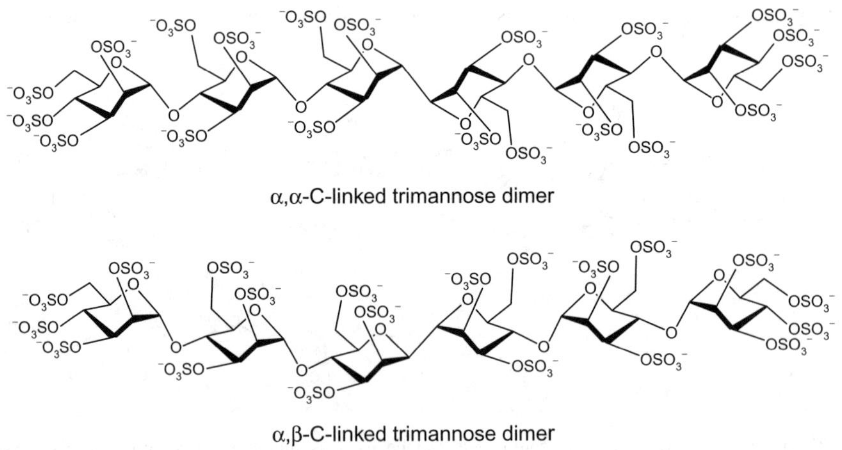

α,α-C-linked trimannose dimer

α,β-C-linked trimannose dimer

FIG. 33. Synthesized sulfated trimannose C—C-linked dimers.

growth and metastasis.[290] The sulfated trimannose dimers successfully attenuated metastasis in melanoma (B1–BL6) cells that express high levels of heparanase but had no effect on carcinoma (MC-38) cells, which have negligible heparanase expression.[289] They were also capable of inhibiting P-selectin binding *in vivo* and preventing P-selectin-mediated metastases.

The expression of glycans on liposomes enables multivalent presentation, which could potentially display improved binding. The ability of sLe^x-displaying liposomes to inhibit E-selectin was studied, using sterically stabilized liposomes (coated with PEG chains), and inhibition was studied by using a number of different assays.[237] Sterically stabilized liposomes with sLe^x-ligands at the PEG termini were compared to those with sLe^x-embedded in the PEG layer, as well as compared to nonsterically stabilized liposomes. The sterical stabilization was successful in preventing uptake by macrophages (J774 cell line), since previously it had been observed that liposomes expressing glycans without sterical stabilization were extremely likely to be taken up by macrophages.[291] The sLe^x-terminated-PEG, sterically stabilized liposomes were observed to be the best inhibitors and successfully reduced the adhesion of colon carcinoma (HT29) and Lewis lung carcinoma cells by 60–80%.[237]

f. Inhibition by Glycoproteins.—sLe^x-expressing glycoproteins can also act as inhibitors of selectin-mediated processes. The human glycoprotein, chorionic gonadotropin, is a placental hormone expressed at elevated levels during pregnancy, and it expresses a high density of both sLe^a and sLe^x at the cell surface.[241] Inhibition studies involving the glycoprotein resulted in binding improved by greater than 1000-fold as compared to the monovalent tetrasaccharide,[241] suggesting it could be further developed as a therapeutic agent to effectively block tumor-cell metastasis.

g. Peptidomimetics.—In sLe^x-containing glycopeptides, it was observed that binding selectivity and affinities to E- and P-selectins could be mediated by the peptide region. In the E-selectin natural ligand ESL-1, an amino acid sequence (672–681, GN☆LTELESED) was identified that contained multiple glycosylation sites and appeared to be conserved among other E-selectin ligands throughout multiple species (rat, human, chicken).[292] A series of sLe^x–Asn conjugate building blocks were synthesized, which could then be applied in solid-phase synthesis to afford target glycopeptides for studying interactions with selectins.[281] It was established using flow cytometry in a mouse neutrophile cell line (32Dc13) that binding affinities to E-selectin were improved for glycopeptide ligands containing this ESL-1 peptide sequence.[293] Upon substituting sugar residues, an exchange for L-galactose displayed very little change in binding affinity, while with D-arabinose, the binding affinity was decreased twofold; (S)-cyclohexyl lactic acid decreased binding by an order of magnitude (Fig. 34).

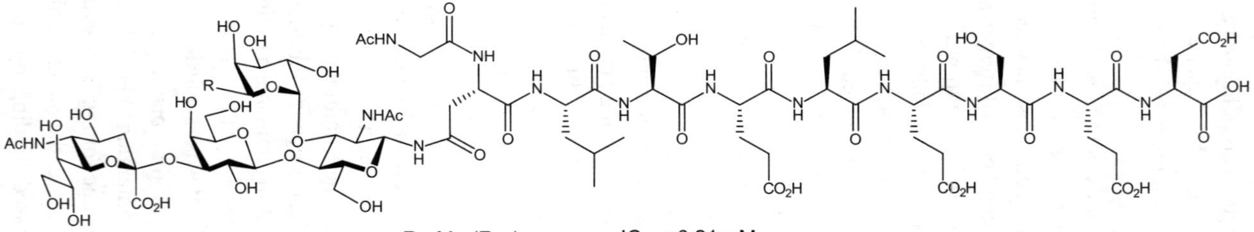

R = Me (Fuc) IC_{50} = 0.21 mM
R = H (Ara) IC_{50} = 0.59 mM
R = CH_2OH (L-Gal) IC_{50} = 0.26 mM

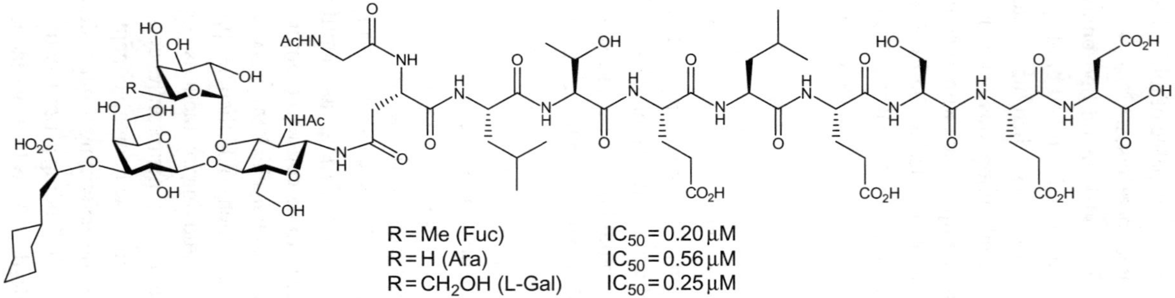

R = Me (Fuc) IC_{50} = 0.20 μM
R = H (Ara) IC_{50} = 0.56 μM
R = CH_2OH (L-Gal) IC_{50} = 0.25 μM

FIG. 34. Synthesized sLe^x-glycopeptides and (*S*)-cyclohexyl lactic acid Le^x analogues.

A library of peptide–phage conjugates was screened in an effort to identify inhibitors of lectins involved in metastasis[294]; initial attempts were not particularly successful, probably because of the inherently low binding affinities involved in lectin–carbohydrate interactions. In an alternative approach, antibodies were generated against E-selectin-binding glycans, and then the peptides were screened for their affinity to the elicited antibodies.[294] The best binding with E-selectin was observed for the peptide sequence IELLQAR, which was also able to bind anti-Le^a Ab, L-selectin, and P-selectin in a Ca(II)-dependent manner.[294] Both synthetic peptides and peptide-containing phages effectively inhibited E-selectin binding to sLe^x-, sLe^a-, and sLe^x-expressing human lung-tumor (HL-60) and melanoma (B16) cells. In an *in vitro* model, intravenous injection of the peptide 20 min prior to inoculation with either sLe^x-expressing cancer cell line successfully attenuated lung colonization in a mouse model.

V. Conclusions

The galectins and selectins have both been implicated in a number of processes related to more-aggressive cancer phenotypes, such as improved adhesion, neoangiogenesis, and immune-cell evasion. The development of inhibitors against lectins has displayed promising *in vitro* and *in vivo* properties, resulting in the attenuation of tumor metastases in experimental models. In addition, cotherapies involving a lectin inhibitor in combination with a cytotoxic agent or anticancer vaccine can improve the immune responses observed in conventional therapeutic approaches, which is a very promising avenue for future exploration. Furthermore, a better understanding of the role that multivalency plays in binding affinities may help in the generation of more effective inhibitors.

References

1. J. C. Aub, A. Lankester, and C. Tieslau, Reactions of normal and tumor cell surfaces to enzymes. I. Wheat-germ lipase and associated mucopolysaccharides, *Proc. Natl. Acad. Sci. U.S.A.*, 50 (1963) 613–619.
2. Y. S. Kim and G. R. Deng, Aberrant expression of carbohydrate antigens in cancer: The role of genetic and epigenetic regulation, *Gastroenterology*, 135 (2008) 305–309.
3. (a) N. Sharon and H. Lis, Lectins as cell recognition molecules, *Science*, 246 (1989) 227–234; (b) K. Drickamer, Increasing diversity of animal lectin structures, *Curr. Opin. Struct. Biol.*, 5 (1995) 612–616; (c) K. Drickamer and M. E. Taylor, Biology of animal lectins, *Annu. Rev. Cell Biol.*, 9 (1993) 237–264.

4. E. Gorelik, U. Galili, and A. Raz, On the role of cell surface carbohydrates and their binding proteins (lectins) in tumor metastasis, *Cancer Metastasis Rev.*, 20 (2001) 245–277.
5. (a) M. J. Bevan and M. Cohn, Cytotoxic effects of antigen- and mitogen-induced T cells on various targets, *J. Immunol.*, 114 (1975) 559–565; (b) T. W. Tao and M. M. Burger, Non-metastasizing variants selected from metastasizing melanoma cells, *Nature*, 270 (1977) 437–438; (c) J. W. Dennis, Different metastatic phenotypes in two genetic classes of wheat germ agglutinin-resistant tumor cell mutants, *Cancer Res.*, 46 (1986) 4594–4600; (d) A. Raz, W. L. McLellan, I. R. Hart, C. D. Bucana, L. C. Hoyer, B. A. Sela, P. Dragsten, and I. J. Fidler, Cell surface properties of B16 melanoma variants with differing metastatic potential, *Cancer Res.*, 40 (1980) 1645–1651.
6. R. D. Astronomo and D. R. Burton, Carbohydrate vaccines: Developing sweet solutions to sticky situations? *Nat. Rev. Drug Discov.*, 9 (2010) 308–324.
7. (a) D. R. Bundle and N. M. Young, Carbohydrate-protein interactions in antibodies and lectins, *Curr. Opin. Struct. Biol.*, 2 (1992) 666–673; (b) B. C. Braden and R. J. Poljak, Structural features of the reactions between antibodies and protein antigens, *FASEB J.*, 9 (1995) 9–16.
8. (a) R. U. Lemieux, L. T. J. Delbaere, H. Beierbeck, and U. Spohr, Involvement of water in host-guest interactions, *Ciba Found. Symp.*, 158 (1991) 231–248; (b) J. P. Carver, Oligosaccharides—How can flexible molecules act as signals? *Pure Appl. Chem.*, 65 (1993) 763–770.
9. (a) K. Sada and H. Yamamura, Effect of lectins on protein kinase activity, *Methods Mol. Med.*, 9 (1998) 423–432; (b) P. Carinci, E. Becchetti, and M. Bodo, Effects of lectins on cytoskeletal organization in mammalian cells, *Methods Mol. Med.*, 9 (1998) 407–421.
10. A. V. Timoshenko, K. Kayser, P. Drings, S. Andre, X. Dong, H. Kaltner, M. Schneller, and H. J. Gabius, Carbohydrate-binding proteins (plant/human lectins and autoantibodies from human serum) as mediators of release of lysozyme, elastase, and myeloperoxidase from human neutrophils, *Res. Exp. Med. (Berl.)*, 195 (1995) 153–162.
11. N. R. Zaccai, K. Maenaka, T. Maenaka, P. R. Crocker, R. Brossmer, S. Kelm, and E. Y. Jones, Structure-guided design of sialic acid-based siglec inhibitors and crystallographic analysis in complex with sialoadhesin, *Structure*, 11 (2003) 557–567.
12. O. Blixt, B. E. Collins, I. M. van den Nieuwenhof, P. R. Crocker, and J. C. Paulson, Sialoside specificity of the siglec family assessed using novel multivalent probes—Identification of potent inhibitors of myelin-associated glycoprotein, *J. Biol. Chem.*, 278 (2003) 31007–31019.
13. (a) P. R. Crocker, Siglecs: Sialic-acid-binding immunoglobulin-like lectins in cell-cell interactions and signalling, *Curr. Opin. Struct. Biol.*, 12 (2002) 609–615; (b) P. R. Crocker and A. Varki, Siglecs, sialic acids and innate immunity, *Trends Immunol.*, 22 (2001) 337–342.
14. S. Kelm, A. Pelz, R. Schauer, M. T. Filbin, S. Tang, M. E. DeBellard, R. L. Schnaar, J. A. Mahoney, A. Hartnell, P. Bradfield, and P. R. Crocker, Sialoadhesin, myelin-associated glycoprotein and CD22 define a new family of sialic acid-dependent adhesion molecules of the immunoglobulin superfamily, *Curr. Biol.*, 4 (1994) 965–972.
15. L. D. Powell and A. Varki, The oligosaccharide binding specificities of CD22β, a sialic acid-specific lectin of B cells, *J. Biol. Chem.*, 269 (1994) 10628–10636.
16. S. Kelm, R. Schauer, J. C. Manuguerra, H. J. Gross, and P. R. Crocker, Modifications of cell surface sialic acids modulate cell adhesion mediated by sialoadhesin and CD22, *Glycoconj. J.*, 11 (1994) 576–585.
17. T. Yamaji, T. Teranishi, M. S. Alphey, P. R. Crocker, and Y. Hashimoto, A small region of the natural killer cell receptor, siglec-7, is responsible for its preferred binding to α2,8-disialyl and branched α2,6-sialyl residues—A comparison with siglec-9, *J. Biol. Chem.*, 277 (2002) 6324–6332.
18. M. S. Alphey, H. Attrill, P. R. Crocker, and D. M. F. van Aalten, High resolution crystal structures of siglec-7—Insights into ligand specificity in the siglec family, *J. Biol. Chem.*, 278 (2003) 3372–3377.
19. (a) A. P. May, R. C. Robinson, M. Vinson, P. R. Crocker, and E. Y. Jones, Crystal structure of the N-terminal domain of sialoadhesin in complex with 3' sialyllactose at 1.85 Å resolution, *Mol. Cell*, 1

(1998) 719–728; (b) S. Tang, Y. J. Shen, M. E. DeBellard, G. Mukhopadhyay, J. L. Salzer, P. R. Crocker, and M. T. Filbin, Myelin-associated glycoprotein interacts with neurons via a sialic acid binding site at Arg118 and a distinct neurite inhibition site, *J. Cell Biol.*, 138 (1997) 1355–1366; (c) P. A. van der Merwe, P. R. Crocker, M. Vinson, A. N. Barclay, R. Schauer, and S. Kelm, Localization of the putative sialic acid-binding site on the immunoglobulin superfamily cell-surface molecule CD22, *J. Biol. Chem.*, 271 (1996) 9273–9280.
20. P. M. Collins, C. T. Oberg, H. Leffler, U. J. Nilsson, and H. Blanchard, Taloside inhibitors of galectin-1 and galectin-3, *Chem. Biol. Drug Des.*, 79 (2012) 339–346.
21. D. N. W. Cooper, Galectinomics: Finding themes in complexity, *Biochim. Biophys. Acta*, 1572 (2002) 209–231.
22. U. V. Sathisha, S. Jayaram, M. A. H. Nayaka, and S. M. Dharmesh, Inhibition of galectin-3 mediated cellular interactions by pectic polysaccharides from dietary sources, *Glycoconjug. J.*, 24 (2007) 497–507.
23. (a) F.-T. Liu and D. K. Hsu, The role of galectin-3 in promotion of the inflammatory response, *Drug News Perspect.*, 20 (2007) 455–460; (b) N. C. Henderson, A. C. Mackinnon, S. L. Farnworth, T. Kipari, C. Haslett, J. P. Iredale, F.-T. Liu, J. Hughes, and T. Sethi, Galectin-3 expression and secretion links macrophages to the promotion of renal fibrosis, *Am. J. Pathol.*, 172 (2008) 288–298; (c) S. Saussez and R. Kiss, Galectin-7, *Cell. Mol. Life Sci.*, 63 (2006) 686–697; (d) G. A. Rabinovich, F. T. Liu, M. Hirashima, and A. Anderson, An emerging role for galectins in tuning the immune response: Lessons from experimental models of inflammatory disease, autoimmunity and cancer, *Scand. J. Immunol.*, 66 (2007) 143–158; (e) H. Lahm, S. Andre, A. Hoeflich, H. Kaltner, H. C. Siebert, B. Sordat, C. W. von der Lieth, E. Wolf, and H. J. Gabius, Tumor galectinology: Insights into the complex network of a family of endogenous lectins, *Glycoconjug. J.*, 20 (2004) 227–238.
24. F. T. Liu and G. A. Rabinovich, Galectins as modulators of tumour progression, *Nat. Rev. Cancer*, 5 (2005) 29–41.
25. I. Camby, M. Le Mercier, F. Lefranc, and R. Kiss, Galectin-1: A small protein with major functions, *Glycobiology*, 16 (2006) 137R–157R.
26. (a) S. H. Barondes, D. N. W. Cooper, M. A. Gitt, and H. Leffler, Galectins—Structure and function of a large family of animal lectins, *J. Biol. Chem.*, 269 (1994) 20807–20810; (b) N. L. Perillo, M. E. Marcus, and L. G. Baum, Galectins: Versatile modulators of cell adhesion, cell proliferation, and cell death, *J. Mol. Med. (Berl.)*, 76 (1998) 402–412; (c) G. A. Rabinovich, Galectins: An evolutionarily conserved family of animal lectins with multifunctional properties; a trip from the gene to clinical therapy, *Cell Death Differ.*, 6 (1999) 711–721; (d) D. N. W. Cooper and S. H. Barondes, God must love galectins; he made so many of them, *Glycobiology*, 9 (1999) 979–984.
27. P. Nangia-Makker, Y. Honjo, R. Sarvis, S. Akahani, V. Hogan, K. J. Pienta, and A. Raz, Galectin-3 induces endothelial cell morphogenesis and angiogenesis, *Am. J. Pathol.*, 156 (2000) 899–909.
28. H. Leffler, S. Carlsson, M. Hedlund, Y. Qian, and F. Poirier, Introduction to galectins, *Glycoconjug. J.*, 19 (2004) 433–440.
29. R. J. Pieters, Inhibition and detection of galectins, *Chembiochem*, 7 (2006) 721–728.
30. C. Cederfur, E. Salomonsson, J. Nilsson, A. Halim, C. T. Oeberg, G. Larson, U. J. Nilsson, and H. Leffler, Different affinity of galectins for human serum glycoproteins: Galectin-3 binds many protease inhibitors and acute phase proteins, *Glycobiology*, 18 (2008) 384–394.
31. N. L. Perillo, K. E. Pace, J. J. Seilhamer, and L. G. Baum, Apoptosis of T cells mediated by galectin-1, *Nature*, 378 (1995) 736–739.
32. R. S. Bresalier, P. S. Yan, J. C. Byrd, R. Lotan, and A. Raz, Expression of the endogenous galactose-binding protein galectin-3 correlates with the malignant potential of tumors in the central nervous system, *Cancer*, 80 (1997) 776–787.
33. (a) H. Inohara, S. Akahani, and A. Raz, Galectin-3 stimulates cell proliferation, *Exp. Cell Res.*, 245 (1998) 294–302; (b) P. Nangia-Makker, S. Nakahara, V. Hogan, and A. Raz, Galectin-3 in apoptosis, a

novel therapeutic target, *J. Bioenerg. Biomembr.*, 39 (2007) 79–84; (c) L. Wang, H. Inohara, K. J. Pienta, and A. Raz, Galectin-3 is a nuclear matrix protein which binds RNA, *Biochem. Biophys. Res. Commun.*, 217 (1995) 292–303; (d) S. F. Dagher, J. L. Wang, and R. J. Patterson, Identification of galectin-3 as a factor in pre-mRNA splicing, *Proc. Natl. Acad. Sci. U.S.A.*, 92 (1995) 1213–1217.
34. (a) Y. Bourne, B. Bolgiano, D. L. Liao, G. Strecker, P. Cantau, O. Herzberg, T. Feizi, and C. Cambillau, Cross-linking of mammalian lectin (galectin-1) by complex biantennary saccharides, *Nat. Struct. Biol.*, 1 (1994) 863–870; (b) D. I. Liao, G. Kapadia, H. Ahmed, G. R. Vasta, and O. Herzberg, Structure of S-lectin, a developmentally regulated vertebrate β-galactoside-binding protein, *Proc. Natl. Acad. Sci. U.S.A.*, 91 (1994) 1428–1432; (c) Y. D. Lobsanov, M. A. Gitt, H. Leffler, S. H. Barondes, and J. M. Rini, X-ray crystal structure of the human dimeric S-Lac lectin, L-14-II, in complex with lactose at 2.9-Å resolution, *J. Biol. Chem.*, 268 (1993) 27034–27038; (d) D. D. Leonidas, B. L. Elbert, Z. Zhou, H. Leffler, S. J. Ackerman, and K. R. Acharya, Crystal structure of human Charcot-Leyden crystal protein, an eosinophil lysophospholipase, identifies it as a new member of the carbohydrate-binding family of galectins, *Structure*, 3 (1995) 1379–1393; (e) D. D. Leonidas, E. H. Vatzaki, H. Vorum, J. E. Celis, P. Madsen, and K. R. Acharya, Structural basis for the recognition of carbohydrates by human galectin-7, *Biochemistry*, 37 (1998) 13930–13940.
35. A. Bergh, H. Leffler, A. Sundin, U. J. Nilsson, and N. Kann, Cobalt-mediated solid phase synthesis of 3-*O*-alkynylbenzyl galactosides and their evaluation as galectin inhibitors, *Tetrahedron*, 62 (2006) 8309–8317.
36. S. H. Barondes, V. Castronovo, D. N. W. Cooper, R. D. Cummings, K. Drickamer, T. Feizi, M. A. Gitt, J. Hirabayashi, C. Hughes, K. Kasai, H. Leffler, F. T. Liu, R. Lotan, A. M. Mercurio, M. Monsigny, S. Pillai, F. Poirer, A. Raz, P. W. J. Rigby, J. M. Rini, and J. L. Wang, Galectins: A family of animal β-galactoside-binding lectins, *Cell*, 76 (1994) 597–598.
37. G. A. Rabinovich, A. Ariel, R. Hershkovitz, J. Hirabayashi, K. I. Kasai, and O. Lider, Specific inhibition of T-cell adhesion to extracellular matrix and proinflammatory cytokine secretion by human recombinant galectin-1, *Immunology*, 97 (1999) 100–106.
38. M. S. Sujatha and P. V. Balaji, Identification of common structural features of binding sites in galactose-specific proteins, *Proteins*, 55 (2004) 44–65.
39. S. Fort, H. S. Kim, and O. Hindsgaul, Screening for galectin-3 inhibitors from synthetic lacto-*N*-biose libraries using microscale affinity chromatography coupled to mass spectrometry, *J. Org. Chem.*, 71 (2006) 7146–7154.
40. C. T. Oberg, H. Blanchard, H. Leffler, and U. J. Nilsson, Protein subtype-targeting through ligand epimerization: Talose-selectivity of galectin-4 and galectin-8, *Bioorg. Med. Chem. Lett.*, 18 (2008) 3691–3694.
41. C. P. Sparrow, H. Leffler, and S. H. Barondes, Multiple soluble β-galactoside-binding lectins from human lung, *J. Biol. Chem.*, 262 (1987) 7383–7390.
42. H. Leffler and S. H. Barondes, Specificity of binding of three soluble rat lung lectins to substituted and unsubstituted mammalian β-galactosides, *J. Biol. Chem.*, 261 (1986) 119–126.
43. K. S. Lau, E. A. Partridge, A. Grigorian, C. I. Silvescu, V. N. Reinhold, M. Demetriou, and J. W. Dennis, Complex N-glycan number and degree of branching cooperate to regulate cell proliferation and differentiation, *Cell*, 129 (2007) 123–134.
44. N. Ahmad, H. J. Gabius, H. Kaltner, S. Andre, I. Kuwabara, F. T. Liu, S. Oscarson, T. Norberg, and C. F. Brewer, Thermodynamic binding studies of cell surface carbohydrate epitopes to galectins-1,-3, and -7: Evidence for differential binding specificities, *Can. J. Chem.*, 80 (2002) 1096–1104.
45. P. Sorme, B. Kahl-Knutson, U. Wellmar, U. J. Nilsson, and H. Leffler, Fluorescence polarization to study galectin-ligand interactions, *Methods Enzymol.*, 362 (2003) 504–512.
46. J. Hirabayashi, T. Hashidate, Y. Arata, N. Nishi, T. Nakamura, M. Hirashima, T. Urashima, T. Oka, M. Futai, W. E. G. Muller, F. Yagi, and K. Kasai, Oligosaccharide specificity of galectins: A search by frontal affinity chromatography, *Biochim. Biophys. Acta*, 1572 (2002) 232–254.

47. T. F. Tedder, D. A. Steeber, A. Chen, and P. Engel, The selectins: Vascular adhesion molecules, *FASEB J.*, 9 (1995) 866–873.
48. A. Varki, Selectin ligands, *Proc. Natl. Acad. Sci. U.S.A.*, 91 (1994) 7390–7397.
49. M. Nguyen, N. A. Strubel, and J. Bischoff, A role for sialyl Lewis-X/A glycoconjugates in capillary morphogenesis, *Nature*, 365 (1993) 267–269.
50. A. E. Koch, M. M. Halloran, C. J. Haskell, M. R. Shah, and P. J. Polverini, Angiogenesis mediated by soluble forms of E-selectin and vascular cell adhesion molecule-1, *Nature*, 376 (1995) 517–519.
51. R. P. McEver, Selectin-carbohydrate interactions during inflammation and metastasis, *Glycoconjug. J.*, 14 (1997) 585–591.
52. J. L. Stevenson, S. H. Choi, and A. Varki, Differential metastasis inhibition by clinically relevant levels of heparins—Correlation with selectin inhibition, not antithrombotic activity, *Clin. Cancer Res.*, 11 (2005) 7003–7011.
53. M. P. Skinner, C. M. Lucas, G. F. Burns, C. N. Chesterman, and M. C. Berndt, GMP-140 binding to neutrophils is inhibited by sulfated glycans, *J. Biol. Chem.*, 266 (1991) 5371–5374.
54. P. P. Wilkins, K. L. Moore, R. P. McEver, and R. D. Cummings, Tyrosine sulfation of P-selectin glycoprotein ligand-1 is required for high affinity binding to P-selectin, *J. Biol. Chem.*, 270 (1995) 22677–22680.
55. (a) T. Pouyani and B. Seed, PSGL-1 recognition of P-selectin is controlled by a tyrosine sulfation consensus at the PSGL-1 amino terminus, *Cell*, 83 (1995) 333–343; (b) D. Sako, K. M. Comess, K. M. Barone, R. T. Camphausen, D. A. Cumming, and G. D. Shaw, A sulfated peptide segment at the amino terminus of PSGL-1 is critical for P-selectin binding, *Cell*, 83 (1995) 323–331.
56. K. L. Moore, S. F. Eaton, D. E. Lyons, H. S. Lichenstein, R. D. Cummings, and R. P. McEver, The P-selectin glycoprotein ligand from human neutrophils displays sialylated, fucosylated, *O*-linked poly-*N*-acetyllactosamine, *J. Biol. Chem.*, 269 (1994) 23318–23327.
57. W. S. Somers, J. Tang, G. D. Shaw, and R. T. Camphausen, Insights into the molecular basis of leukocyte tethering and rolling revealed by structures of P- and E-selectin bound to sLe^x and PSGL-1, *Cell*, 103 (2000) 467–479.
58. (a) C. T. Yuen, A. M. Lawson, W. G. Chai, M. Larkin, M. S. Stoll, A. C. Stuart, F. X. Sullivan, T. J. Ahern, and T. Feizi, Novel sulfated ligands for the cell adhesion molecule E-selectin revealed by the neoglycolipid technology among O-linked oligosaccharides on an ovarian cystadenoma glycoprotein, *Biochemistry*, 31 (1992) 9126–9131; (b) C. Galustian, A. M. Lawson, S. Komba, H. Ishida, M. Kiso, and T. Feizi, Sialyl-lewis^x sequence 6-*O*-sulfated at *N*-acetylglucosamine rather than at galactose is the preferred ligand for L-selectin and de-*N*-acetylation of the sialic acid enhances the binding strength, *Biochem. Biophys. Res. Commun.*, 240 (1997) 748–751; (c) K. E. Norgardsumnicht, N. M. Varki, and A. Varki, Calcium-dependent heparin-like ligands for L-selectin in nonlymphoid endothelial cells, *Science*, 261 (1993) 480–483; (d) Y. Imai, L. A. Lasky, and S. D. Rosen, Sulphation requirement for glycam-1, an endothelial ligand for L-selectin, *Nature*, 361 (1993) 555–557; (e) S. Baumhueter, M. S. Singer, W. Henzel, S. Hemmerich, M. Renz, S. D. Rosen, and L. A. Lasky, Binding of L-selectin to the vascular sialomucin CD34, *Science*, 262 (1993) 436–438.
59. C. Musselli, P. O. Livingston, and G. Ragupathi, Keyhole limpet hemocyanin conjugate vaccines against cancer: The Memorial Sloan Kettering experience, *J. Cancer Res. Clin. Oncol.*, 127 (2001) R20–R26.
60. P. O. Livingston, Construction of cancer vaccines with carbohydrate and protein (peptide) tumor antigens, *Curr. Opin. Immunol.*, 4 (1992) 624–629.
61. S. H. Itzkowitz, E. J. Bloom, W. A. Kokal, G. Modin, S. Hakomori, and Y. S. Kim, Sialosyl-Tn—A novel mucin antigen associated with prognosis in colorectal cancer patients, *Cancer*, 66 (1990) 1960–1966.
62. (a) S. H. Itzkowitz, M. Yuan, C. K. Montgomery, T. Kjeldsen, H. K. Takahashi, W. L. Bigbee, and Y. S. Kim, Expression of Tn, sialosyl-Tn, and T antigens in human colon cancer, *Cancer Res.*, 49

(1989) 197–204; (b) G. F. Springer, T and Tn, general carcinoma auto-antigens, *Science*, 224 (1984) 1198–1206.
63. S. Dziadek and H. Kunz, Synthesis of tumor-associated glycopeptide antigens for the development of tumor-selective vaccines, *Chem. Rec.*, 3 (2004) 308–321.
64. S. Hakomori, Glycosylation defining cancer malignancy: New wine in an old bottle, *Proc. Natl. Acad. Sci. U.S.A.*, 99 (2002) 10231–10233.
65. H. Oguchi, T. Toyokuni, B. Dean, H. Ito, E. Otsuji, V. L. Jones, K. K. Sadozai, and S. Hakomori, Effect of lactose derivatives on metastatic potential of B16 melanoma cells, *Cancer Commun.*, 2 (1990) 311–316.
66. J. W. Dennis, N-linked oligosaccharide processing and tumor cell biology, *Semin. Cancer Biol.*, 2 (1991) 411–420.
67. J. W. Dennis, S. Laferte, C. Waghorne, M. L. Breitman, and R. S. Kerbel, β1-6 Branching of Asn-linked oligosaccharides is directly associated with metastasis, *Science*, 236 (1987) 582–585.
68. Y. J. Kim and A. Varki, Perspectives on the significance of altered glycosylation of glycoproteins in cancer, *Glycoconjug. J.*, 14 (1997) 569–576.
69. (a) S. L. Zhang, H. S. Zhang, C. Cordon-Cardo, G. Ragupathi, and P. O. Livingston, Selection of tumor antigens as targets for immune attack using immunohistochemistry: Protein antigens, *Clin. Cancer Res.*, 4 (1998) 2669–2676; (b) J. L. Zhu, Q. Wan, G. Ragupathi, C. M. George, P. O. Livingston, and S. J. Danishefsky, Biologics through chemistry: Total synthesis of a proposed dual-acting vaccine targeting ovarian cancer orchestration of oligosaccharide and polypeptide domains, *J. Am. Chem. Soc.*, 131 (2009) 4151–4158.
70. (a) S. L. Zhang, L. A. Walberg, S. Ogata, S. H. Itzkowitz, R. R. Koganty, M. Reddish, S. S. Gandhi, B. M. Longenecker, K. O. Lloyd, and P. O. Livingston, Immune sera and monoclonal antibodies define two configurations for the sialyl Tn tumor antigen, *Cancer Res.*, 55 (1995) 3364–3368; (b) H. Nakada, Y. Numata, M. Inoue, N. Tanaka, H. Kitagawa, I. Funakoshi, S. Fukui, and I. Yamashina, Elucidation of an essential structure recognized by an anti-GalNAcα-Ser(Thr) monoclonal antibody (MLS 128), *J. Biol. Chem.*, 266 (1991) 12402–12405; (c) N. Tanaka, H. Nakada, M. Inoue, and I. Yamashina, Binding characteristics of an anti-Siaα2-6GalNAcα-Ser/Thr (sialyl Tn) monoclonal antibody (MLS 132), *Eur. J. Biochem.*, 263 (1999) 27–32.
71. S. L. Zhang, H. S. Zhang, C. Cordon-Cardo, V. E. Reuter, A. K. Singhal, K. O. Lloyd, and P. O. Livingston, Selection of tumor antigens as targets for immune attack using immunohistochemistry. 2. Blood group-related antigens, *Int. J. Cancer*, 73 (1997) 50–56.
72. S. L. Zhang, C. Cordon-Cardo, H. S. Zhang, V. E. Reuter, S. Adluri, W. B. Hamilton, K. O. Lloyd, and P. O. Livingston, Selection of tumor antigens as targets for immune attack using immunohistochemistry. 1. Focus on gangliosides, *Int. J. Cancer*, 73 (1997) 42–49.
73. (a) J. G. Steele, C. Rowlatt, J. K. Sandall, and L. M. Franks, Cell surface properties of high- and low-metastatic cell lines selected from a spontaneous mouse lung carcinoma, *Int. J. Cancer*, 32 (1983) 769–779; (b) M. Fogel, P. Altevogt, and V. Schirrmacher, Metastatic potential severely altered by changes in tumor cell adhesiveness and cell-surface sialylation, *J. Exp. Med.*, 157 (1983) 371–376; (c) P. Altevogt, M. Fogel, R. Cheingsong-Popov, J. Dennis, P. Robinson, and V. Schirrmacher, Different patterns of lectin binding and cell surface sialylation detected on related high- and low-metastatic tumor lines, *Cancer Res.*, 43 (1983) 5138–5144; (d) G. Yogeeswaran and P. L. Salk, Metastatic potential is positively correlated with cell surface sialylation of cultured murine tumor cell lines, *Science*, 212 (1981) 1514–1516.
74. A. Passaniti and G. W. Hart, Cell surface sialylation and tumor metastasis—Metastatic potential of B16 melanoma variants correlates with their relative numbers of specific penultimate oligosaccharide structures, *J. Biol. Chem.*, 263 (1988) 7591–7603.
75. G. Picco, S. Julien, I. Brockhausen, R. Beatson, A. Antonopoulos, S. Haslam, U. Mandel, A. Dell, S. Pinder, J. Taylor-Papadimitriou, and J. Burchell, Over-expression of ST3Gal-I promotes mammary tumorigenesis, *Glycobiology*, 20 (2010) 1241–1250.

76. J. W. Dennis, S. Laferte, M. Fukuda, A. Dell, and J. P. Carver, Asn-linked oligosaccharides in lectin-resistant tumor-cell mutants with varying metastatic potential, *Eur. J. Biochem.*, 161 (1986) 359–373.
77. (a) J. W. Dennis, K. Kosh, D. M. Bryce, and M. L. Breitman, Oncogenes conferring metastatic potential induce increased branching of Asn-linked oligosaccharides in rat2 fibroblasts, *Oncogene*, 4 (1989) 853–860; (b) P. J. Seberger and W. G. Chaney, Control of metastasis by Asn-linked, β1-6 branched oligosaccharides in mouse mammary cancer cells, *Glycobiology*, 9 (1999) 235–241.
78. V. V. Glinsky, G. V. Glinsky, K. Rittenhouse-Olson, M. E. Huflejt, O. V. Glinskii, S. L. Deutscher, and T. P. Quinn, The role of Thomsen-Friedenreich antigen in adhesion of human breast and prostate cancer cells to the endothelium, *Cancer Res.*, 61 (2001) 4851–4857.
79. (a) J. Kanitakis, I. Al-Rifai, M. Faure, and A. Claudy, Differential expression of the cancer associated antigens T (Thomsen-Friedenreich) and Tn to the skin in primary and metastatic carcinomas, *J. Clin. Pathol.*, 51 (1998) 588–592; (b) G. F. Springer, P. R. Desai, M. Ghazizadeh, and H. Tegtmeyer, T/Tn pancarcinoma autoantigens—Fundamental, diagnostic, and prognostic aspects, *Cancer Detect. Prev.*, 19 (1995) 173–182; (c) T. Janssen, M. Petein, R. VanVelthoven, P. VanLeer, M. Fourmarier, J. P. Vanegas, A. Danguy, C. Schulman, J. L. Pasteels, and R. Kiss, Differential histochemical peanut agglutinin stain in benign and malignant human prostate tumors: Relationship with prostatic specific antigen immunostain and nuclear DNA content, *Hum. Pathol.*, 27 (1996) 1341–1347.
80. H. J. Gabius, R. Engelhardt, S. Rehm, and F. Cramer, Biochemical characterization of endogenous carbohydrate-binding proteins from spontaneous murine rhabdomyosarcoma, mammary adenocarcinoma, and ovarian teratoma, *J. Natl. Cancer Inst.*, 73 (1984) 1349–1357.
81. S. Nakamori, M. Kameyama, S. Imaoka, H. Furukawa, O. Ishikawa, Y. Sasaki, T. Kabuto, T. Iwanaga, Y. Matsushita, and T. Irimura, Increased expression of sialyl Lewis^x antigen correlates with poor survival in patients with colorectal carcinoma: Clinicopathological and immunohistochemical study, *Cancer Res.*, 53 (1993) 3632–3637.
82. (a) S. I. Hakomori, Aberrant glycosylation in tumors and tumor-associated carbohydrate antigens, *Adv. Cancer Res.*, 52 (1989) 257–331; (b) S. Hakomori, Tumor malignancy defined by aberrant glycosylation and sphingo(glyco)lipid metabolism, *Cancer Res.*, 56 (1996) 5309–5318.
83. M. Miyake, T. Taki, S. Hitomi, and S. Hakomori, Correlation of expression of H/Le^y/Le^b antigens with survival in patients with carcinoma of the lung, *N. Engl. J. Med.*, 327 (1992) 14–18.
84. Y. Izumi, Y. Taniuchi, T. Tsuji, C. W. Smith, S. Nakamori, I. J. Fidler, and T. Irimura, Characterization of human colon carcinoma variant cells selected for sialyl Le^x carbohydrate antigen: Liver colonization and adhesion to vascular endothelial cells, *Exp. Cell Res.*, 216 (1995) 215–221.
85. N. Yamada, Y. S. Chung, S. Takatsuka, Y. Arimoto, T. Sawada, T. Dohi, and M. Sowa, Increased sialyl Lewis A expression and fucosyltransferase activity with acquisition of a high metastatic capacity in a colon cancer cell line, *Br. J. Cancer*, 76 (1997) 582–587.
86. S. Julien, A. Ivetic, A. Grigoriadis, D. Qize, B. Burford, D. Sproviero, G. Picco, C. Gillett, S. L. Papp, L. Schaffer, A. Tutt, J. Taylor-Papadimitriou, S. E. Pinder, and J. M. Burchell, Selectin ligand sialyl-Lewis X antigen drives metastasis of hormone-dependent breast cancers, *Cancer Res.*, 71 (2011) 7683–7693.
87. (a) G. Walz, A. Aruffo, W. Kolanus, M. Bevilacqua, and B. Seed, Recognition by ELAM-1 of the sialyl-Le^x determinant on myeloid and tumor cells, *Science*, 250 (1990) 1132–1135; (b) A. Takada, K. Ohmori, N. Takahashi, K. Tsuyuoka, A. Yago, K. Zenita, A. Hasegawa, and R. Kannagi, Adhesion of human cancer cells to vascular endothelium mediated by a carbohydrate antigen, sialyl Lewis A, *Biochem. Biophys. Res. Commun.*, 179 (1991) 713–719.
88. T. Narita, H. Funahashi, Y. Satoh, T. Watanabe, J. Sakamoto, and H. Takagi, Association of expression of blood group-related carbohydrate antigens with prognosis in breast cancer, *Cancer*, 71 (1993) 3044–3053.
89. H. Sakahara, K. Endo, K. Nakajima, T. Nakashima, M. Koizumi, H. Ohta, A. Hidaka, S. Kohno, Y. Nakano, A. Naito, T. Suzuki, and K. Torizuka, Serum CA 19-9 concentrations and computed tomography findings in patients with pancreatic carcinoma, *Cancer*, 57 (1986) 1324–1326.

90. (a) K. Steplewska-Mazur, A. Gabriel, W. Zajecki, M. Wylezol, and M. Gluck, Breast cancer progression and expression of blood group-related tumor-associated antigens, *Hybridoma*, 19 (2000) 129–133; (b) E. L. Berg, M. K. Robinson, O. Mansson, E. C. Butcher, and J. L. Magnani, A carbohydrate domain common to both sialyl Le[a] and sialyl Le[x] is recognized by the endothelial cell leukocyte adhesion molecule ELAM-1, *J. Biol. Chem.*, 266 (1991) 14869–14872.
91. N. Rubinstein, M. Alvarez, N. W. Zwirner, M. A. Toscano, J. M. Ilarregui, A. Bravo, J. Mordoh, L. Fainboim, O. L. Podhajcer, and G. A. Rabinovich, Targeted inhibition of galectin-1 gene expression in tumor cells results in heightened T cell-mediated rejection: A potential mechanism of tumor-immune privilege, *Cancer Cell*, 5 (2004) 241–251.
92. N. D. S. Rambaruth and M. V. Dwek, Cell surface glycan-lectin interactions in tumor metastasis, *Acta Histochem.*, 113 (2011) 591–600.
93. R. Takamiya, K. Ohtsubo, S. Takamatsu, N. Taniguchi, and T. Angata, The interaction between siglec-15 and tumor-associated sialyl-Tn antigen enhances TGF-β secretion from monocytes/macrophages through the DAP12-Syk pathway, *Glycobiology*, 23 (2013) 178–187.
94. B. J. Swanson, K. M. McDermott, P. K. Singh, J. P. Eggers, P. R. Crocker, and M. A. Hollingsworth, MUC1 is a counter-receptor for myelin-associated glycoprotein (siglec-4a) and their interaction contributes to adhesion in pancreatic cancer perineural invasion, *Cancer Res.*, 67 (2007) 10222–10229.
95. (a) N. Mitra, K. Banda, T. K. Altheide, L. Schaffer, T. L. Johnson-Pais, J. Beuten, R. J. Leach, T. Angata, N. Varki, and A. Varki, Siglec12, a human-specific segregating (pseudo)gene, encodes a signaling molecule expressed in prostate carcinomas, *J. Biol. Chem.*, 286 (2011) 23003–23011; (b) M. Ohta, A. Ishida, M. Toda, K. Akita, M. Inoue, K. Yamashita, M. Watanabe, T. Murata, T. Usui, and H. Nakada, Immunomodulation of monocyte-derived dendritic cells through ligation of tumor-produced mucins to siglec-9, *Biochem. Biophys. Res. Commun.*, 402 (2010) 663–669.
96. (a) A. Gillenwater, X. C. Xu, A. K. El-Naggar, G. L. Clayman, and R. Lotan, Expression of galectins in head and neck squamous cell carcinoma, *Head Neck*, 18 (1996) 422–432; (b) H. L. Schoeppner, A. Raz, S. B. Ho, and R. S. Bresalier, Expression of an endogenous galactose-binding lectin correlates with neoplastic progression in the colon, *Cancer*, 75 (1995) 2818–2826; (c) J. J. Shen, M. D. Person, J. J. Zhu, J. L. Abbruzzese, and D. H. Li, Protein expression profiles in pancreatic adenocarcinoma compared with normal pancreatic tissue and tissue affected by pancreatitis as detected by two-dimensional gel electrophoresis and mass spectrometry, *Cancer Res.*, 64 (2004) 9018–9026.
97. A. F. Chambers, A. C. Groom, and I. C. MacDonald, Dissemination and growth of cancer cells in metastatic sites, *Nat. Rev. Cancer*, 2 (2002) 563–572.
98. (a) V. Castronovo, F. A. van den Brule, P. Jackers, N. Clausse, F. T. Liu, C. Gillet, and M. E. Sobel, Decreased expression of galectin-3 is associated with progression of human breast cancer, *J. Pathol.*, 179 (1996) 43–48; (b) F. A. van den Brule, A. Berchuck, R. C. Bast, F. T. Liu, C. Gillet, M. E. Sobel, and V. Castronovo, Differential expression of the 67-kD laminin receptor and 31-kD human laminin-binding protein in human ovarian carcinomas, *Eur. J. Cancer*, 30A (1994) 1096–1099; (c) F. A. van den Brule, C. Buicu, A. Berchuck, R. C. Bast, M. Deprez, F. T. Liu, D. N. W. Cooper, C. Pieters, M. E. Sobel, and V. Castronovo, Expression of the 67-kD laminin receptor, galectin-1, and galectin-3 in advanced human uterine adenocarcinoma, *Hum. Pathol.*, 27 (1996) 1185–1191.
99. S. Nakahara, N. Oka, and A. Raz, On the role of galectin-3 in cancer apoptosis, *Apoptosis*, 10 (2005) 267–275.
100. R. S. Bresalier, N. Mazurek, L. R. Sternberg, J. C. Byrd, C. K. Yunker, P. Nangia-Makker, and A. Raz, Metastasis of human colon cancer is altered by modifying expression of the β-galactoside-binding protein galectin 3, *Gastroenterology*, 115 (1998) 287–296.
101. P. Nangia-Makker, E. Thompson, C. Hogan, J. Ochieng, and A. Raz, Induction of tumorigenicity by galectin-3 in a nontumorigenic human breast carcinoma cell line, *Int. J. Oncol.*, 7 (1995) 1079–1087.
102. N. LeMarer and R. C. Hughes, Effects of the carbohydrate-binding protein galectin-3 on the invasiveness of human breast carcinoma cells, *J. Cell. Physiol.*, 168 (1996) 51–58.

103. (a) N. L. Perillo, C. H. Uittenbogaart, J. T. Nguyen, and L. G. Baum, Galectin-1, an endogenous lectin produced by thymic epithelial cells, induces apoptosis of human thymocytes, *J. Exp. Med.*, 185 (1997) 1851–1858; (b) J. Wada, K. Ota, A. Kumar, E. I. Wallner, and Y. S. Kanwar, Developmental regulation, expression, and apoptotic potential of galectin-9, a β-galactoside binding lectin, *J. Clin. Invest.*, 99 (1997) 2452–2461.
104. (a) R. Y. Yang, D. K. Hsu, and F. T. Liu, Expression of galectin-3 modulates T-cell growth and apoptosis, *Proc. Natl. Acad. Sci. U.S.A.*, 93 (1996) 6737–6742; (b) S. Akahani, P. Nangia-Makker, H. Inohara, H. R. C. Kim, and A. Raz, Galectin-3: A novel antiapoptotic molecule with a functional BH1 (NWGR) domain of Bcl-2 family, *Cancer Res.*, 57 (1997) 5272–5276.
105. J. B. Lowe, Glycosylation, immunity, and autoimmunity, *Cell*, 104 (2001) 809–812.
106. (a) C. D. Chung, V. P. Patel, M. Moran, L. A. Lewis, and M. C. Miceli, Galectin-1 induces partial TCR ζ-chain phosphorylation and antagonizes processive TCR signal transduction, *J. Immunol.*, 165 (2000) 3722–3729; (b) G. Rabinovich, G. Daly, H. Dreja, H. Tailor, C. M. Riera, J. Hirabayashi, and A. Chernajovsky, Recombinant galectin-1 and its genetic delivery suppress collagen-induced arthritis via T cell apoptosis, *J. Exp. Med.*, 190 (1999) 385–397.
107. N. S. Brown and R. Bicknell, Hypoxia and oxidative stress in breast cancer—Oxidative stress: Its effects on the growth, metastatic potential and response to therapy of breast cancer, *Breast Cancer Res.*, 3 (2001) 323–327.
108. X. Y. Zhao, T. T. Chen, L. Xia, M. Guo, Y. Xu, F. Yue, Y. Jiang, G. Q. Chen, and K. W. Zhao, Hypoxia inducible factor-1 mediates expression of galectin-1: The potential role in migration/invasion of colorectal cancer cells, *Carcinogenesis*, 31 (2010) 1367–1375.
109. D. Hanahan and J. Folkman, Patterns and emerging mechanisms of the angiogenic switch during tumorigenesis, *Cell*, 86 (1996) 353–364.
110. A. Marchetti, N. Tinari, F. Buttitta, A. Chella, C. A. Angeletti, R. Sacco, F. Mucilli, A. Ullrich, and S. Iacobelli, Expression of 90K (Mac-2 BP) correlates with distant metastasis and predicts survival in stage I non-small cell lung cancer patients, *Cancer Res.*, 62 (2002) 2535–2539.
111. N. Tinari, I. Kuwabara, M. E. Huflejt, P. F. Shen, S. Iacobelli, and F. T. Liu, Glycoprotein 90K/Mac-2BP interacts with galectin-1 and mediates galectin-1-induced cell aggregation, *Int. J. Cancer*, 91 (2001) 167–172.
112. (a) P. A. Baeuerle and O. Gires, Epcam (CD326) finding its role in cancer, *Br. J. Cancer*, 96 (2007) 417–423; (b) P. Klingbeil, R. Marhaba, T. Jung, R. Kirmse, T. Ludwig, and M. Zoeller, CD44 variant isoforms promote metastasis formation by a tumor cell-matrix cross-talk that supports adhesion and apoptosis resistance, *Mol. Cancer Res.*, 7 (2009) 168–179.
113. (a) Q. Zhou, M. E. Munger, R. G. Veenstra, B. J. Weigel, M. Hirashima, D. H. Munn, W. J. Murphy, M. Azuma, A. C. Anderson, V. K. Kuchroo, and B. R. Blazar, Coexpression of Tim-3 and PD-1 identifies a $CD8^+$ T-cell exhaustion phenotype in mice with disseminated acute myelogenous leukemia, *Blood*, 117 (2011) 4501–4510; (b) K. Nagahara, T. Arikawa, S. Oomizu, K. Kontani, A. Nobumoto, H. Tateno, K. Watanabe, T. Niki, S. Katoh, M. Miyake, S.-I. Nagahata, J. Hirabayashi, V. K. Kuchroo, A. Yamauchi, and M. Hirashima, Galectin-9 increases Tim-3^+ dendritic cells and $CD8^+$ T cells and enhances antitumor immunity via galectin-9-Tim-3 interactions, *J. Immunol.*, 181 (2008) 7660–7669.
114. F. van den Brule, S. Califice, F. Garnier, P. L. Fernandez, A. Berchuck, and V. Castronovo, Galectin-1 accumulation in the ovary carcinoma peritumoral stroma is induced by ovary carcinoma cells and affects both cancer cell proliferation and adhesion to laminin-1 and fibronectin, *Lab. Invest.*, 83 (2003) 377–386.
115. B. N. Stillman, P. S. Mischel, and L. G. Baum, New roles for galectins in brain tumors—From prognostic markers to therapeutic targets, *Brain Pathol.*, 15 (2005) 124–132.
116. (a) H. Inohara, S. Akahani, K. Koths, and A. Raz, Interactions between galectin-3 and Mac-2-binding protein mediate cell-cell adhesion, *Cancer Res.*, 56 (1996) 4530–4534; (b) H. Inohara and A. Raz,

Functional evidence that cell surface galectin-3 mediates homotypic cell adhesion, *Cancer Res.*, 55 (1995) 3267–3271.
117. (a) B. Rotblat, H. Niv, S. Andre, H. Kaltner, H. J. Gabius, and Y. Kloog, Galectin-1(L11A) predicted from a computed galectin-1 farnesyl-binding pocket selectively inhibits Ras-GTP, *Cancer Res.*, 64 (2004) 3112–3118; (b) A. Paz, R. Haklai, G. Elad-Sfadia, E. Ballan, and Y. Kloog, Galectin-1 binds oncogenic H-Ras to mediate Ras membrane anchorage and cell transformation, *Oncogene*, 20 (2001) 7486–7493; (c) R. Shalom-Feuerstein, T. Cooks, A. Raz, and Y. Kloog, Galectin-3 regulates a molecular switch from N-Ras to K-Ras usage in human breast carcinoma cells, *Cancer Res.*, 65 (2005) 7292–7300.
118. H. R. C. Kim, H. M. Lin, H. Biliran, and A. Raz, Cell cycle arrest and inhibition of anoikis by galectin-3 in human breast epithelial cells, *Cancer Res.*, 59 (1999) 4148–4154.
119. P. Brodt, L. Fallavollita, R. S. Bresalier, S. Meterissian, C. R. Norton, and B. A. Wolitzky, Liver endothelial E-selectin mediates carcinoma cell adhesion and promotes liver metastasis, *Int. J. Cancer*, 71 (1997) 612–619.
120. L. Borsig, R. Wong, J. Feramisco, D. R. Nadeau, N. M. Varki, and A. Varki, Heparin and cancer revisited: Mechanistic connections involving platelets, P-selectin, carcinoma mucins, and tumor metastasis, *Proc. Natl. Acad. Sci. U.S.A.*, 98 (2001) 3352–3357.
121. Y. J. Kim, L. Borsig, N. M. Varki, and A. Varki, P-selectin deficiency attenuates tumor growth and metastasis, *Proc. Natl. Acad. Sci. U.S.A.*, 95 (1998) 9325–9330.
122. A. Fryer, Y. C. Huang, G. Rao, D. Jacoby, E. Mancilla, R. Whorton, C. A. Piantadosi, T. Kennedy, and J. Hoidal, Selective O-desulfation produces nonanticoagulant heparin that retains pharmacological activity in the lung, *J. Pharmacol. Exp. Ther.*, 282 (1997) 208–219.
123. G. C. Jayson, M. Lyon, C. Paraskeva, J. E. Turnbull, J. A. Deakin, and J. T. Gallagher, Heparan sulfate undergoes specific structural changes during the progression from human colon adenoma to carcinoma *in vitro*, *J. Biol. Chem.*, 273 (1998) 51–57.
124. N. M. Varki and A. Varki, Heparin inhibition of selectin-mediated interactions during the hematogenous phase of carcinoma metastasis: Rationale for clinical studies in humans, *Semin. Thromb. Hemost.*, 28 (2002) 53–66.
125. O. J. T. McCarty, S. A. Mousa, P. F. Bray, and K. Konstantopoulos, Immobilized platelets support human colon carcinoma cell tethering, rolling, and firm adhesion under dynamic flow conditions, *Blood*, 96 (2000) 1789–1797.
126. (a) J. Polgar, J. Matuskova, and D. D. Wagner, The P-selectin, tissue factor, coagulation triad, *J. Thromb. Haemost.*, 3 (2005) 1590–1596; (b) I. Hrachovinova, B. Cambien, A. Hafezi-Moghadam, J. Kappelmayer, R. T. Camphausen, A. Widom, L. J. Xia, H. H. Kazazian, R. G. Schaub, R. P. McEver, and D. D. Wagner, Interaction of P-selectin and PSGL-1 generates microparticles that correct hemostasis in a mouse model of hemophilia A, *Nat. Med.*, 9 (2003) 1020–1025.
127. (a) L. A. Lasky, Selectins: Interpreters of cell-specific carbohydrate information during inflammation, *Science*, 258 (1992) 964–969; (b) P. M. Sass, The involvement of selectins in cell adhesion, tumor progression, and metastasis, *Cancer Invest.*, 16 (1998) 322–328.
128. (a) Y. S. Kim, J. Gum, and I. Brockhausen, Mucin glycoproteins in neoplasia, *Glycoconjug. J.*, 13 (1996) 693–707; (b) D. H. Dube and C. R. Bertozzi, Glycans in cancer and inflammation. Potential for therapeutics and diagnostics, *Nat. Rev. Drug Discov.*, 4 (2005) 477–488.
129. M. Martin-Satue, R. Marrugat, J. A. Cancelas, and J. Blanco, Enhanced expression of α(1,3)fucosyltransferase genes correlates with E-selectin-mediated adhesion and metastatic potential of human lung adenocarcinoma cells, *Cancer Res.*, 58 (1998) 1544–1550.
130. F. R. Liu, C. G. Jiang, Y. S. Li, J. B. Li, and F. Li, Cimetidine inhibits the adhesion of gastric cancer cells expressing high levels of sialyl Lewis X in human vascular endothelial cells by blocking E-selectin expression, *Int. J. Mol. Med.*, 27 (2011) 537–544.
131. (a) K. Shimodaira, J. Nakayama, N. Nakamura, O. Hasebe, T. Katsuyama, and M. Fukuda, Carcinoma-associated expression of core 2 β-1,6-N-acetylglucosaminyltransferase gene in human

colorectal cancer: Role of O-glycans in tumor progression, *Cancer Res.*, 57 (1997) 5201–5206; (b) J. Renkonen, T. Paavonen, and R. Renkonen, Endothelial and epithelial expression of sialyl Lewis[x] and sialyl Lewis[a] in lesions of breast carcinoma, *Int. J. Cancer*, 74 (1997) 296–300.
132. (a) R. Sawada, S. Tsuboi, and M. Fukuda, Differential E-selectin-dependent adhesion efficiency in sublines of a human colon cancer exhibiting distinct metastatic potentials, *J. Biol. Chem.*, 269 (1994) 1425–1431; (b) Y. Ikeda, M. Mori, K. Kajiyama, Y. Haraguchi, O. Sasaki, and K. Sugimachi, Immunohistochemical expression of sialyl Tn, sialyl Lewis[a], sialyl Lewis[a-b-], and sialyl Lewis[x] in primary tumor and metastatic lymph nodes in human gastric cancer, *J. Surg. Oncol.*, 62 (1996) 171–176.
133. C. Ohyama, S. Tsuboi, and M. Fukuda, Dual roles of sialyl Lewis X oligosaccharides in tumor metastasis and rejection by natural killer cells, *EMBO J.*, 18 (1999) 1516–1525.
134. H. Zhang, S. L. Zhang, N. K. V. Cheung, G. Ragupathi, and P. O. Livingston, Antibodies against GD2 ganglioside can eradicate syngeneic cancer micrometastases, *Cancer Res.*, 58 (1998) 2844–2849.
135. T. Kishimoto, H. Ishikura, C. Kimura, T. Takahashi, H. Kato, and T. Yoshiki, Phenotypes correlating to metastatic properties of pancreas adenocarcinoma *in vivo*: The importance of surface sialyl Lewis[a] antigen, *Int. J. Cancer*, 69 (1996) 290–294.
136. J. Beuth, H. L. Ko, V. Schirrmacher, G. Uhlenbruck, and G. Pulverer, Inhibition of liver tumor cell colonization in two animal tumor models by lectin blocking with D-galactose or arabinogalactan, *Clin. Exp. Metastasis*, 6 (1988) 115–120.
137. R. I. Zuberi, D. K. Hsu, O. Kalayci, H. Y. Chen, H. K. Sheldon, L. Yu, J. R. Apgar, T. Kawakami, C. M. Lilly, and F. T. Liu, Critical role for galectin-3 in airway inflammation and bronchial hyperresponsiveness in a murine model of asthma, *Am. J. Pathol.*, 165 (2004) 2045–2053.
138. (a) G. A. Rabinovich, N. Rubinstein, and M. A. Toscano, Role of galectins in inflammatory and immunomodulatory processes, *Biochim. Biophys. Acta*, 1572 (2002) 274–284; (b) W. Peng, H. Y. Wang, Y. Miyahara, G. Peng, and R.-F. Wang, Tumor-associated galectin 3 modulates the function of tumor-reactive T cells, *Cancer Res.*, 68 (2008) 7228–7236.
139. M.-H. Wu, H.-C. Hong, T.-M. Hong, W.-F. Chiang, Y.-T. Jin, and Y.-L. Chen, Targeting galectin-1 in carcinoma-associated fibroblasts inhibits oral squamous cell carcinoma metastasis by downregulating MCP-1/CCL2 expression, *Clin. Cancer Res.*, 17 (2011) 1306–1316.
140. Y. H. Shi, B. A. He, K. M. Kuchenbecker, L. You, Z. D. Xu, I. Mikami, A. Yagui-Beltran, G. Clement, Y. C. Lin, J. Okamoto, D. T. Bravo, and D. M. Jablons, Inhibition of *Wnt-2* and galectin-3 synergistically destabilizes β-catenin and induces apoptosis in human colorectal cancer cells, *Int. J. Cancer*, 121 (2007) 1175–1181.
141. K. A. Stannard, P. M. Collins, K. Ito, E. M. Sullivan, S. A. Scott, E. Gabutero, I. D. Grice, P. Low, U. J. Nilsson, H. Leffler, H. Blanchard, and S. J. Ralph, Galectin inhibitory disaccharides promote tumour immunity in a breast cancer model, *Cancer Lett.*, 299 (2010) 95–110.
142. J. Dumic, S. Dabelic, and M. Flogel, Curcumin—A potent inhibitor of galectin-3 expression, *Food Technol. Biotechnol.*, 40 (2002) 281–287.
143. L. Meromsky, R. Lotan, and A. Raz, Implications of endogenous tumor cell surface lectins as mediators of cellular interactions and lung colonization, *Cancer Res.*, 46 (1986) 5270–5275.
144. U. Jeschke, U. Karsten, I. Wiest, S. Schulze, C. Kuhn, K. Friese, and H. Walzel, Binding of galectin-1 (gal-1) to the Thomsen-Friedenreich (TF) antigen on trophoblast cells and inhibition of proliferation of trophoblast tumor cells in vitro by gal-1 or an anti-TF antibody, *Histochem. Cell Biol.*, 126 (2006) 437–444.
145. C. M. John, H. Leffler, B. Kahl-Knutsson, I. Svensson, and G. A. Jarvis, Truncated galectin-3 inhibits tumor growth and metastasis in orthotopic nude mouse model of human breast cancer, *Clin. Cancer Res.*, 9 (2003) 2374–2383.
146. T. Fukumori, Y. Takenaka, T. Yoshii, H. R. C. Kim, V. Hogan, H. Inohara, S. Kagawa, and A. Raz, CD29 and CD7 mediate galectin-3-induced type II T-cell apoptosis, *Cancer Res.*, 63 (2003) 8302–8311.

147. Y. Wang, P. Nangia-Makker, V. Balan, V. Hogan, and A. Raz, Calpain activation through galectin-3 inhibition sensitizes prostate cancer cells to cisplatin treatment, *Cell Death Dis.*, 1 (2010) e101.
148. (a) K. E. Pace, C. Lee, P. L. Stewart, and L. G. Baum, Restricted receptor segregation into membrane microdomains occurs on human T cells during apoptosis induced by galectin-1, *J. Immunol.*, 163 (1999) 3801–3811; (b) S. Karmakar, S. R. Stowell, R. D. Cummings, and R. P. McEver, Galectin-1 signaling in leukocytes requires expression of complex-type N-glycans, *Glycobiology*, 18 (2008) 770–778; (c) H. J. Allen, H. Ahmed, and K. L. Matta, Binding of synthetic sulfated ligands by human splenic galectin 1, a β-galactoside-binding lectin, *Glycoconjug. J.*, 15 (1998) 691–695.
149. F. Cedeno-Laurent, M. J. Opperman, S. R. Barthel, D. Hays, T. Schatton, Q. Zhan, X. He, K. L. Matta, J. G. Supko, M. H. Frank, G. F. Murphy, and C. J. Dimitroff, Metabolic inhibition of galectin-1-binding carbohydrates accentuates antitumor immunity, *J. Invest. Dermatol.*, 132 (2012) 410–420.
150. A. Banh, J. Zhang, H. Cao, D. M. Bouley, S. Kwok, C. Kong, A. J. Giaccia, A. C. Koong, and Q.-T. Le, Tumor galectin-1 mediates tumor growth and metastasis through regulation of T-cell apoptosis, *Cancer Res.*, 71 (2011) 4423–4431.
151. H. C. Hang, C. Yu, K. G. Ten Hagen, E. Tian, K. A. Winans, L. A. Tabak, and C. R. Bertozzi, Small molecule inhibitors of mucin-type O-linked glycosylation from a uridine-based library, *Chem. Biol.*, 11 (2004) 337–345.
152. J. Tejler, H. Leffler, and U. J. Nilsson, Synthesis of O-galactosyl aldoximes as potent LacNAc-mimetic galectin-3 inhibitors, *Bioorg. Med. Chem. Lett.*, 15 (2005) 2343–2345.
153. P. Nangia-Makker, V. Hogan, Y. Honjo, S. Baccarini, L. Tait, R. Bresalier, and A. Raz, Inhibition of human cancer cell growth and metastasis in nude mice by oral intake of modified citrus pectin, *J. Natl. Cancer Inst.*, 94 (2002) 1854–1862.
154. X. Gao, Y. Zhi, T. Zhang, H. Xue, X. Wang, A. D. Foday, G. Tai, and Y. Zhou, Analysis of the neutral polysaccharide fraction of MCP and its inhibitory activity on galectin-3, *Glycoconjug. J.*, 29 (2012) 159–165.
155. H. Inohara and A. Raz, Effects of natural complex carbohydrate (citrus pectin) on murine melanoma cell properties related to galectin-3 functions, *Glycoconjug. J.*, 11 (1994) 527–532.
156. D. Platt and A. Raz, Modulation of the lung colonization of B16-F1 melanoma cells by citrus pectin, *J. Natl. Cancer Inst.*, 84 (1992) 438–442.
157. K. J. Pienta, H. Naik, A. Akhtar, K. Yamazaki, T. S. Replogle, J. Lehr, T. L. Donat, L. Tait, V. Hogan, and A. Raz, Inhibition of spontaneous metastasis in a rat prostate cancer model by oral administration of modified citrus pectin, *J. Natl. Cancer Inst.*, 87 (1995) 348–353.
158. C. L. Jackson, T. M. Dreaden, L. K. Theobald, N. M. Tran, T. L. Beal, M. Eid, M. Y. Gao, R. B. Shirley, M. T. Stoffel, M. V. Kumar, and D. Mohnen, Pectin induces apoptosis in human prostate cancer cells: Correlation of apoptotic function with pectin structure, *Glycobiology*, 17 (2007) 805–819.
159. D. Chauhan, G. L. Li, K. Podar, T. Hideshima, P. Neri, D. L. He, N. Mitsiades, P. Richardson, Y. Chang, J. Schindler, B. Carver, and K. C. Anderson, A novel carbohydrate-based therapeutic GCS-100 overcomes bortezomib resistance and enhances dexamethasone-induced apoptosis in multiple myeloma cells, *Cancer Res.*, 65 (2005) 8350–8358.
160. A. P. Gunning, R. J. M. Bongaerts, and V. J. Morris, Recognition of galactan components of pectin by galectin-3, *FASEB J.*, 23 (2009) 415–424.
161. B. M. Yapo, P. Lerouge, J.-F. Thibault, and M.-C. Ralet, Pectins from citrus peel cell walls contain homogalacturonans homogenous with respect to molar mass, rhamnogalacturonan I and rhamnogalacturonan II, *Carbohydr. Polym.*, 69 (2007) 426–435.
162. (a) T. Fukumori, N. Oka, Y. Takenaka, P. Nangia-Makker, E. Elsamman, T. Kasai, N. Shono, H. Kanayama, J. Ellerhorst, R. Lotan, and A. Raz, Galectin-3 regulates mitochondrial stability and antiapoptotic function in response to anticancer drug in prostate cancer, *Cancer Res.*, 66 (2006) 3114–3119; (b) F. Yu, R. L. Finley, A. Raz, and H. R. C. Kim, Galectin-3 translocates to the perinuclear membranes and inhibits cytochrome *c* release from the mitochondria—A role for synexin

in galectin-3 translocation, *J. Biol. Chem.*, 277 (2002) 15819–15827; (c) V. V. Glinsky, Intravascular cell-to-cell adhesive interactions and bone metastasis, *Cancer Metastasis Rev.*, 25 (2006) 531–540.
163. K. D. Johnson, O. V. Glinskii, V. V. Mossine, J. R. Turk, T. P. Mawhinney, D. C. Anthony, C. J. Henry, V. H. Huxley, G. V. Glinsky, K. J. Pienta, A. Raz, and V. V. Glinsky, Galectin-3 as a potential therapeutic target in tumors arising from malignant endothelia, *Neoplasia*, 9 (2007) 662–670.
164. V. V. Glinsky, G. Kiriakova, O. V. Glinskii, V. V. Mossine, T. P. Mawhinney, J. R. Turk, A. B. Glinskii, V. H. Huxley, J. E. Price, and G. V. Glinsky, Synthetic galectin-3 inhibitor increases metastatic cancer cell sensitivity to taxol-induced apoptosis *in vitro* and *in vivo*, *Neoplasia*, 11 (2009) 901–909.
165. T. Nomura, M. Yamasaki, Y. Nomura, and H. Mimata, Expression of the inhibitors of apoptosis proteins in cisplatin-resistant prostate cancer cells, *Oncol. Rep.*, 14 (2005) 993–997.
166. P. Nangia-Makker, S. Baccarini, and A. Raz, Carbohydrate-recognition and angiogenesis, *Cancer Metastasis Rev.*, 19 (2000) 51–57.
167. A. Hayashi, A. C. Gillen, and J. R. Lott, Effects of daily oral administration of quercetin chalcone and modified citrus pectin on implanted colon-25 tumor growth in Balb-c mice, *Altern. Med. Rev.*, 5 (2000) 546–552.
168. B. Hagmar, W. Ryd, and H. Skomedal, Arabinogalactan blockade of experimental metastases to liver by murine hepatoma, *Invasion Metastasis*, 11 (1991) 348–355.
169. J. Beuth, H. L. Ko, K. Oette, G. Pulverer, K. Roszkowski, and G. Uhlenbruck, Inhibition of liver metastasis in mice by blocking hepatocyte lectins with arabinogalactan infusions and D-galactose, *J. Cancer Res. Clin. Oncol.*, 113 (1987) 51–55.
170. I. Iurisci, A. Cumashi, A. A. Sherman, Y. E. Tsvetkov, N. Tinari, E. Piccolo, M. D'Egidio, V. Adamo, C. Natoli, G. A. Rabinovich, S. Iacobelli, N. E. Nifantiev, and C. I. Behalf, Synthetic inhibitors of galectin-1 and -3 selectively modulate homotypic cell aggregation and tumor cell apoptosis, *Anticancer Res.*, 29 (2009) 403–410.
171. P. Sorme, B. Kahl-Knutsson, U. Wellmar, B. G. Magnusson, H. Leffler, and U. J. Nilsson, Design and synthesis of galectin inhibitors, *Methods Enzymol.*, 363 (2003) 157–169.
172. K. Henrick, S. Bawumia, E. A. M. Barboni, B. Mehul, and R. C. Hughes, Evidence for subsites in the galectins involved in sugar binding at the nonreducing end of the central galactose of oligosaccharide ligands: Sequence analysis, homology modeling and mutagenesis studies of hamster galectin-3, *Glycobiology*, 8 (1998) 45–57.
173. J. Seetharaman, A. Kanigsberg, R. Slaaby, H. Leffler, S. H. Barondes, and J. M. Rini, X-ray crystal structure of the human galectin-3 carbohydrate recognition domain at 2.1-Å resolution, *J. Biol. Chem.*, 273 (1998) 13047–13052.
174. B. A. Salameh, H. Leffler, and U. J. Nilsson, 3-(1,2,3-Triazol-1-yl)-1-thio-galactosides as small, efficient, and hydrolytically stable inhibitors of galectin-3, *Bioorg. Med. Chem. Lett.*, 15 (2005) 3344–3346.
175. P. Sorme, Y. N. Qian, P. G. Nyholm, H. Leffler, and U. J. Nilsson, Low micromolar inhibitors of galectin-3 based on 3 '-derivatization of *N*-acetyllactosamine, *Chembiochem*, 3 (2002) 183–189.
176. C. T. Oberg, H. Leffler, and U. J. Nilsson, Inhibition of galectins with small molecules, *Chimia*, 65 (2011) 18–23.
177. J. C. Ma and D. A. Dougherty, The cation-π interaction, *Chem. Rev.*, 97 (1997) 1303–1324.
178. (a) V. V. Rostovtsev, L. G. Green, V. V. Fokin, and K. B. Sharpless, A stepwise Huisgen cycloaddition process: Copper(I)-catalyzed regioselective "ligation" of azides and terminal alkynes, *Angew. Chem. Int. Ed. Engl.*, 41 (2002) 2596–2599; (b) C. W. Tornoe, C. Christensen, and M. Meldal, Peptidotriazoles on solid phase: [1,2,3]-Triazoles by regiospecific copper(I)-catalyzed 1,3-dipolar cycloadditions of terminal alkynes to azides, *J. Org. Chem.*, 67 (2002) 3057–3064.
179. D. Giguere, R. Patnam, M. A. Bellefleur, C. St-Pierre, S. Sato, and R. Roy, Carbohydrate triazoles and isoxazoles as inhibitors of galectins-1 and -3, *Chem. Commun. (Camb.)* (2006) 2379–2381.

180. B. A. Salameh, A. Sundin, H. Leffler, and U. J. Nilsson, Thioureido N-acetyllactosamine derivatives as potent galectin-7 and 9N inhibitors, *Bioorg. Med. Chem.*, 14 (2006) 1215–1220.
181. I. Cumpstey, E. Salomonsson, A. Sundin, H. Leffler, and U. J. Nilsson, Studies of arginine-arene interactions through synthesis and evaluation of a series of galectin-binding aromatic lactose esters, *Chembiochem*, 8 (2007) 1389–1398.
182. G. Masuyer, T. Jabeen, C. T. Oberg, H. Leffler, U. J. Nilsson, and K. R. Acharya, Inhibition mechanism of human galectin-7 by a novel galactose-benzylphosphate inhibitor, *FEBS J.*, 279 (2012) 193–202.
183. D. Giguere, S. Sato, C. St-Pierre, S. Sirois, and R. Roy, Aryl O- and S-galactosides and lactosides as specific inhibitors of human galectins-1 and -3: Role of electrostatic potential at O-3, *Bioorg. Med. Chem. Lett.*, 16 (2006) 1668–1672.
184. I. Cumpstey, S. Carlsson, H. Leffler, and U. J. Nilsson, Synthesis of a phenyl thio-β-D-galactopyranoside library from 1,5-difluoro-2,4-dinitrobenzene: Discovery of efficient and selective monosaccharide inhibitors of galectin-7, *Org. Biomol. Chem.*, 3 (2005) 1922–1932.
185. S. Carlsson, C. T. Oberg, M. C. Carlsson, A. Sundin, U. J. Niisson, D. Smith, R. D. Cummings, J. Almkvist, A. Karlsson, and H. Leffler, Affinity of galectin-8 and its carbohydrate recognition domains for ligands in solution and at the cell surface, *Glycobiology*, 17 (2007) 663–676.
186. E. Cauet, M. Rooman, R. Wintjens, J. Lievin, and C. Biot, Histidine-aromatic interactions in proteins and protein-ligand complexes: Quantum chemical study of X-ray and model structures, *J. Chem. Theory Comput.*, 1 (2005) 472–483.
187. K. Hotta, T. Funahashi, Y. Matsukawa, M. Takahashi, H. Nishizawa, K. Kishida, M. Matsuda, H. Kuriyama, S. Kihara, T. Nakamura, Y. Tochino, N. L. Bodkin, B. C. Hansen, and Y. Matsuzawa, Galectin-12, an adipose-expressed galectin-like molecule possessing apoptosis-inducing activity, *J. Biol. Chem.*, 276 (2001) 34089–34097.
188. J. Tejler, B. Salameh, H. Leffler, and U. J. Nilsson, Fragment based development of triazole-substituted O-galactosyl aldoximes with fragment-induced affinity and selectivity for galectin-3, *Org. Biomol. Chem.*, 7 (2009) 3982–3990.
189. J. Tejler, F. Skogman, H. Leffler, and U. J. Nilsson, Synthesis of galactose-mimicking $1H$-(1,2,3-triazol-1-yl)-mannosides as selective galectin-3 and 9N inhibitors, *Carbohydr. Res.*, 342 (2007) 1869–1875.
190. (a) C. Fradin, D. Poulain, and T. Jouault, β-1,2-linked oligomannosides from *Candida albicans* bind to a 32-kilodalton macrophage membrane protein homologous to the mammalian lectin galectin-3, *Infect. Immun.*, 68 (2000) 4391–4398; (b) G. J. Swaminathan, D. D. Leonidas, M. P. Savage, S. J. Ackerman, and K. R. Acharya, Selective recognition of mannose by the human eosinophil Charcot-Leyden crystal protein (galectin-10): A crystallographic study at 1.8 Å resolution, *Biochemistry*, 38 (1999) 13837–13843.
191. C. T. Oberg, H. Leffler, and U. J. Nilsson, Arginine binding motifs: Design and synthesis of galactose-derived arginine tweezers as galectin-3 inhibitors, *J. Med. Chem.*, 51 (2008) 2297–2301.
192. P. Sorme, P. Arnoux, B. Kahl-Knutsson, H. Leffler, J. M. Rini, and U. J. Nilsson, Structural and thermodynamic studies on cation–π interactions in lectin-ligand complexes: High-affinity galectin-3 inhibitors through fine-tuning of an arginine-arene interaction, *J. Am. Chem. Soc.*, 127 (2005) 1737–1743.
193. D. Giguere, M.-A. Bonin, P. Cloutier, R. Patnam, C. St-Pierre, S. Sato, and R. Roy, Synthesis of stable and selective inhibitors of human galectins-1 and-3, *Bioorg. Med. Chem.*, 16 (2008) 7811–7823.
194. (a) H. Ahmed, H. J. Allen, A. Sharma, and K. L. Matta, Human splenic galaptin: Carbohydrate-binding specificity and characterization of the combining site, *Biochemistry*, 29 (1990) 5315–5319; (b) R. T. Lee, Y. Ichikawa, H. J. Allen, and Y. C. Lee, Binding characteristics of galactoside-binding lectin (galaptin) from human spleen, *J. Biol. Chem.*, 265 (1990) 7864–7871.

195. G. A. Rabinovich, A. Cumashi, G. A. Bianco, D. Ciavardelli, I. Iurisci, M. D'Egidio, E. Piccolo, N. Tinari, N. Nifantiev, and S. Iacobelli, Synthetic lactulose amines: Novel class of anticancer agents that induce tumor-cell apoptosis and inhibit galectin-mediated homotypic cell aggregation and endothelial cell morphogenesis, *Glycobiology*, 16 (2006) 210–220.
196. L. Ingrassia, P. Nshimyumukiza, J. Dewelle, F. Lefranc, L. Wlodarczak, S. Thomas, G. Dielie, C. Chiron, C. Zedde, P. Tisnes, R. van Soest, J. C. Braekman, F. Darro, and R. Kiss, A lactosylated steroid contributes in vivo therapeutic benefits in experimental models of mouse lymphoma and human glioblastoma, *J. Med. Chem.*, 49 (2006) 1800–1807.
197. G. V. Glinsky, J. E. Price, V. V. Glinsky, V. V. Mossine, G. Kiriakova, and J. B. Metcalf, Inhibition of human breast cancer metastasis in nude mice by synthetic glycoamines, *Cancer Res.*, 56 (1996) 5319–5324.
198. I. Cumpstey, E. Salomonsson, A. Sundin, H. Leffler, and U. J. Nilsson, Double affinity amplification of galectin-ligand interactions through arginine-arene interactions: Synthetic, thermodynamic, and computational studies with aromatic diamido thiodigalactosides, *Chemistry*, 14 (2008) 4233–4245.
199. I. Cumpstey, A. Sundin, H. Leffler, and U. J. Nilsson, C_2-Symmetrical thiodigalactoside bis-benzamido derivatives as high-affinity inhibitors of galectin-3: Efficient lectin inhibition through double arginine-arene interactions, *Angew. Chem. Int. Ed. Engl.*, 44 (2005) 5110–5112.
200. K. Ito and S. J. Ralph, Inhibiting galectin-1 reduces murine lung metastasis with increased $CD4^+$ and $CD8^+$ T cells and reduced cancer cell adherence, *Clin. Exp. Metastasis*, 29 (2012) 561–572.
201. K. Ito, S. A. Scott, S. Cutler, L.-F. Dong, J. Neuzil, H. Blanchard, and S. J. Ralph, Thiodigalactoside inhibits murine cancers by concurrently blocking effects of galectin-1 on immune dysregulation, angiogenesis and protection against oxidative stress, *Angiogenesis*, 14 (2011) 293–307.
202. (a) S. A. Scott, A. Bugarcic, and H. Blanchard, Characterisation of oxidized recombinant human galectin-1, *Protein Pept. Lett.*, 16 (2009) 1249–1255; (b) Y. Inagaki, Y. Sohma, H. Horie, R. Nozawa, and T. Kadoya, Oxidized galectin-1 promotes axonal regeneration in peripheral nerves but does not possess lectin properties, *Eur. J. Biochem.*, 267 (2000) 2955–2964.
203. T. Delaine, I. Cumpstey, L. Ingragsia, M. Le Mercier, P. Okcchukwu, H. Leffler, R. Kiss, and U. J. Nilsson, Galectin-inhibitory thiodigalactoside ester derivatives have antimigratory effects in cultured lung and prostate cancer cells, *J. Med. Chem.*, 51 (2008) 8109–8114.
204. C.-l. Lin, E. E. Whang, D. B. Donner, X. Jiang, B. D. Price, A. M. Carothers, T. Delaine, H. Leffler, U. J. Nilsson, V. Nose, F. D. Moore, Jr., and D. T. Ruan, Galectin-3 targeted therapy with a small molecule inhibitor activates apoptosis and enhances both chemosensitivity and radiosensitivity in papillary thyroid cancer, *Mol. Cancer Res.*, 7 (2009) 1655–1662.
205. J. Tejler, E. Tullberg, T. Frejd, H. Leffler, and U. J. Nilsson, Synthesis of multivalent lactose derivatives by 1,3-dipolar cycloadditions: Selective galectin-1 inhibition, *Carbohydr. Res.*, 341 (2006) 1353–1362.
206. A. Nelson, J. M. Belitsky, S. Vidal, C. S. Joiner, L. G. Baum, and J. F. Stoddart, A self-assembled multivalent pseudopolyrotaxane for binding galectin-1, *J. Am. Chem. Soc.*, 126 (2004) 11914–11922.
207. J. U. Baenziger, The role of glycosylation in protein recognition, *Am. J. Pathol.*, 121 (1985) 382–391.
208. D. Tyrrell, P. James, N. Rao, C. Foxall, S. Abbas, F. Dasgupta, M. Nashed, A. Hasegawa, M. Kiso, D. Asa, J. Kidd, and B. K. Brandley, Structural requirements for the carbohydrate ligand of E-selectin, *Proc. Natl. Acad. Sci. U.S.A.*, 88 (1991) 10372–10376.
209. S. Andre, R. J. Pieters, I. Vrasidas, H. Kaltner, L. Kuwabara, F. T. Liu, R. M. J. Liskamp, and H. J. Gabius, Wedgelike glycodendrimers as inhibitors of binding of mammalian galectins to glycoproteins, lactose maxiclusters, and cell surface glycoconjugates, *Chembiochem*, 2 (2001) 822–830.

210. I. Vrasidas, S. Andre, P. Valentini, C. Bock, M. Lensch, H. Kaltner, R. M. J. Liskamp, H. J. Gabius, and R. J. Pieters, Rigidified multivalent lactose molecules and their interactions with mammalian galectins; a route to selective inhibitors, *Org. Biomol. Chem.*, 1 (2003) 803–810.
211. R. J. Pieters, Interference with lectin binding and bacterial adhesion by multivalent carbohydrates and peptidic carbohydrate mimics, *Trends Glycosci. Glycotechnol.*, 16 (2004) 243–254.
212. E. Salomonsson, A. Larumbe, J. Tejler, E. Tullberg, H. Rydberg, A. Sundin, A. Khabut, T. Frejd, Y. D. Lobsanov, J. M. Rini, U. J. Nilsson, and H. Leffler, Monovalent interactions of galectin-1, *Biochemistry*, 49 (2010) 9518–9532.
213. N. Ahmad, H. J. Gabius, S. Andre, H. Kaltner, S. Sabesan, R. Roy, B. C. Liu, F. Macaluso, and C. F. Brewer, Galectin-3 precipitates as a pentamer with synthetic multivalent carbohydrates and forms heterogeneous cross-linked complexes, *J. Biol. Chem.*, 279 (2004) 10841–10847.
214. J. Kopitz, C. von Reitzenstein, S. Andre, H. Kaltner, J. Uhl, V. Ehemann, M. Cantz, and H. J. Gabius, Negative regulation of neuroblastoma cell growth by carbohydrate-dependent surface binding of galectin-1 and functional divergence from galectin-3, *J. Biol. Chem.*, 276 (2001) 35917–35923.
215. (a) N. Ahmad, H. J. Gabius, S. Sabesan, S. Oscarson, and C. F. Brewer, Thermodynamic binding studies of bivalent oligosaccharides to galectin-1, galectin-3, and the carbohydrate recognition domain of galectin-3, *Glycobiology*, 14 (2004) 817–825; (b) K. Bachhawat-Sikder, C. J. Thomas, and A. Surolia, Thermodynamic analysis of the binding of galactose and poly-*N*-acetyllactosamine derivatives to human galectin-3, *FEBS Lett.*, 500 (2001) 75–79.
216. S. Andre, H. Kaltner, T. Furuike, S. I. Nishimura, and H. J. Gabius, Persubstituted cyclodextrin-based glycoclusters as inhibitors of protein-carbohydrate recognition using purified plant and mammalian lectins and wild-type and lectin-gene-transfected tumor cells as targets, *Bioconjug. Chem.*, 15 (2004) 87–98.
217. (a) D. A. Fulton and J. F. Stoddart, Neoglycoconjugates based on cyclodextrins and calixarenes, *Bioconjug. Chem.*, 12 (2001) 655–672; (b) O. Renaudet and P. Dumy, Chemoselectively template-assembled glycoconjugates as mimics for multivalent presentation of carbohydrates, *Org. Lett.*, 5 (2003) 243–246; (c) L. Baldini, A. Casnati, F. Sansone, and R. Ungaro, Calixarene-based multivalent ligands, *Chem. Soc. Rev.*, 36 (2007) 254–266; (d) S. K. Pandey, X. Zheng, J. Morgan, J. R. Missert, T.-H. Liu, M. Shibata, D. A. Bellnier, A. R. Oseroff, B. W. Henderson, T. J. Dougherty, and R. K. Pandey, Purpurinimide carbohydrate conjugates: Effect of the position of the carbohydrate moiety in photosensitizing efficacy, *Mol. Pharm.*, 4 (2007) 448–464.
218. S. Andre, F. Sansone, H. Kaltner, A. Casnati, J. Kopitz, H.-J. Gabius, and R. Ungaro, Calix[*n*]arene-based glycoclusters: Bioactivity of thiourea-linked galactose/lactose moieties as inhibitors of binding of medically relevant lectins to a glycoprotein and cell-surface glycoconjugates and selectivity among human adhesion/growth-regulatory galectins, *Chembiochem*, 9 (2008) 1649–1661.
219. J. M. Garcia Fernandez and C. Ortiz Mellet, Chemistry and developments of *N*-thiocarbonyl carbohydrate derivatives: Sugar isothiocyanates, thioamides, thioureas, thiocarbamates, and their conjugates, *Adv. Carbohydr. Chem. Biochem.*, 55 (2000) 35–135.
220. B. Monzavi-Karbassi, G. Cunto-Amesty, P. Luo, and T. Kieber-Emmons, Peptide mimotopes as surrogate antigens of carbohydrates in vaccine discovery, *Trends Biotechnol.*, 20 (2002) 207–214.
221. S. Andre, C. J. Arnusch, I. Kuwabara, R. Russwurm, H. Kaltner, H. J. Gabius, and R. J. Pieters, Identification of peptide ligands for malignancy- and growth-regulating galectins using random phage-display and designed combinatorial peptide libraries, *Bioorg. Med. Chem.*, 13 (2005) 563–573.
222. J. Zou, V. V. Glinsky, L. A. Landon, L. Matthews, and S. L. Deutscher, Peptides specific to the galectin-3 carbohydrate recognition domain inhibit metastasis-associated cancer cell adhesion, *Carcinogenesis*, 26 (2005) 309–318.
223. C. E. P. Maljaars, S. Andre, K. M. Halkes, H.-J. Gabius, and J. P. Kamerling, Assessing the inhibitory potency of galectin ligands identified from combinatorial (glyco)peptide libraries using surface plasmon resonance spectroscopy, *Anal. Biochem.*, 378 (2008) 190–196.

224. G. V. Glinsky, V. V. Mossine, J. E. Price, D. Bielenberg, V. V. Glinsky, H. N. Ananthaswamy, and M. S. Feather, Inhibition of colony formation in agarose of metastatic human breast carcinoma and melanoma cells by synthetic glycoamine analogs, *Clin. Exp. Metastasis*, 14 (1996) 253–267.
225. V. V. Glinsky, G. V. Glinsky, O. V. Glinskii, V. H. Huxley, J. R. Turk, V. V. Mossine, S. L. Deutscher, K. J. Pienta, and T. P. Quinn, Intravascular metastatic cancer cell homotypic aggregation at the sites of primary attachment to the endothelium, *Cancer Res.*, 63 (2003) 3805–3811.
226. O. V. Glinskii, S. Sud, V. V. Mossine, T. P. Mawhinney, D. C. Anthony, G. V. Glinsky, K. J. Pienta, and V. V. Glinsky, Inhibition of prostate cancer bone metastasis by synthetic TF antigen mimic/galectin-3 inhibitor lactulose-L-leucine, *Neoplasia*, 14 (2012) 65–73.
227. K. R. Oldenburg, D. Loganathan, I. J. Goldstein, P. G. Schultz, and M. A. Gallop, Peptide ligands for a sugar-binding protein isolated from a random peptide library, *Proc. Natl. Acad. Sci. U.S.A.*, 89 (1992) 5393–5397.
228. C. J. Arnusch, S. Andre, P. Valentini, M. Lensch, R. Russwurm, H. C. Siebert, M. J. E. Fischer, H. J. Gabius, and R. J. Pieters, Interference of the galactose-dependent binding of lectins by novel pentapeptide ligands, *Bioorg. Med. Chem. Lett.*, 14 (2004) 1437–1440.
229. E. Weber, A. Hetenyi, B. Vaczi, E. Szolnoki, R. Fajka-Boja, V. Tubak, E. Monostori, and T. A. Martinek, Galectin-1-asialofetuin interaction is inhibited by peptides containing the Tyr-Xxx-Tyr motif acting on the glycoprotein, *Chembiochem*, 11 (2010) 228–234.
230. A. W. Kahsai, J. Cui, H. U. Kaniskan, P. P. Garner, and G. Fenteany, Analogs of tetrahydroisoquinoline natural products that inhibit cell migration and target galectin-3 outside of its carbohydrate-binding site, *J. Biol. Chem.*, 283 (2008) 24534–24545.
231. R. P. M. Dings, E. S. Van Laar, M. Loren, J. Webber, Y. Zhang, S. J. Waters, J. R. MacDonald, and K. H. Mayo, Inhibiting tumor growth by targeting tumor vasculature with galectin-1 antagonist anginex conjugated to the cytotoxic acylfulvene, 6-hydroxypropylacylfulvene, *Bioconjug. Chem.*, 21 (2010) 20–27.
232. V. L. J. L. Thijssen, R. Postel, R. J. M. G. E. Brandwijk, R. P. M. Dings, I. Nesmelova, S. Satijn, N. Verhofstad, Y. Nakabeppu, L. G. Baum, J. Bakkers, K. H. Mayo, F. Poirier, and A. W. Griffioen, Galectin-1 is essential in tumor angiogenesis and is a target for antiangiogenesis therapy, *Proc. Natl. Acad. Sci. U.S.A.*, 103 (2006) 15975–15980.
233. K. H. Mayo, R. P. M. Dings, C. Flader, I. Nesmelova, B. Hargittai, D. W. J. van der Schaft, L. I. van Eijk, D. Walek, J. Haseman, T. R. Hoye, and A. W. Griffioen, Design of a partial peptide mimetic of anginex with antiangiogenic and anticancer activity, *J. Biol. Chem.*, 278 (2003) 45746–45752.
234. R. P. M. Dings and K. H. Mayo, A journey in structure-based drug discovery: From designed peptides to protein surface topomimetics as antibiotic and antiangiogenic agents, *Acc. Chem. Res.*, 40 (2007) 1057–1065.
235. R. P. M. Dings, X. Chen, D. M. E. I. Hellebrekers, L. I. van Eijk, Y. Zhang, T. R. Hoye, A. W. Griffioen, and K. H. Mayo, Design of nonpeptidic topomimetics of antiangiogenic proteins with antitumor activities, *J. Natl. Cancer Inst.*, 98 (2006) 932–936.
236. R. P. M. Dings, M. C. Miller, I. Nesmelova, L. Astorgues-Xerri, N. Kumar, M. Serova, X. Chen, E. Raymond, T. R. Hoye, and K. H. Mayo, Antitumor agent calixarene 0118 targets human galectin-1 as an allosteric inhibitor of carbohydrate binding, *J. Med. Chem.*, 55 (2012) 5121–5129.
237. R. Zeisig, R. Stahn, K. Wenzel, D. Behrens, and I. Fichtner, Effect of sialyl Lewis X-glycoliposomes on the inhibition of E-selectin-mediated tumour cell adhesion in vitro, *Biochim. Biophys. Acta*, 1660 (2004) 31–40.
238. (a) L. K. Needham and R. L. Schnaar, The HNK-1 reactive sulfoglucuronyl glycolipids are ligands for L-selectin and P-selectin but not E-selectin, *Proc. Natl. Acad. Sci. U.S.A.*, 90 (1993) 1359–1363; (b) A. Aruffo, W. Kolanus, G. Walz, P. Fredman, and B. Seed, CD62/P-selectin recognition of myeloid and tumor cell sulfatides, *Cell*, 67 (1991) 35–44.

239. M. Wei, G. H. Tai, Y. G. Gao, N. Li, B. Q. Huang, Y. F. Zhou, S. Hao, and X. L. Zeng, Modified heparin inhibits P-selectin-mediated cell adhesion of human colon carcinoma cells to immobilized platelets under dynamic flow, *J. Biol. Chem.*, 279 (2004) 29202–29210.
240. A. K. Sarkar, T. A. Fritz, W. H. Taylor, and J. D. Esko, Disaccharide uptake and priming in animal cells: Inhibition of sialyl Lewis X by acetylated Galβ1-4GlcNAcβ-*O*-naphthalenemethanol, *Proc. Natl. Acad. Sci. U.S.A.*, 92 (1995) 3323–3327.
241. R. Stahn, S. Goletz, R. Wilmanowski, X. Y. Wang, V. Briese, K. Friese, and U. Jeschke, Human chorionic gonadotropin (hCG) as inhibitior of E-selectin-mediated cell adhesion, *Anticancer Res.*, 25 (2005) 1811–1816.
242. Z. H. Chen, Y. J. Jing, B. H. Song, Y. L. Han, and Y. H. Chu, Chemically modified heparin inhibits in vitro L-selectin-mediated human ovarian carcinoma cell adhesion, *Int. J. Gynecol. Cancer*, 19 (2009) 540–546.
243. C. G. Jiang, J. B. Li, F. R. Liu, T. Wu, M. Yu, and H. M. Xu, Andrographolide inhibits the adhesion of gastric cancer cells to endothelial cells by blocking E-selectin expression, *Anticancer Res.*, 27 (2007) 2439–2447.
244. (a) T. Nubel, W. Dippold, H. Kleinert, B. Kaina, and G. Fritz, Lovastatin inhibits Rho-regulated expression of E-selectin by TNF-α and attenuates tumor cell adhesion, *FASEB J.*, 17 (2003) 140–161; (b) L. L. Yue, H. X. Cui, C. C. Li, Y. Lin, Y. X. Sun, Y. C. Niu, X. C. Wen, and J. C. Liu, A polysaccharide from *Agaricus blazei* attenuates tumor cell adhesion via inhibiting E-selectin expression, *Carbohydr. Polym.*, 88 (2012) 1326–1333.
245. A. M. Khatib, L. Fallavollita, E. V. Wancewicz, B. P. Monia, and P. Brodt, Inhibition of hepatic endothelial E-selectin expression by C-Raf antisense oligonucleotides blocks colorectal carcinoma liver metastasis, *Cancer Res.*, 62 (2002) 5393–5398.
246. K. Kobayashi, S. Matsumoto, T. Morishima, T. Kawabe, and T. Okamoto, Cimetidine inhibits cancer cell adhesion to endothelial cells and prevents metastasis by blocking E-selectin expression, *Cancer Res.*, 60 (2000) 3978–3984.
247. D. L. Morris and W. J. Adams, Cimetidine and colorectal cancer—Old drug, new use, *Nat. Med.*, 1 (1995) 1243–1244.
248. J. Kawase, S. Ozawa, K. Kobayashi, Y. Imaeda, S. Umemoto, S. Matsumoto, and M. Ueda, Increase in E-selectin expression in umbilical vein endothelial cells by anticancer drugs and inhibition by cimetidine, *Oncol. Rep.*, 22 (2009) 1293–1297.
249. T. Nubel, W. Dippold, B. Kaina, and G. Fritz, Ionizing radiation-induced E-selectin gene expression and tumor cell adhesion is inhibited by lovastatin and *all-trans* retinoic acid, *Carcinogenesis*, 25 (2004) 1335–1344.
250. G. Mannori, D. Santoro, L. Carter, C. Corless, R. M. Nelson, and M. P. Bevilacqua, Inhibition of colon carcinoma cell lung colony formation by a soluble form of E-selectin, *Am. J. Pathol.*, 151 (1997) 233–243.
251. (a) L. Descheny, M. E. Gainers, B. Walcheck, and C. J. Dimitroff, Ameliorating skin-homing receptors on malignant T cells with a fluorosugar analog of *N*-acetylglucosamine: P-selectin ligand is a more sensitive target than E-selectin ligand, *J. Invest. Dermatol.*, 126 (2006) 2065–2073; (b) C. J. Dimitroff, T. S. Kupper, and R. Sackstein, Prevention of leukocyte migration to inflamed skin with a novel fluorosugar modifier of cutaneous lymphocyte-associated antigen, *J. Clin. Invest.*, 112 (2003) 1008–1018.
252. (a) M. M. Fuster, J. R. Brown, L. C. Wang, and J. D. Esko, A disaccharide precursor of sialyl Lewis X inhibits metastatic potential of tumor cells, *Cancer Res.*, 63 (2003) 2775–2781; (b) A. K. Sarkar, K. S. Rostand, R. K. Jain, K. L. Matta, and J. D. Esko, Fucosylation of disaccharide precursors of sialyl Lewis^x inhibit selectin-mediated cell adhesion, *J. Biol. Chem.*, 272 (1997) 25608–25616; (c) A. K. Sarkar, J. R. Brown, and J. D. Esko, Synthesis and glycan priming activity of acetylated disaccharides, *Carbohydr. Res.*, 329 (2000) 287–300; (d) J. R. Brown, M. M. Fuster, T. Whisenant,

and J. D. Esko, Expression patterns of α2,3-sialyltransferases and α1,3-fucosyltransferases determine the mode of sialyl Lewis X inhibition by disaccharide decoys, *J. Biol. Chem.*, 278 (2003) 23352–23359.
253. J. R. Brown, F. Yang, A. Sinha, B. Ramakrishnan, Y. Tor, P. K. Qasba, and J. D. Esko, Deoxygenated disaccharide analogs as specific inhibitors of β1-4-galactosyltransferase 1 and selectin-mediated tumor metastasis, *J. Biol. Chem.*, 284 (2009) 4952–4959.
254. B. W. Weston, K. M. Hiller, J. P. Mayben, G. A. Manousos, K. M. Bendt, R. Liu, and J. C. Cusack, Expression of human α(1,3)fucosyltransferase antisense sequences inhibits selectin-mediated adhesion and liver metastasis of colon carcinoma cells, *Cancer Res.*, 59 (1999) 2127–2135.
255. S. Mathieu, M. Prorok, A. M. Benoliel, R. Uch, C. Langlet, P. Bongrand, R. Gerolami, and A. El-Battari, Transgene expression of α(1,2)-fucosyltransferase-I (FUT1) in tumor cells selectively inhibits sialyl-Lewis X expression and binding to E-selectin without affecting synthesis of sialyl-Lewis A or binding to P-selectin, *Am. J. Pathol.*, 164 (2004) 371–383.
256. P. A. Prieto, R. D. Larsen, M. Cho, H. N. Rivera, A. Shilatifard, J. B. Lowe, R. D. Cummings, and D. F. Smith, Expression of human *H*-type α1,2-fucosyltransferase encoding for blood group H(O) antigen in chinese hamster ovary cells—Evidence for preferential fucosylation and truncation of polylactosamine sequences, *J. Biol. Chem.*, 272 (1997) 2089–2097.
257. M. Aubert, L. Panicot, C. Crotte, P. Gibier, D. Lombardo, M. O. Sadoulet, and E. Mas, Restoration of α(1,2) fucosyltransferase activity decreases adhesive and metastatic properties of human pancreatic cancer cells, *Cancer Res.*, 60 (2000) 1449–1456.
258. (a) R. K. Jain, Delivery of molecular medicine to solid tumors, *Science*, 271 (1996) 1079–1080; (b) J. A. Phillips, S. J. Craig, D. Bayley, R. A. Christian, R. Geary, and P. L. Nicklin, Pharmacokinetics, metabolism, and elimination of a 20-mer phosphorothioate oligodeoxynucleotide (CGP 69846A) after intravenous and subcutaneous administration, *Biochem. Pharmacol.*, 54 (1997) 657–668.
259. Y. I. Kawamura, R. Kawashima, R. Fukunaga, K. Hirai, N. Toyama-Sorimachi, M. Tokuhara, T. Shimizu, and T. Dohi, Introduction of Sd[a] carbohydrate antigen in gastrointestinal cancer cells eliminates selectin and inhibits metastasis, *Cancer Res.*, 65 (2005) 6220–6227.
260. T. Dohi, A. Nishikawa, I. Ishizuka, M. Totani, K. Yamaguchi, K. Nakagawa, O. Saitoh, S. Ohshiba, and M. Oshima, Substrate specificity and distribution of UDP-GalNAc: Sialylparagloboside *N*-acetylgalactosaminyl transferase in the human stomach, *Biochem. J.*, 288 (1992) 161–165.
261. G. S. Kansas, Selectins and their ligands: Current concepts and controversies, *Blood*, 88 (1996) 3259–3287.
262. J. Y. Luo, M. Kato, H. M. Wang, M. Bernfield, and J. Bischoff, Heparan sulfate and chondroitin sulfate proteoglycans inhibit E-selectin binding to endothelial cells, *J. Cell. Biochem.*, 80 (2001) 522–531.
263. A. Koenig, K. Norgard-Sumnicht, R. Linhardt, and A. Varki, Differential interactions of heparin and heparan sulfate glycosaminoglycans with the selectins—Implications for the use of unfractionated and low molecular weight heparins as therapeutic agents, *J. Clin. Invest.*, 101 (1998) 877–889.
264. S. A. Mousa and L. J. Petersen, Anti-cancer properties of low-molecular-weight heparin: Preclinical evidence, *Thromb. Haemost.*, 102 (2009) 258–267.
265. X. Xie, A. S. Rivier, A. Zakrzewicz, M. Bernimoulin, X. L. Zeng, H. P. Wessel, M. Schapira, and O. Spertini, Inhibition of selectin-mediated cell adhesion and prevention of acute inflammation by nonanticoagulant sulfated saccharides—Studies with carboxyl-reduced and sulfated heparin and with trestatin a sulfate, *J. Biol. Chem.*, 275 (2000) 34818–34825.
266. N. Hostettler, A. Naggi, G. Torri, R. Ishai-Michaeli, B. Casu, I. Vlodavsky, and L. Borsig, P-selectin- and heparanase-dependent antimetastatic activity of non-anticoagulant heparins, *FASEB J.*, 21 (2007) 3562–3572.
267. J. L. Stevenson, A. Varki, and L. Borsig, Heparin attenuates metastasis mainly due to inhibition of P- and L-selectin, but non-anticoagulant heparins can have additional effects, *Thromb. Res.*, 120 (2007) S107–S111.

268. (a) K. A. Valentine, R. D. Hull, and G. F. Pineo, Low-molecular-weight heparin therapy and mortality, *Semin. Thromb. Hemost.*, 23 (1997) 173–178; (b) D. L. Ornstein and L. R. Zacharski, The use of heparin for treating human malignancies, *Haemostasis*, 29 (1999) 48–60; (c) L. R. Zacharski, D. L. Ornstein, and A. C. Mamourian, Low-molecular-weight heparin and cancer, *Semin. Thromb. Hemost.*, 26 (2000) 69–77.
269. Y. G. Gao, M. Wei, S. Zheng, X. Q. Ba, S. Hao, and X. L. Zeng, Chemically modified heparin inhibits the in vitro adhesion of nonsmall cell lung cancer cells to P-selectin, *J. Cancer Res. Clin. Oncol.*, 132 (2006) 257–264.
270. J. H. Im, W. L. Fu, H. Wang, S. K. Bhatia, D. A. Hammer, M. A. Kowalska, and R. J. Muschel, Coagulation facilitates tumor cell spreading in the pulmonary vasculature during early metastatic colony formation, *Cancer Res.*, 64 (2004) 8613–8619.
271. M. Salmivirta, F. Safaiyan, K. Prydz, M. S. Andresen, M. Aryan, and S. O. Kolset, Differentiation-associated modulation of heparan sulfate structure and function in CaCo-2 colon carcinoma cells, *Glycobiology*, 8 (1998) 1029–1036.
272. M. Sobel, W. R. Fish, N. Toma, S. Luo, K. Bird, K. Mori, S. Kusumoto, S. D. Blystone, and Y. Suda, Heparin modulates integrin function in human platelets, *J. Vasc. Surg.*, 33 (2001) 587–594.
273. M. Wei, Y. Gao, M. H. Tian, N. Li, S. Hao, and X. L. Zeng, Selectively desulfated heparin inhibits P-selectin-mediated adhesion of human melanoma cells, *Cancer Lett.*, 229 (2005) 123–126.
274. J. Fritzsche, S. Alban, R. J. Ludwig, S. Rubant, W. H. Boehncke, G. Schumacher, and G. Bendas, The influence of various structural parameters of semisynthetic sulfated polysaccharides on the P-selectin inhibitory capacity, *Biochem. Pharmacol.*, 72 (2006) 474–485.
275. D. H. Mi, Y. G. Gao, S. Zheng, X. Q. Ba, and X. L. Zeng, Inhibitory effects of chemically modified heparin on the P-selectin-mediated adhesion of breast cancer cells *in vitro*, *Mol. Med. Rep.*, 2 (2009) 301–306.
276. E. O. Kozlowski, M. S. G. Pavao, and L. Borsig, Ascidian dermatan sulfates attenuate metastasis, inflammation and thrombosis by inhibition of P-selectin, *J. Thromb. Haemost.*, 9 (2011) 1807–1815.
277. R. F. Wang, J. F. Huang, M. Wei, and X. L. Zeng, The synergy of 6-*O*-sulfation and *N*- or 3-*O*-sulfation of chitosan is required for efficient inhibition of P-selectin-mediated human melanoma A375 cell adhesion, *Biosci. Biotechnol. Biochem.*, 74 (2010) 1697–1700.
278. (a) G. S. Jacob, C. Kirmaier, S. Z. Abbas, S. C. Howard, C. N. Steininger, J. K. Welply, and P. Scudder, Binding of sialyl Lewis X to E-selectin as measured by fluorescence polarization, *Biochemistry*, 34 (1995) 1210–1217; (b) M. E. Beauharnois, K. C. Lindquist, D. Marathe, P. Vanderslice, J. Xia, K. L. Matta, and S. Neelamegham, Affinity and kinetics of sialyl Lewis-X and core-2 based oligosaccharides binding to L- and P-selectin, *Biochemistry*, 44 (2005) 9507–9519.
279. (a) M. J. Bamford, M. Bird, P. M. Gore, D. S. Holmes, R. Priest, J. C. Prodger, and V. Saez, Synthesis and biological activity of conformationally constrained sialyl Lewis X analogues with reduced carbohydrate character, *Bioorg. Med. Chem. Lett.*, 6 (1996) 239–244; (b) K. Scheffler, J. R. Brisson, R. Weisemann, J. L. Magnani, W. T. Wong, B. Ernst, and T. Peters, Application of homonuclear 3D NMR experiments and 1D analogs to study the conformation of sialyl Lewis^X bound to E-selectin, *J. Biomol. NMR*, 9 (1997) 423–436; (c) R. Banteli and B. Ernst, Synthesis of sialyl Lewis^x mimics, modifications of the 6-position of galactose, *Bioorg. Med. Chem. Lett.*, 11 (2001) 459–462; (d) W. Stahl, U. Sprengard, G. Kretzschmar, and H. Kunz, Synthesis of deoxy sialyl Lewis^x analogs, potential selectin antagonists, *Angew. Chem. Int. Ed. Engl.*, 33 (1994) 2096–2098.
280. R. W. Denton, X. Cheng, K. A. Tony, A. Dilhas, J. Jose Hernandez, A. Canales, J. Jimenez-Barbero, and D. R. Mootoo, C-disaccharides as probes for carbohydrate recognition—Investigation of the conformational requirements for binding of disaccharide mimetics of sialyl Lewis X, *Eur. J. Org. Chem.*, 2007 (2007) 645–654.
281. C. Filser, D. Kowalczyk, C. Jones, M. K. Wild, U. Ipe, D. Vestweber, and H. Kunz, Synthetic glycopeptides from the E-selectin ligand 1 with varied sialyl Lewis^x structure as cell-adhesion inhibitors of E-selectin, *Angew. Chem. Int. Ed. Engl.*, 46 (2007) 2108–2111.

282. H. C. Kolb and B. Ernst, Development of tools for the design of selectin antagonists, *Chemistry*, 3 (1997) 1571–1578.
283. T. Shodai, J. Suzuki, S. Kudo, S. Itoh, M. Terada, S. Fujita, H. Shimazu, and T. Tsuji, Inhibition of P-selectin-mediated cell adhesion by a sulfated derivative of sialic acid, *Biochem. Biophys. Res. Commun.*, 312 (2003) 787–793.
284. T. Ikami, N. Tsuruta, H. Inagaki, T. Kakigami, Y. Matsumoto, N. Tomiya, T. Jomori, T. Usui, Y. Suzuki, H. Tanaka, D. Miyamoto, H. Ishida, A. Hasegawa, and M. Kiso, Synthetic studies on selectin ligands/inhibitors. Synthesis and biological evaluation of sulfated and phosphorylated β-D-galacto- and lactopyranosides containing fatty-alkyl residues of different carbon chain lengths, *Chem. Pharm. Bull.*, 46 (1998) 797–806.
285. J. Perez-Castells, J. J. Hernandez-Gay, R. W. Denton, K. A. Tony, D. R. Mootoo, and J. Jimenez-Barbero, The conformational behaviour and P-selectin inhibition of fluorine-containing sialyl LeX glycomimetics, *Org. Biomol. Chem.*, 5 (2007) 1087–1092.
286. V. Gouge-Ibert, C. Pierry, F. Poulain, A. L. Serre, C. Largeau, V. Escriou, D. Scherman, P. Jubault, J. C. Quirion, and E. Leclerc, Synthesis of fluorinated C-mannopeptides as sialyl Lewis[x] mimics for E- and P-selectin inhibition, *Bioorg. Med. Chem. Lett.*, 20 (2010) 1957–1960.
287. A. Fukami, K. Iijima, M. Hayashi, K. Komiyama, and S. Omura, Macrosphelide B suppressed metastasis through inhibition of adhesion of sLe[x]/E-selectin molecules, *Biochem. Biophys. Res. Commun.*, 291 (2002) 1065–1070.
288. S. Ushiyama, T. M. Laue, K. L. Moore, H. P. Erickson, and R. P. McEver, Structural and functional characterization of monomeric soluble P-selectin and comparison with membrane P-selectin, *J. Biol. Chem.*, 268 (1993) 15229–15237.
289. L. Borsig, I. Vlodavsky, R. Ishai-Michaeli, G. Torri, and E. Vismara, Sulfated hexasaccharides attenuate metastasis by inhibition of P-selectin and heparanase, *Neoplasia*, 13 (2011) 445–452.
290. E. Edovitsky, M. Elkin, E. Zcharia, T. Peretz, and I. Vlodavsky, Heparanase gene silencing, tumor invasiveness, angiogenesis, and metastasis, *J. Natl. Cancer Inst.*, 96 (2004) 1219–1230.
291. G. A. Koning, H. W. M. Morselt, A. Gorter, T. M. Allen, S. Zalipsky, J. Kamps, and G. L. Scherphof, Pharmacokinetics of differently designed immunoliposome formulations in rats with or without hepatic colon cancer metastases, *Pharm. Res.*, 18 (2001) 1291–1298.
292. (a) M. Steegmaier, A. Levinovitz, S. Isenmann, E. Borges, M. Lenter, H. P. Kocher, B. Kleuser, and D. Vestweber, The E-selectin ligand ESL-1 is a variant of a receptor for fibroblast growth factor, *Nature*, 373 (1995) 615–620; (b) Z. Mourelatos, J. O. Gonatas, E. Cinato, and N. K. Gonatas, Cloning and sequence analysis of the human MG160, a fibroblast growth factor and E-selectin binding membrane sialoglycoprotein of the Golgi apparatus, *DNA Cell Biol.*, 15 (1996) 1121–1128.
293. M. Rosch, H. Herzner, W. Dippold, M. Wild, D. Vestweber, and H. Kunz, Synthetic inhibitors of cell adhesion: A glycopeptide from E-selectin ligand 1 (ESL-1) with the arabino sialyl Lewis[x] structure, *Angew. Chem. Int. Ed. Engl.*, 40 (2001) 3836–3839.
294. M. N. Fukuda, C. Ohyama, K. Lowitz, O. Matsuo, R. Pasqualini, E. Ruoslahti, and M. Fukuda, A peptide mimic of E-selectin ligand inhibits sialyl Lewis X-dependent lung colonization of tumor cells, *Cancer Res.*, 60 (2000) 450–456.

BACTERIAL CELL-ENVELOPE GLYCOCONJUGATES

Paul Messner[a], Christina Schäffer[a], and Paul Kosma[b]

[a]Department of NanoBiotechnology, NanoGlycobiology Unit, University of Natural Resources and Life Sciences, Vienna, Austria
[b]Department of Chemistry, University of Natural Resources and Life Sciences, Vienna, Austria

I. Introduction	210
1. Outline	210
2. Background—Bacterial Protein Glycosylation	211
II. Surface-Layer Glycoproteins	217
1. Bacterial S-Layer Glycoproteins	218
2. Archaeal S-Layer Glycoproteins	229
III. Nonclassical Secondary Cell-Envelope Polysaccharides	231
1. Background	231
2. The Nonclassical Group of SCWPs	232
IV. Structural Analysis	241
1. Isolation of Cell-Envelope Polysaccharides and Glycopeptides	241
2. Degradation Reactions	242
3. Structure Elucidation by Nuclear Magnetic Resonance Spectroscopy	243
4. Mass Spectrometry	245
V. Cell-Envelope Glycan Biosynthesis	246
1. Genetic Basis for S-Layer Glycoprotein Biosynthesis	246
2. Nucleotide Sugar Biosynthesis	247
3. Multispecific Glycosyltransferases	249
4. Proposed Pathway for S-Layer Glycoprotein Biosynthesis	250
5. SCWP Glycosylation Gene Clusters	252
VI. Glycan Engineering and Applications	254
1. The S-Layer Glycobiology Toolbox	254
2. Glycosylation Engineering	255
VII. Concluding Remarks	257
References	257

ABBREVIATIONS

AFM, atomic force microscopy; Bac, "bacillosamine," 2,4-diacetamido-2,4,6-trideoxy-D-glucose; CD, circular dichroism; COSY, correlated spectroscopy; CP, capsular polysaccharide; EM, electron microscopy; ESI, electrospray ionization; GC, gas chromatography; HF, hydrofluoric acid; HMBC, heteronuclear multiple-bond correlation; HPAEC, high-performance anion-exchange chromatography; HR MAS, high-resolution magic-angle spinning; Legionaminic acid, 5,7-diacetamido-3,5,7,9-tetradeoxy-D-*glycero*-D-*galacto*-nonulosonic acid; LPS, lipopolysaccharide; MALDI, matrix-assisted laser-desorption ionization; MS, mass spectrometry; NMR, nuclear magnetic resonance; NOESY, nuclear Overhauser effect spectroscopy; PA, polyacrylamide; Pse, pseudaminic acid 5,7-diacetamido-3,5,7,9-tetradeoxy-L-*glycero*-L-*manno*-nonulosonic acid; PAD, pulsed amperometric detection; PG, peptidoglycan; Quinovose, 6-deoxy-D-glucose; RP, reverse phase; S, S-layer (surface layer); SCWP, secondary cell-wall polymer; SDS, sodium dodecyl sulfate; SLH, S-layer homology domain; TMS, trimethylsilyl; TOCSY, total correlated spectroscopy; TOF, time of flight

I. INTRODUCTION

1. Outline

Nature has equipped prokaryotes (namely, bacteria and archaea) from almost all phylogenetic branches with a considerable repertoire of components from its "glycodiversification toolbox." Especially, bacterial-cell envelopes present an amazing repertoire of glycoconjugates, including (i) surface-located compounds, such as lipopolysaccharides, capsular polysaccharides, lipooligosaccharides, and glycoproteins; (ii) cell-wall-associated compounds categorized as "classical" and "nonclassical" secondary cell-wall polymers (SCWP); as well as (iii) secreted exopolysaccharides.

This article focuses on a special class of bacterial cell-surface glycoproteins, namely, S (surface)-layer glycoproteins and their inherent SCWP cell-envelope anchor. S-Layer glycoproteins have the unique feature of self-assembling into 2-D crystalline lattices, providing a display matrix for glycans having periodicity at the nanometer scale. This feature, in combination with the high degree of glycan variability which exceeds by far that of their eukaryotic counterparts, makes them attractive targets for glycoengineering aimed at (nano)biotechnology and biomedical applications.

2. Background—Bacterial Protein Glycosylation

Glycosylation of proteins is a modification ubiquitous in the natural world; it involves the co- or postsecretional addition of sugar residues to the respective polypeptides and serves to expand the diversity of the proteome.[1-3] A glycoprotein is a glycoconjugate in which a protein carries one or more glycans covalently attached to the polypeptide backbone, usually via an N- or O-glycosylic linkage. Many different carbohydrate components, together with a variety of covalent glycan–protein linkages, have been identified in glycoproteins originating from organisms of all three domains of life, namely, the Eukaryotes,[4] the *Bacteria*,[5-13] and the *Archaea*.[14-17] S-Layer glycoproteins are the focus of this article and are described in detail herein. A detailed list of other prokaryotic glycoproteins, including glycosylated enzymes, toxins, plasma and outer membrane proteins, antigens, aggregation factors, and other cellular glycoproteins, has been given by Messner and Schäffer.[18]

To understand fully the potential of bacterial protein glycosylation, it is reasonable to recapitulate details about the eukaryotic glycoproteins, which have been well investigated. Eukaryotic glycoproteins have been implicated in a multitude of cellular processes, including immune response, intracellular targeting, intercellular recognition, and protein folding and stability.[4,19-22] The common classes of glycans found in or on eukaryotic cells are primarily defined according to the nature of the linkage to the aglycone (protein or lipid). In N-linked glycans, a sugar chain is linked covalently to an asparagine residue of a polypeptide chain, commonly involving a GlcNAc residue and the consensus peptide sequence Asn-X-Ser/Thr.[3] Eukaryotic N-glycans share a common pentasaccharide-core region and can be generally divided into three main classes: oligomannosidic (or high-mannose) type, complex type, and hybrid type. In O-linked glycans, an *N*-acetylgalactosamine (GalNAc) residue is usually linked to the hydroxyl group of a serine or threonine residue, and it can be extended into a variety of different structural-core classes. In addition, several other types of O-linkages have been identified in a variety of eukaryotic glycoproteins (such as mannose–serine/threonine, xylose–serine, glucose–tyrosine, galactose–hydroxylysine, and L-arabinose–hydroxyproline).[4,21]

For a long time, reports on prokaryotic glycoproteins were restricted to a few archaeal and bacterial examples,[23,24] and thus, initially, the presence of glycoproteins in prokaryotes was considered a controversial matter. With the investigations of halobacterial S-layer glycoproteins conducted during the late 1970s and 1980s, the occurrence of glycosylated proteins in prokaryotes became generally accepted.[15,25,26] These first prokaryotic glycoproteins were shown to have similar types of glycosidic linkages as those found in eukaryotes. However, as the S-layer glycoproteome of more bacterial species was unraveled, it became evident that this class of bacterial cell-surface glycoconjugates displays a high structural variability which exceeds that of eukaryotic glycoproteins.[18] Recent work on bacterial and archaeal S-layer glycoproteins is discussed in greater detail in Section II.

Even though S-layer glycoproteins are a major class of prokaryotic glycoproteins, glycosylation of prokaryotic proteins is not limited to the S-layers. About 15 years ago, Sandercock and colleagues coined the term "non-S-layer glycoproteins" to subsume all other representatives of this heterogeneous group of bacterial glycoproteins.[27] One of the very first reports in this context was the partial description of an enterococcal (formerly streptococcal) *N*-acetylmuramoylhydrolase[28] and also the characterization of the TibA glycoprotein from an enterotoxigenic (ETEC) *Escherichia coli* that acts as an adhesin and plays an important role in the virulence of this organism.[29] Other important members of "non-S-layer glycoproteins" are such cell-surface appendages as flagella and pili.[30] Among the first fully described species were the flagella of the archaeon *Halobacterium salinarum* (formerly *Hbt. halobium*)[31] and the flagellin of the nitrogen-fixing bacterium *Azospirillum brasilense*.[32]

During the past decade, high-resolution mass-spectrometric investigations revealed the existence of uncommon sialic acid-like sugars (5,7-diacetamido-3,5,7,9-tetradeoxy-nonulosonate derivatives), such as pseudaminic acid or legionaminic acid, as major components of flagella[33–37] and pili[38] of different pathogenic bacteria. These can also incorporate the previously mentioned sialic acid-like sugars into other virulence-associated cell-surface glycoconjugates such as lipopolysaccharides (LPS)[39,40] and capsular polysaccharides.[41] The sialic acid-like sugars are unique to microorganisms and may exhibit configurational differences as compared to sialic acid. Legionaminic acid (5,7-diacetamido-3,5,7,9-tetradeoxy-D-*glycero*-D-*galacto*-nonulosonic acid), for instance, has the same absolute configuration as sialic acid. In this context, the properties of the sialic acid-like glycoconjugates mimicking host structures are of utmost importance for the host immune defense.

The seminal observation of the late 1990s that *Campylobacter jejuni* is able to synthesize a large number of periplasmic glycoproteins strongly supported the already proposed concept concerning the existence of prokaryotic glycoproteins.[42] *C. jejuni* is a pathogen of the human gut mucosa that is one of the main causes of bacterial gastroenteritis worldwide. A striking finding was the identification of two glycosylation loci, which encode, respectively, both the O-linked and N-linked protein glycosylation pathways. The O-linked flagellin glycosylation system comprises a glycosylation cluster of approximately 50 genes, including genes encoding the flagellin structural proteins FlaA and FlaB. Although multiple orthologues of the genes for sialic acid biosynthesis were identified, recent structural analysis of the flagellins from *C. jejuni* and the related species *Campylobacter coli* has shown that flagellins are not modified by sialic acid, but by several monosaccharide analogues of the related sugar, pseudaminic acid.[43] The *Campylobacter* flagella are extensively modified, with approximately 10% of the total FlaA glycoprotein mass consisting of O-linked glycans, and each filament containing approximately 20,000 protein subunits.[44]

The glycosylation system for the N-linked protein was originally thought to be involved in the biosynthesis of lipopolysaccharide (LPS), but later it was shown to be involved in the glycosylation of multiple *C. jejuni* proteins.[42] Structural investigations revealed that the N-linked glycan consists of the heptasaccharide α-GalNAc-(1$\rightarrow$4)-α-GalNAc-(1$\rightarrow$4)[β-Glc-(1$\rightarrow$3)]-α-GalNAc-(1$\rightarrow$4)-α-GalNAc-(1$\rightarrow$4)-α-GalNAc-(1$\rightarrow$3)-β-Bac-(1$\rightarrow$N)-Asn, where Bac is bacillosamine (2,4-diacetamido-2,4,6-trideoxy-D-glucose).[45] The key enzyme for protein glycosylation is the *N*-oligosaccharyltransferase (OTase) PglB, which has significant amino acid similarity to the eukaryotic oligosaccharyltransferase complex STT3.[46] Glycosylation in *C. jejuni* involves the transfer of the heptasaccharide from an undecaprenyl pyrophosphate donor to the asparagine side chain of proteins at the bacterial periplasmic membrane to form N-linked glycoproteins.[9] The *in vitro* biosynthesis machinery responsible for this elaborate protein modification follows a similar overall progression, whereby an oligosaccharide is assembled in a stepwise fashion on a polyisoprenyl pyrophosphate carrier and is then ultimately transferred to a protein.[22] A major scientific and biotechnological breakthrough could be achieved by the transfer of the complete *C. jejuni* N-glycosylation machinery into *E. coli*.[47] This allowed for the first time the recombinant production of glycoproteins in prokaryotes. In a series of excellent papers, the optimization of both the glycosylation process and the production conditions for biosynthesis of "humanized glycoproteins" in bacteria has been worked out.[48–53]

Similar considerations were also applied for the characterization of other prokaryotic glycoproteins, such as type IV pili.[54] The constituting pilins are *O*-glycosylated in such pathogenic bacteria as *Neisseria meningitidis* and *N. gonorrhoeae*, where they are decorated with short glycans, up to three sugars in length. The final demonstration that PglL is the O-OTase in *N. meningitidis*, responsible for the transfer of the glycan to pilin, was provided by Faridmoayer and colleagues, who showed the PglL-dependent glycosylation of pilin by reconstituting the process in *E. coli*.[55] In *N. gonorrhoeae*, the glycosylation pathway to O-linked protein is responsible for the synthesis of the trisaccharide on undecaprenyl diphosphate and subsequent *en bloc* transfer of the glycan by PglO, the respective O-OTase, to serine residues of select periplasmic proteins. The linkage sugar nucleotide UDP-2,4-diacetamido-2,4,6-trideoxy-α-D-glucose, which is the first sugar in glycan biosynthesis, was produced enzymatically, and the stereochemistry as uridine diphosphate N-diacetylbacillosamine (UDP-Bac-2,4-diNAc) was assigned by NMR.[56]

The biological role of many prokaryotic glycoproteins requires further exploration. During the past decade, it became evident that glycosylation plays a vital role in pathogenicity and host invasion. In addition, roles in adhesion, protection against proteolytic cleavage, protein assembly, solubility, antigenic variation, and protective immunity have been suggested.[8,57–62]

a. General Architecture of Bacterial Cell Envelopes.—In the course of evolution, prokaryotic organisms have developed a considerable diversity in their supramolecular architectures. The different cell-wall structures, particularly those facing the surrounding environment, obviously reflect evolutionary adaptations of the organisms to a broad spectrum of selection criteria in the natural habitat.[63] Among them are S-layers which, although not a universal feature, have been identified as common features of many organisms belonging to the prokaryotic domains *Bacteria* (Section II.1) and *Archaea* (Section II.2). These monomolecular arrays of (glyco) proteinaceous subunits present as the outermost component of the cell envelope were detected in a large number of different species of nearly every phylogenetic group of bacteria[64–74] and represent an almost universal feature of archaeal cell envelopes.[17,75–77] Despite the fact that considerable variations exist in the structure and chemistry of prokaryotic envelopes (Fig. 1), S-layers have apparently coevolved with these diverse structures. In Gram-positive bacteria and Gram-positive archaea (Fig. 1, left panel), the regular arrays assemble on the surface of the rigid wall matrix, which is mainly composed of peptidoglycan or pseudomurein, respectively.[63] In Gram-negative bacteria (Fig. 1, middle panel), the S-layer is presumably attached to the LPS of the outer membrane, and in most archaea (Fig. 1, right panel), S-layers are attached or inserted to the plasma membrane (for review, see Ref. 74).

The location and ultrastructure of S-layers of a great variety of organisms have been investigated by electron microscopy (EM) (for reviews, see Refs. 67,78–80) and atomic force microscopy (AFM) (for reviews, see Refs. 81–84). On the prokaryotic cell surface, the S-layer subunits can be aligned in lattices with oblique (p1, p2), square (p4), or hexagonal (p3, p6) symmetry. While hexagonal symmetry is predominant among archaea, in bacterial S-layers, any type of lattice has been observed.[69,74]

It is an intrinsic property to rapidly self-assemble into two-dimensional (2-D), regularly ordered protein crystals, thus preventing the formation of 3-D crystals suitable for X-ray diffraction analysis.[85,86] Although S-layer proteins are usually insoluble in aqueous systems, we observed that this was not the case with the glycosylated S-layer protein of *Aneurinibacillus thermoaerophilus* (*Ane.*

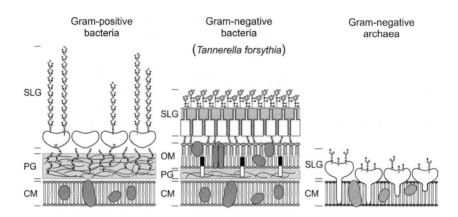

FIG. 1. Scheme of cell envelopes of prokaryotic organisms, showing representative cellular glycan structures. Abbreviations: CM, cytoplasmic membrane; PG, peptidoglycan; OM, outer membrane; SLG, glycosylated S-layer protein. , different S-layer glycans; , , S-layer proteins; , secondary cell-wall glycan (SCWP); , peptidoglycan strand; , bacterial membrane phospholipid; , archaeal membrane tetraetherlipid; , outer membrane lipid A.

thermoaerophilus, formerly *Bacillus thermoaerophilus*).[87] This S-layer glycoprotein was completely water-soluble and was thus potentially suited for 3-D crystallization experiments. Later on, however, the S-layer protein SbsC from *Geobacillus stearothermophilus* (formerly *Bacillus stearothermophilus*) ATCC 12980^T became the first one for which different N- or C-terminally truncated S-layer protein forms were recombinantly produced[88] and systematically surveyed for their self-assembling and recrystallization properties by a dissection approach.[89] Native and heavy-atom derivative data confirmed the results of the secondary-structure prediction, which indicated that the N-terminal region comprising the first 257 amino acids is mainly organized as α-helices, whereas the middle and C-terminal parts of SbsC consist of loops and β-sheets.[90,91] The crystal-structure model of truncated rSbsC revealed a novel fold, consisting of six separate domains, which are connected by short flexible linkers. Instead of a dissection approach, Baranova *et al.* used a nanobody-aided crystallization system and provided the first 3-D crystal structure of the nontruncated S-layer protein SbsB from *Lysinibacillus sphaericus* (formerly *Bacillus sphaericus*) CCM 2177 with a resolution of 2.4 Å.[92] SbsB consists of a seven-domain protein, formed by an amino-terminal cell-wall attachment domain and six consecutive immunoglobulin-like domains, that organize into a χ-shaped disk-like monomeric crystallization unit stabilized by interdomain Ca^{2+} ion coordination.

Independent of the 3-D crystallization efforts, secondary-structure predictions have been made for different S-layer proteins (as from *Aeromonas salmonicida*, *Campylobacter fetus*, and *G. stearothermophilus* PV72/p2; for review, see Ref. 93) by comparison of circular dichroism (CD) spectra under native and denaturing conditions. Furthermore, equilibrium unfolding profiles induced thermally or by guanidinium hydrochloride, monitored by intrinsic fluorescence and CD spectroscopy, revealed that the N-terminal portion of truncated rSbsB from *G. stearothermophilus* PV72/p2 is an α-helical protein. The C-terminal form was characterized as a β-sheet protein with typical multidomain unfolding.[89] The first tertiary-structure prediction for SbsB, based on molecular dynamics simulations using the mean force method, has been published.[94] By using a combination of the structural information and a Monte Carlo method with a coarse-grained model, the functional protein self-assembly of SbsB from *G. stearothermophilus* PV72/p2 was studied.[95] It was found that only a few amino acids and mainly hydrophobic ones, located on the surface of the monomer, are responsible for the formation of a highly ordered, anisotropic protein lattice. The broad potential for application of S-layers in nanobiotechnology is based on the specific intrinsic features of the monomolecular arrays composed of identical protein or glycoprotein subunits. Many applications depend on the capability of isolated subunits to recrystallize into monomolecular arrays in suspension or on suitable surfaces (as on polymers, metals, or silicon wafers) or interfaces (such as lipid films, liposomes, or emulsomes).[96]

b. Biological Functions.—Only little is known about S-layer glycoprotein structure–function relationships. Most of the presumptive functions of S-layer glycoprotein glycans relate to the functions attributed to the 2-D crystalline S-layer protein portions. Considering that S-layer-carrying organisms are ubiquitous in the biosphere, the supramolecular concept of a closed, isoporous, protein meshwork has the potential to fulfill a broad spectrum of functions. When bacteria are no longer exposed to selection pressures in the natural environment, S-layers can be lost, indicating that the considerable biosynthetic effort is only required in natural habitats. In functional terms, (glycosylated) S-layers are generally part of complex envelope structures (Fig. 1) and consequently should not be considered as isolated layers. S-Layer functions are generally enabled through the prominent location at the cell surface and include, for instance, isoporous and/or protective coating of the cell, surface recognition and cell adhesion to substrates, receptor–substrate interactions, templated fine-grain mineralization, as well as mediation of pathogenicity-related phenomena.[74,97–100] Many of these assigned functions are still hypothetical and are not based on firm experimental data.

Concerning the latter function, S-layer variation might have evolved as an important strategy of pathogens—but also of nonpathogens—to respond to changing environmental conditions.[101–105] With regard to an interplay of glycosylation and pathogenicity, specific data have recently been accumulated for the periodontal pathogen *Tannerella forsythia*. For this bacterium, S-layer glycosylation is an important aspect affecting its cellular integrity, lifestyle, and virulence potential.[100] The bacterium possesses a unique Gram-negative cell envelope with a glycosylated S-layer as outermost decoration that is formed by coassembly of the two S-layer glycoproteins TfsA and TfsB.[84] The O-glycosidically linked *T. forsythia* S-layer decasaccharide (see Section II for details) is overall a highly diverse structure containing several rare sugar residues.[100] This S-layer glycan directly impacts the lifestyle of *T. forsythia* because increased biofilm formation of a UDP-*N*-acetylmannosaminuronic acid dehydrogenase mutant could be correlated with the presence of truncated S-layer glycans in which the "acidic" branch of the decasaccharide is missing.[106] It is also tempting to speculate that the terminal 5-acetimidoyl-7-glycolyl pseudaminic acid (Pse5Am7Gc) residue of the S-layer glycans participates in the bacterium–host cross talk, although the relevance of the modification of the pseudaminic acid remains yet unclear. This notion is supported by the fact that members of this class of sialic acid-like sugars have been found in many Gram-negative bacterial species as constituents of important cell-surface glycoconjugates, such as LPS,[40] capsules,[41] pili,[38] and flagella,[33,36] all of which are important mediators of pathogenicity and possibly influence bacterial adhesion, invasion, and immune evasion.[107]

While the S-layer has been shown to be a virulence factor and to delay the bacterium's recognition by the innate immune system of the host,[84,108] the contribution of glycosylation to modulating the host immunity is currently unraveling. It was already shown that surface glycosylation of *Tannerella* has a role in restraining the Th17-mediated neutrophil infiltration in the gingival tissues.[109]

II. Surface-Layer Glycoproteins

During the past 35 years, considerable progress has been made in research on prokaryotic glycoproteins, providing a wealth of data on structures, biological properties, and functions of the different cellular and external prokaryotic glycoproteins (compare with Section I). The first data on prokaryotic S-layer glycoproteins date back to the mid-1970s, when reports on halobacteria[23] and on thermophilic clostridia[24] were published. Since then a considerable body of knowledge has accumulated about bacterial (for recent reviews, see Refs. 18,110,111) and archaeal S-layer glycoproteins (for reviews, see Refs. 16,17,72,112–114), indicating that glycosylation is the major modification of S-layer proteins; however, prokaryotic S-layer proteins are not necessarily glycosylated. Glycosylation generally contributes to an enormous diversification potential of the prokaryotic cell surface, which may be advantageous for the survival of the organisms in their natural, competitive habitats.

1. Bacterial S-Layer Glycoproteins

A major class among prokaryotic glycoconjugates are the bacterial S-layer glycoproteins from members of the *Bacillaceae* (for reviews, see Refs. 5,11,18,58,74,111, 115,116). Among the *Bacillaceae* species investigated are highly thermophilic *G. stearothermophilus* strains[117] and thermophilic clostridia,[24] as well as the less-thermophilic *Ane. thermoaerophilus*[118] and *Anoxybacillus tepidamans* (*Ano. tepidamans*, formerly *Geobacillus tepidamans*) strains[119,119a]; all of these bacteria are isolates from extraction plants of Austrian beet-sugar factories. The mesophilic Gram-positive bacteria investigated include *Paenibacillus alvei* (formerly *Bacillus alvei*), a type culture organism originally isolated from honeybee hives, and different *Lactobacillus buchneri* strains, including a dairy factory isolate[120] and a silage isolate.[121]

In these organisms, up to 20% of the total protein-synthesis effort may be devoted to the production of S-layer glycoproteins, and this indicates that S-layer protein synthesis is under the control of very strong promoters.[122] The degree of glycosylation of S-layer proteins generally varies between 2% and 10% (w/w), yielding overall apparent molecular masses of the constituting protomers between 45 and 200 kDa, according to SDS-PA gels. Most of these S-layer glycoproteins have the unique property of assembling into two-dimensional crystalline arrays on the supporting peptidoglycan layer, yielding a complete coverage of the bacterial cells during all stages of the bacterial growth cycle, with the glycan moieties protruding from the cell surface into the exterior environment (Fig. 2). This feature resembles the coating of Gram-negative bacterial cells by lipopolysaccharides.[123,124] While in the past most of the structural investigations on S-layer glycoproteins have been performed on selected Gram-positive bacilli, clostridia and lactobacilli,[18,111,116] investigations on Gram-negative organisms were only recently initiated. Glycosylated S-layer proteins have been unambiguously identified on the human oral pathogen *T. forsythia*,[106,125] and comparable glycosylated proteins, though not yet explicitly identified as S-layer proteins, on the closely related human commensal *Bacteroides* fragilis.[126,127] Thus far, only the structure of the *T. forsythia* S-layer oligosaccharide has been completely elucidated,[106] but preliminary analyses revealed that the *B. fragilis* glycan is structurally comparable.[127]

a. Core Structures.—Despite a highly diverse glycan composition, most S-layer glycoproteins of the Gram-positive bacteria thus far analyzed follow a general

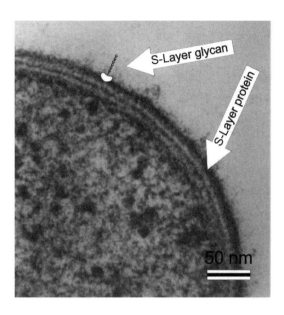

FIG. 2. Electron micrograph of a freeze-substituted and cross-sectioned cell of *Ane. thermoaerophilus* DSM 10155. With permission from Springer Science + Business Media, compare with Ref. 128, Fig. 1b.

tripartite building plan.[11,112,117,118] This plan comprises a short oligosaccharide, termed "adapter region," linking the entire S-layer glycan to the S-layer polypeptide backbone. To date, in bacterial S-layer glycoproteins from Gram-positive as well as Gram-negative species, only O-glycosidic linkages have been identified.[111,128] This includes linkages to serine, threonine,[4] and tyrosine residues.[130] Typically, S-layer proteins are multiply glycosylated. It is noteworthy that in all structures investigated the linkage sugar (namely, Gal, GalNAc, or Glc) is in the β-anomeric configuration.

(i) Core Oligosaccharides of Elongated S-Layer Glycans. In the long-chain S-layer glycans of *Bacillaceae*, some variation in the composition of the adapter region has been observed. The most common structure, which is present in *G. stearothermophilus* NRS 2004/3a,[131] *Ano. tepidamans* GS5-97,[132] *Ane. thermoaerophilus* (DSM 10155),[133] *P. alvei* (CCM 2051^T),[134,135] *Thermoanaerobacter thermohydrosulfuricus* L111-69, and L110-69 (DSM 568) (formerly *Clostridium thermohydrosulfuricum*),[136] comprises one to three α-(1 → 3)-linked L-rhamnopyranose residues α-(1 → 3)-linked to the linkage sugars β-D-Gal*p* or β-D-Gal*p*NAc, respectively (Fig. 3).

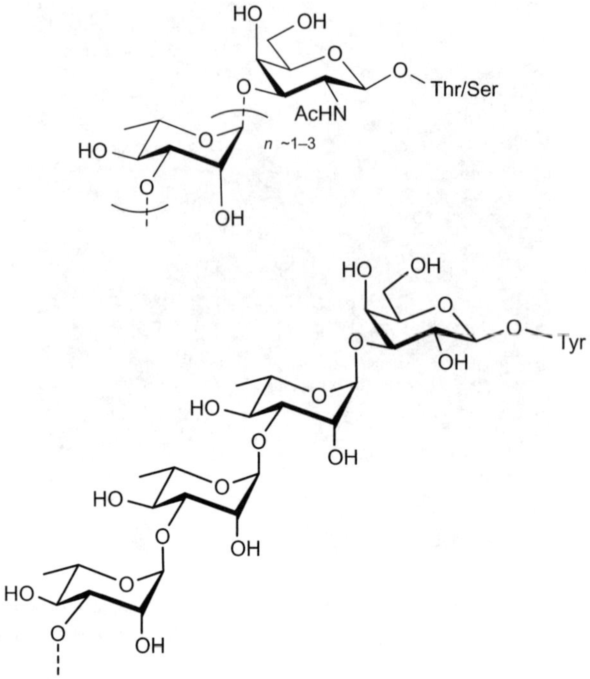

FIG. 3. Rhamnosyl core oligosaccharides.[131–136]

In the *P. alvei* CCM 2051 S-layer glycan core, the distal L-Rha*p* residue is additionally decorated with a β-(1→4)-linked GroA-(2→O)-P(=O)OH-(O→4)-β-D-Man*p*NAc side chain (Fig. 4).[135]

A slightly different adapter region oligosaccharide has been found in the S-layer glycan of *Ane. thermoaerophilus* L420-91^T (DSM 10154), where instead of α-(1→3)-linked L-Rha*p* residues, the adapter glycan is made of one to three α-(1→3)-linked D-Rha*p* residues that are α-(1→3)-linked to a β-D-Gal*p*NAc residue (Fig. 5).[137]

If the oligosaccharide linker is missing, the carbohydrate chain can be bound directly to the glycosylated amino acid via the first repeating unit, as in the S-layer glycan of *Thermoanaerobacterium thermosaccharolyticum* D120-70 (formerly *Clostridium thermosaccharolyticum*).[138,139] The anomeric configuration of this linkage sugar may remain unchanged (no core) or be inverted (pseudo-core).[115] In *Thb. thermohydrosulfuricus* S102-70, a short glycan consisting only of a tyrosine-linked hexasaccharide with β-D-Glc*p* as the linkage sugar has been observed.[140]

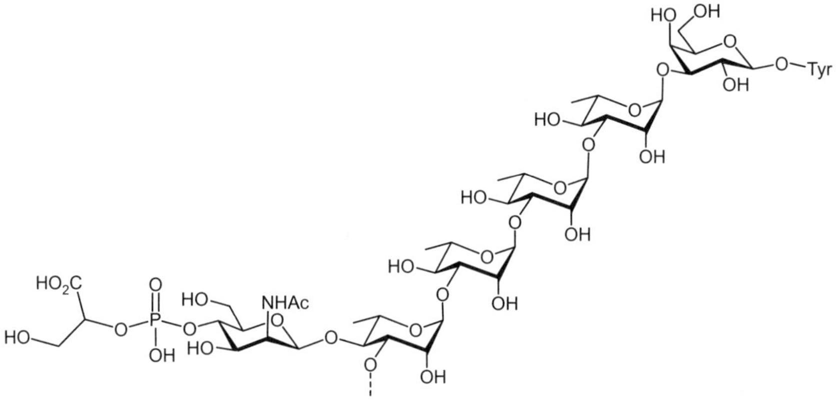

FIG. 4. Core structure of *P. alvei* CCM 2051 S-layer glycan.[135]

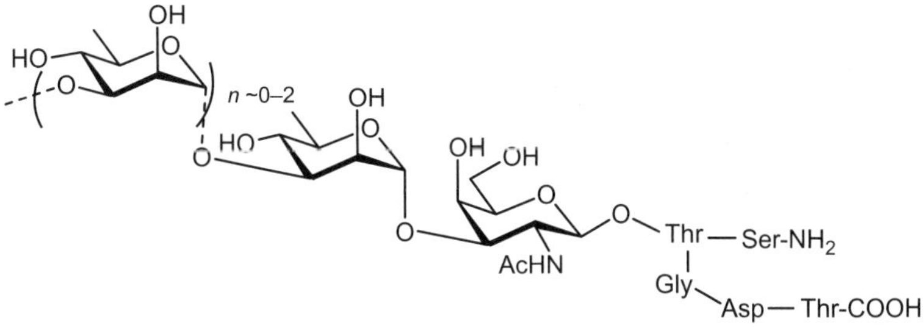

FIG. 5. Core structure of *Ane. thermoaerophilus* L420-91^T.[137]

(ii) S-Layer Glycans Without Core Oligosaccharides. A divergence from the tripartite building plan of the long-chain S-layer glycans from the *Bacillaceae* previously described was observed for the S-layer glycoproteins of Gram-positive *Lb. buchneri* strains. In *Lb. buchneri* 41021/251 and CD034, the S-layer glycans are short D-gluco homooligomers that are O-glycosidically linked to distinct serine residues within the specific amino acid sequence SSASSASSA (glycosylation sites are underlined) for glycosylation of the respective S-layer proteins (Fig. 6).[120,141]

Interestingly, the same *O*-glycosylation motif has also been found on other lactobacillar proteins of the strains investigated, which led to the first proposal of a general protein *O*-glycosylation system in lactobacilli.[141]

Also for the *Bacteroidales*, to which the Gram-negative species *B. fragilis* and *T. forsythia* are affiliated, the glycoprotein glycans so far known are short

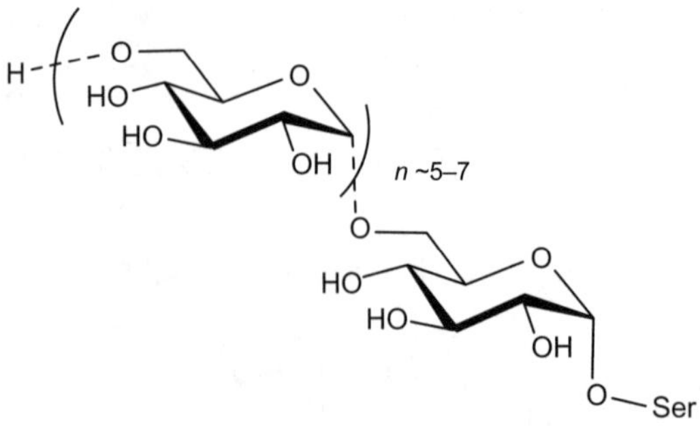

FIG. 6. S-Layer glycan of *Lb. buchneri*.[120]

oligosaccharides without repeats.[106,127] For these glycans, a phylum-wide protein *O*-glycosylation system targeting the amino acid motif D-(S/T)-(A/I/L/V/M/T) has been proposed.[106,127]

b. **S-Layer Glycans.**—At present, about 40 different S-layer glycoprotein glycan structures have been fully or at least partially elucidated. The structures observed and glycosidic linkage types exceed by far the display found in eukaryotes and broaden the spectrum of components found in bacterial polysaccharides, as previously reviewed in this series.[142] Most of the S-layer glycan chains of Gram-positive bacteria so far known are linear or branched homo- or heterosaccharides, which comprise 20–50 identical repeating units. In selected lactic acid bacteria (such as *Lb. buchneri* strains) as well as in the Gram-negative bacterium *T. forsythia* currently analyzed, S-layer glycans are short oligosaccharides without repeats comparable to archaeal glycans. The monosaccharide constituents of the S-layer glycan chains include a wide range of neutral hexoses, 6-deoxyhexoses, amino sugars, and uronic acids.[7] This spectrum of carbohydrates is further extended by rare sugars, including α-D-Rha*p*, α-D-Fuc*p*, α-D-Fuc*p*NAc, β-D-Qui*p*NAc, D-*glycero*-β-D-*manno*-Hep*p*, β-D-Man*p*NAcA, or Pse5Am7Gc. Some of them are otherwise typical constituents of LPS O-antigens of Gram-negative bacteria.[123] O-Glycosidic linkage regions observed on S-layer glycoproteins involve tyrosine, serine, and threonine residues (for reviews, see Refs. 74,116). As is known from eukaryotic glycoproteins,[4] there is no specific glycosylation sequence present for *O*-glycosylation of proteins.[143]

Despite being highly diverse on first analyses, most S-layer glycoproteins from the Gram-positive bacteria analyzed possess the tripartite structure, comprising the elongated glycan chain that is made of a distinct number of repeats, an adapter region, and the glycosidic linkage to the S-layer proteins (see Section II.1.a).

In the majority of glycans, L-rhamnose is the prevailing carbohydrate residue, as in the L-rhamnan found in the S-layer glycan of *G. stearothermophilus* NRS 2004/3a (Fig. 7).[131]

L-Rhamnose linked to mannopyranosyl residues has also been detected frequently as a component of linear and branched repeating units in glycans of *Thb. thermohydrosulfuricus* strains L111-69, L110-69 (DSM 568)[136] and S120-70 (Fig. 8) as well as *Thm. thermosaccharolyticum* D120-70 (Fig. 9),[139] respectively.

The backbone of the *Thm. thermosaccharolyticum* D120-70 glycan comprises a →4)-α-L-Rhap-(1→3)-β-D-Manp-(1→4)-β-L-Rhap-(1→3)-α-D-Glcp-(1→ tetrasaccharide with branching α-D-Galp-(1→2)- and β-D-Glcp-(1→6)- side-chain residues.[139]

The disaccharide L-rhamnose α-(1→2)-linked to D-fucose constitutes the repeating unit in the glycan of *Ano. tepidamans* GS5-97^T (Fig. 10),[132] whereas

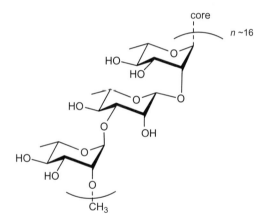

FIG. 7. S-Layer glycan structure of *G. stearothermophilus* NRS 2004/3a.[131]

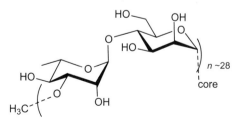

FIG. 8. S-Layer glycan structure of *Thb. thermohydrosulfuricus* strains L111-69 and L110-69.[136]

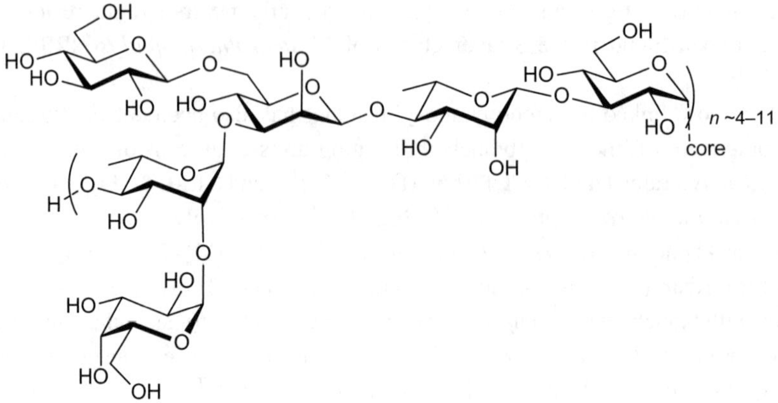

FIG. 9. S-Layer glycan structure of *Thm. thermosaccharolyticum* strain D120-70.[139]

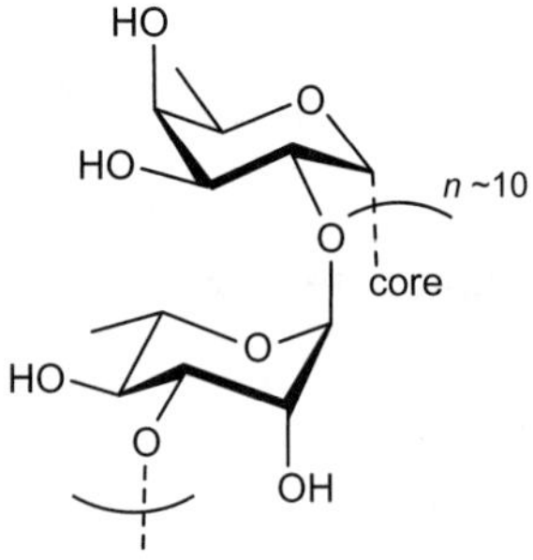

FIG. 10. S-Layer glycan structure of *Ano. tepidamans* GS5-97^T.[132]

Ane. thermoaerophilus DSM 10155 harbors L-rhamnose α-(1 → 3)-linked to D-*glycero*-β-D-*manno*-heptose—otherwise a common constituent of bacterial LPS in Gram-negative bacteria (Fig. 11).[133]

The glycan of *Ane. thermoaerophilus* strains L420-91^T and GS4-97, respectively, contains a branched hexasaccharide repeating unit of α-(1 → 2)-linked 3-acetamido-3-deoxy-D-fucose residues connected to a D-rhamnosyl tetrasaccharide backbone (Fig. 12).[137,144]

N-Acetyl-β-D-mannosamine, N-acetyl-D-glucosamine, and N-acetyl-D-galactosamine residues are present in S-layer glycans from *P. alvei* CCM 2051[134] (Fig. 13) and *Thb. thermohydrosufuricus* strains L77-66 (DSM 569) and L92-71, respectively (Fig. 14).[145]

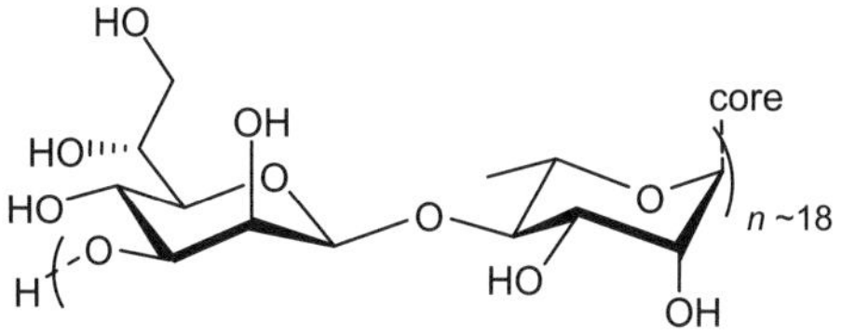

FIG. 11. S-Layer glycan structure of *Ane. thermoaerophilus* DSM 10155.[133]

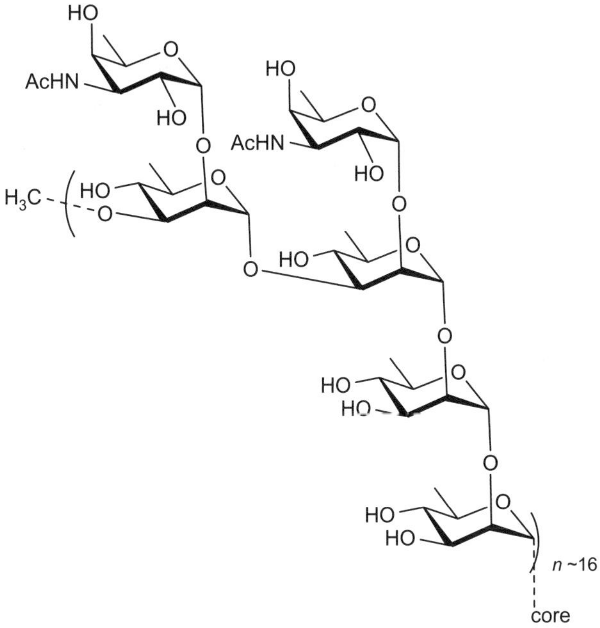

FIG. 12. S-Layer glycan structure of *Ane. thermoaerophilus* strains L420-91^T and GS4-97.[137,144]

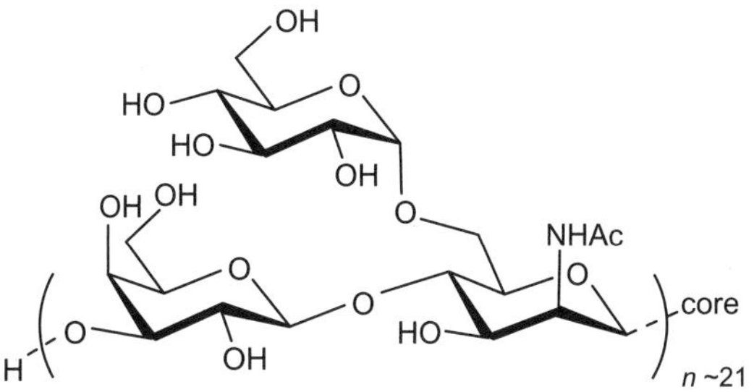

FIG. 13. S-Layer glycan structure of *P. alvei* CCM 2051^T.[134]

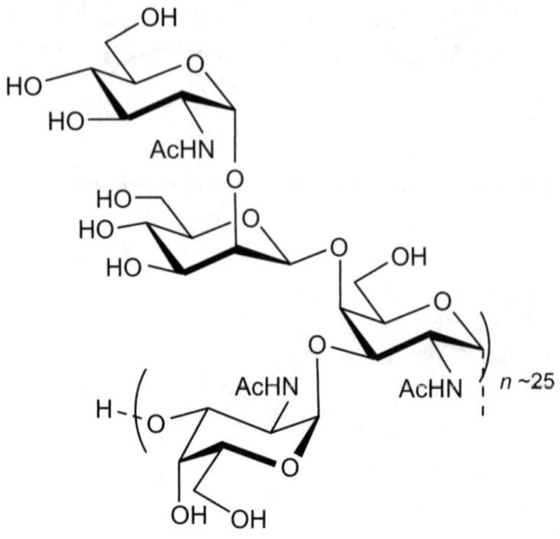

FIG. 14. S-Layer glycan structure of *Thb. thermohydrosufuricus* L77-66 (DSM 569).[145]

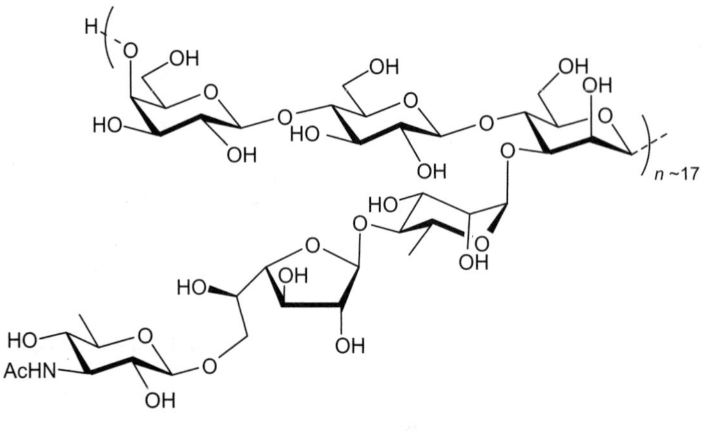

FIG. 15. S-Layer glycan structure of *Thm. thermosaccharolyticum* E207-71.[146]

A complex branched hexasaccharide repeating-unit containing the trisaccharide backbone →4)-β-D-Gal$p$-(1→4)-β-D-Glc$p$-(1→4)-β-D-Man$p$-(1→ has been identified in *Thm. thermosaccharolyticum* E207-71 (Fig. 15).[146] The repeating unit is extended at position 3 of the mannose residues by a branching trisaccharide sidechain β-D-Qui$p$3NAc-(1→6)-β-D-Gal$f$-(1→4)-α-D-Rha$p$.

The overall structure of these glycans, in combination with the broad spectrum of constituent sugars, lets us interpret the highly variable S-layer glycoproteins as the

Gram-positive equivalents of lipopolysaccharides, with the protein component replacing lipid A of Gram-negative bacteria.[123]

As already mentioned, an exception among the *Bacillaceae* is the S-layer glycan of *Thb. thermohydrosulfuricus* S102-70, which consists only of a hexasaccharide β-D-Gal*f*-(1→3)-α-D-Gal*p*-(1→2)-α-L-Rha*p*-(1→3)-α-D-Man*p*-(1→3)-α-L-Rha*p*-(1→3)-β-D-Glc linked to specific tyrosine residues of the S-layer proteins. The nonreducing end of the chain possesses a (less common) β-D-galactofuranose residue as the terminating sugar (Fig. 16).[140] The reason for this truncation is not known.

The other short-chain S-layer glycans from Gram-positive strains are those of *Lb. buchneri* strains.[120,141] In contrast to the previously mentioned S-layer glycans in the other bacilli, the D-glucose residue linking the S-layer glycan to the S-layer protein is in the α configuration (Figure 6). This observation might reflect a different biosynthetic pathway (see Section V).

For *T. forsythia*, S-layer glycosylation is an important factor in defining the bacterium's lifestyle. This S-layer glycan constitutes a complex and so far unique oligosaccharide, having the structure 4-*O*-Me-β-Man*p*NAc6CONH$_2$-(1→3)-[Pse5Am7Gc-(2→4)]-β-Man*p*NAcA-(1→4)-[4-*O*-Me-α-Gal*p*-(1→2)]-α-Fuc*p*-(1→4)-[α-Xyl*p*-(1→3)]-β-Glc*p*A-(1→3)-[β-Dig*p*-(1→2)]-α-Gal*p*, which is O-glycosidically linked to multiple serine and threonine residues within the *Bacteroidales* protein glycosylation motif of the two S-layer proteins constituting the S-layer (Fig. 17).[106] As with the anomeric configuration of the linkage sugar in S-layer

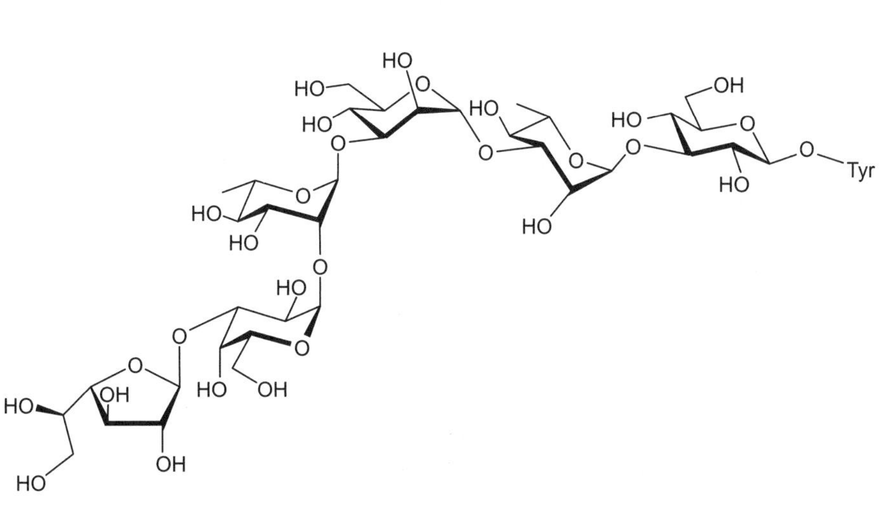

FIG. 16. S-Layer glycan structure of *Thb. thermohydrosulfuricus* S102-70.[140]

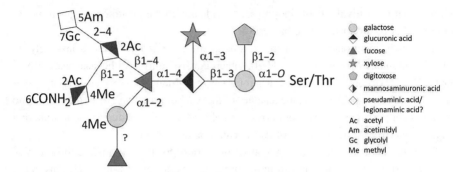

FIG. 17. Schematic structure of the *T. forsythia* S-layer glycan.[100,106] Adapted from the open access journal *Biomolecules*; © 2012 by the authors, licensee MDPI, Basel, Switzerland; http:/creativecommons.org/licenses/by/3.0. (For color version of this figure, see the Color Insert in this volume and the online version of this chapter.)

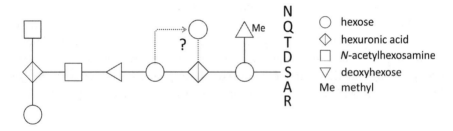

FIG. 18. Schematic putative structure of the *B. fragilis* glycan.[127]

glycans of lactobacilli, the linkage sugar galactose of the *T. forsythia* glycoproteins is also in the α configuration.

Mass-spectrometric analysis of native *B. fragilis* glycoproteins after subjection to β-elimination revealed the glycan to be an oligosaccharide consisting of nine sugar residues (Fig. 18).[127] Notably, the proposed structure resembles that of the *T. forsythia* O-glycan in several aspects. In both glycans, attachment to the protein occurs via a hexose residue that is succeeded by a hexuronic acid. Additionally, a nonpolar constituent branches from the first sugar of the *B. fragilis* O-glycan. Also, both glycans contain a deoxyhexose residue in their linear structure. As in the *T. forsythia* glycan, the deoxyhexose of the *B. fragilis* glycan most likely is a fucose residue, as glycoproteins of *B. fragilis* are readily detected with *Aleuria aurantia* lectin and depend on genes for GDP-fucose biosynthesis.[126]

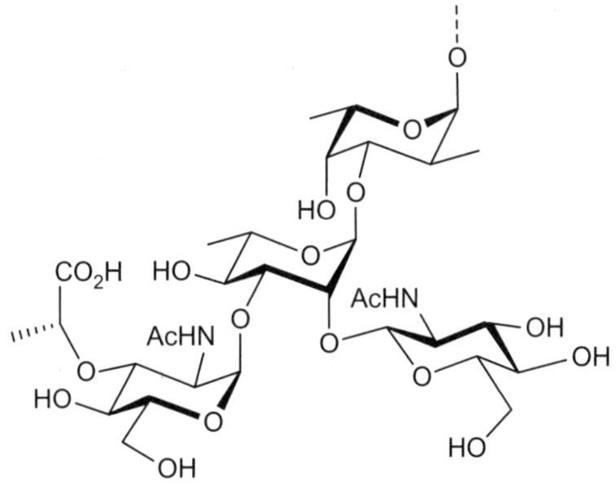

FIG. 19. Terminal end of the S-layer glycan of *Ano. tepidamans* GS5-97^T.[132]

c. Terminal Groups and Noncarbohydrate Substituents.—Frequently, the terminal sugar residue at the nonreducing end of long-chain S-layer glycans is capped with noncarbohydrate substituents (see section II.1.b.). Capping with 2-*O*-methyl groups is observed on the S-layer glycans of *G. stearothermophilus* NRS 2004/3a[131] (Fig. 7) or *Thb. thermohydrosulfuricus* L111-69[136] (Fig. 8), while 3-*O*-methylation has been found in the distal units of the *Ane. thermoaerophilus* L420-91^T (DSM 10154) and GS4-97[137] glycans (Fig. 12), respectively. On the S-layer glycan of *Ano. tepidamans* GS5-97, an unusual capping structure was observed, namely, an α-(1 → 3)-linked *N*-acetylmuramic acid and a β-(1 → 2)-linked GlcNAc residue (Fig. 19).[132]

By analogy to pathways for LPS biosynthesis, these capping structures are assumed to be involved in chain-length determination of the polysaccharides during biosynthesis; however, no conclusive data are yet available.[111] Since not all S-layer glycans are terminated by such structures, alternative mechanisms may exist. In the case of LPS, the O-PS chain length is strictly regulated, since it plays a pivotal role for the protection of the bacterium from defense mechanisms in the host. In the biosynthesis pathway for the Wzy-dependent O-antigen, the integral inner-membrane protein Wzz determines the O-antigen chain length.[147]

2. Archaeal S-Layer Glycoproteins

During the past few years, only a few new structures of archaeal S-layer glycans have been reported, and the initial work on S-layer glycosylation of haloarchaea and methanogens is summarized in several excellent reviews (see Refs.

16,17,113,148,149,149a). There has been, however, considerable progress made in the functional characterization of the carbohydrate-active proteins involved in archaeal S-layer-protein glycosylation. In an impressive series of research papers, Eichler and coworkers reinvestigated the structure of the S-layer glycan of *Haloferax volcanii* and established the biosynthesis pathway for an S-layer pentasaccharide.[150–152] They further discovered that lipid modification resulted in the formation of two distinct populations of S-layer glycoproteins.[153] These observations are consistent with the S-layer glycoprotein being synthesized initially as an integral membrane protein and subsequently undergoing a processing event in which the extracellular portion of the protein is separated from the membrane-spanning domain and transferred to a "waiting" lipid moiety. In a comparative approach, protein N-glycosylation in *Haloarcula marismortui*, a second haloarchaeon also originating from the Dead Sea, was investigated.[154] While both species decorate the respective S-layer glycoprotein with the same N-linked pentasaccharide and employ dolichyl phosphate as the lipid glycan carrier, species-specific differences in the two N-glycosylation pathways exist. *Har. marismortui* first assembles the complete pentasaccharide on dolichyl phosphate and only then transfers the glycan to the target protein. In contrast, *Hfx. volcanii* initially transfers the first four pentasaccharide subunits from a common dolichyl phosphate carrier to the target protein and subsequently delivers the final pentasaccharide from a distinct dolichyl phosphate to the N-linked tetrasaccharide.[154] To cope with life in changing hypersaline environments, *Hfx. volcanii* is able to modulate not only the N-linked glycan structures decorating the S-layer glycoprotein but also the sites of posttranslational modification.[114] This high degree of inherent structural flexibility makes this organism also a potent candidate for glycoengineering.[154]

In the course of structural investigations of glycosylated flagellins from methanogenic archaea, Jarrell and coworkers made the interesting observation that not only the flagellin but also the glycosylated S-layer protein of *Methanococcus voltae* were decorated with the same complex N-linked glycan, where one of the constituting glycan components (β-ManNAcA) having a carbonyl group at C-6 forms an amide bond with the amino group of a threonine residue.[155,156] In other strains, such as *Mco. maripaludis*, glycan structures of even greater complexity, containing also a novel terminal sugar residue (5S)-2-acetamido-2,4-dideoxy-5-O-methyl-α-L-*erythro*-hexos-5-ulo-1,5-pyranose, have been identified in flagellins and potentially in S-layer glycoproteins.[112,157–159] This common usage of identical glycan structures on S-layers and flagella was first reported with the haloarchaeon *Hbt. salinarum*.[31]

Subsequently, it was shown that the oligosaccharide of the S-layer of the thermoacidophilic crenarchaeote *Sulfolobus acidocaldarius* consists of a tri-branched

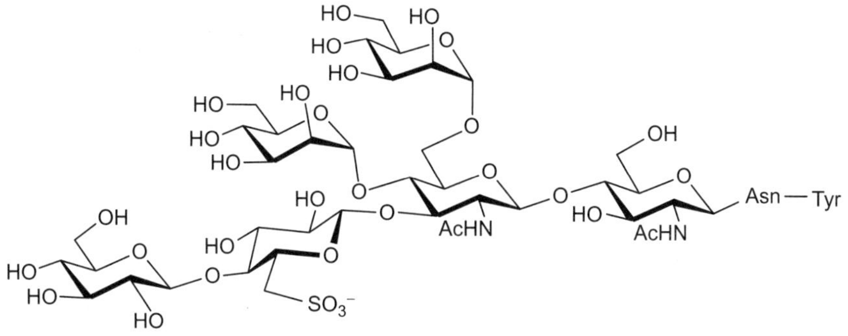

FIG. 20. Structure of a glycopeptide from cytochrome $b_{558/566}$ of *Sl. acidocaldarius*.[161]

hexasaccharide that is *N*-glycosylically linked via a chitobiose core to several sites at the S-layer protein.[160] Furthermore, this glycan contains a 6-sulfoquinovose residue, which was first described in the cytochrome $b_{558/566}$ of *Sl. acidocaldarius* (Fig. 20).[161]

A gene cluster involved in the biosynthesis of sulfoquinovose and the assembly of the S-layer N-glycans was identified.[162] In archaea, *N*-glycosylation relies also on phosphorylated dolichol. The organism *Sl. acidocaldarius*, however, contains an unusually short, highly reduced dolichyl phosphate, which presents a degree of saturation thus far not reported in any other organism.[163]

Regarding anchoring of the archaeal S-layer proteins to the cell envelope, in many strains the S-layer entirely penetrates the cytoplasmic membrane, allowing for a stable positioning of the S-layer protomers in the membrane.[113] This principle is completely different from bacterial S-layer proteins, where carbohydrate interactions obviously play a dominant role in S-layer anchoring, either via secondary cell-wall polymers (SCWPs) as in Gram-positive bacteria[164–167] or via rough lipopolysaccharide, as suggested for Gram-negative bacteria.[168]

III. Nonclassical Secondary Cell-Envelope Polysaccharides

1. Background

As established about three decades ago, cell-wall polysaccharides of Gram-positive organisms can be classified on the basis of their structural characteristics into three distinct groups: (i) teichoic acids,[169] (ii) teichuronic acids,[170] and (iii) other neutral or acidic polysaccharides that cannot be assigned to the two former groups.[171,172] Since all these compounds play a secondary role in cell-wall function, they have been termed "secondary" cell-wall polymers (SCWPs). Biochemical and genetic data accumulated over the past 30 years indicate that the first two groups of these polysaccharides (classical SCWPs) play important roles in normal cell function.[169,173]

In the course of later investigations of bacterial S-layer proteins (see Section II), novel aspects of a group of nonclassical SCWPs have emerged with regard to their structure and function. S-Layers are generally composed of identical (glyco)protein species, forming regular two-dimensional, lattices on the surfaces of bacterial cells. It can be assumed that the stable attachment of S-layer (glyco)proteins to the cell wall is important for the cell, and in this context, nonclassical SCWPs have been identified as mediators for noncovalent attachment of S-layers to the underlying PG meshwork (Fig. 1).[163,165,174,175] Herein we summarize the current knowledge on structural features, possible linkage types to the cell wall, and interactions of nonclassical SCWPs in S-layer-carrying organisms.

2. The Nonclassical Group of SCWPs

Besides the considerable body of knowledge that has accumulated concerning structural, biochemical, and immunological features of teichoic and teichuronic acids from cell walls of Gram-positive bacteria,[169,170] it should be kept in mind that additional cell-wall polysaccharides may be present in these organisms. Our research has focused on glycosylated S-layer proteins (see Section II), which are regarded as ideal model systems for studying the glycosylation of prokaryotic proteins. Improved purification and separation methods, however, eventually led to the clear assignment of a distinct proportion of compounds (originally copurified) as a separate class of SCWPs, and this accounts for a substantial proportion (7–15% by weight) of the PG of the organisms investigated. The SCWP–PG complexes investigated comprise the intact glycan moieties, together usually with proportions of PG of variable size, arising because of random degradation during the preparation procedure (without any acid treatment to prevent undesired removal of acid-labile components).

a. Common and Variable Features of Nonclassical SCWPs from *Bacillaceae*.— Based on compositional and structural data, we suggest that classification of the nonclassical SCWPs from S-layer-carrying *Bacillaceae* into group (iii) as defined by Araki and Ito is the most appropriate.[171] Recent investigations have shown that, in S-layer-carrying organisms, teichoic and teichuronic acids as typical representatives of classical SCWPs are not present. Instead, the (to some extent, structurally comparable) nonclassical SCWPs act as linkers for the random noncovalent attachment of

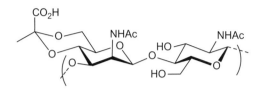

FIG. 21. Repeating unit of the SCWP from *P. alvei* CCM 2051^{T}.[176]

the S-layers to carbon 6 of muramic acid residues of the PG (compare with Fig. 1, left panel). This information can now be summarized as follows:

First group: The structure of the glycan portion of the SCWP–PG complex of *P. alvei* CCM 2051^T was elucidated to be [(Pyr4,6)-β-D-ManpNAc-(1→4)-β-D-GlcpNAc-(1→3)]$_{n\sim 11}$-(Pyr4,6)-β-D-ManpNAc-(1→4)-α-D-GlcpNAc-(1→.[176] Each repeating-unit disaccharide of this SCWP is substituted with 4,6-linked pyruvic acid residues, conferring the overall anionic character of the SCWP (Fig. 21).

Upon prolonged exposure of the polymer sample to acidic conditions, as occurs upon solvation in D$_2$O during NMR experiments, considerable loss of pyruvate residues can occur. The ManNAc–GlcNAc backbone disaccharide motif corresponds to that frequently observed in certain teichoic acids of other bacilli.[171] However, in the nonclassical SCWPs, this motif is repeated several times, thus constituting the entire glycan moiety of those SCWPs. Pyruvic acid-containing SCWPs have also been reported for the S-layer-carrying organisms *L. sphaericus* CCM 2177[177] and *B. anthracis*.[164] So far, however, no information is available on either their full structures or their linkage to the PG layer.

In contrast to anionic polymers, neutral polysaccharides possessing the identical backbone motif are found in the PG of other *Bacillaceae*. The SCWPs of *Thm. thermosaccharolyticum* strains D120-70[138] and E207-71[178] display the commonly encountered →3)-β-D-ManpNAc-(1→4)-β-D-GlcpNAc-(1→4)-β-D-ManpNAc-(1→3)-β-D-GlcpNAc-(1→ motif.[178] Although previously it had been thought that in strain D120-70 this motif is extended by galactose residues,[138] later results show that alternating ManNAc residues are substituted by ribofuranose side chains (C. Schäffer, H. Kählig, R. Christian, and P. Messner, unpublished results), as has also been demonstrated for the SCWP of strain E207-71 (Fig. 22).[178]

Second group: The first SCWP–PG complex of an S-layer-carrying organism for which the structure was completely elucidated was purified from *G. stearothermophilus* NRS 2004/3a.[179,180] The anionic polymer, comprising on average six tetrasaccharide repeating units having the structure →4)-β-D-Manp-2,3-diNAcA-(1→6)-α-D-Glcp-(1→4)-β-D-Manp-2,3-diNAcA-(1→3)-α-D-GlcpNAc(1→, constitutes the SCWP of

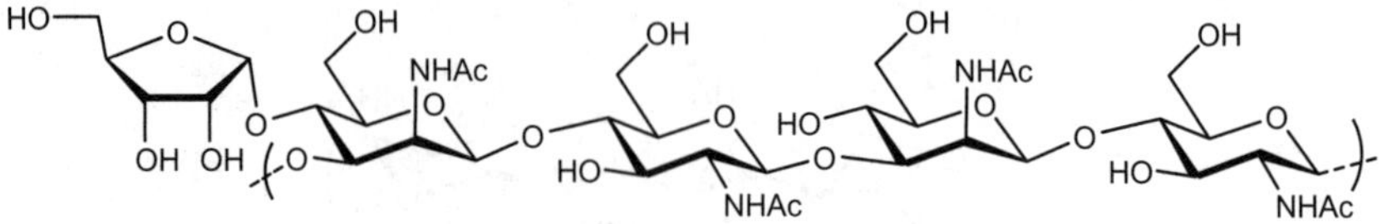

FIG. 22. Structure of the SCWP from *Thm. thermosaccharolyticum* strain E207-71.[178]

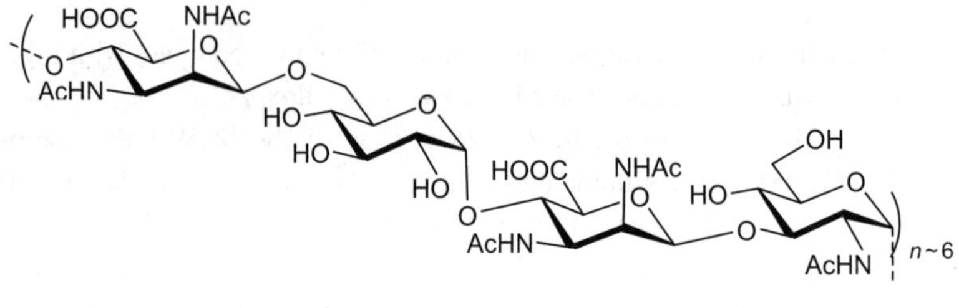

FIG. 23. Structure of the SCWP from *G. stearothermophilus* NRS 2004/3a.[179,180]

many *G. stearothermophilus* wild-type strains (Fig. 23). Preliminary analysis of that SCWP indicated a mass range of 4000–6000. About 20–25% of the muramyl residues are substituted by SCWP glycans.[179,180] Interestingly, analyses of the SCWP of *Ano. tepidamans* GS5-97^T indicate that its backbone structure is reminiscent of that of *G. stearothermophilus* NRS 2004/3a, but with additional modifications of the carboxyl groups by amide formation at the Man*p*-2,3-diNAcA residues, turning the anionic character of the glycan into a neutral one. ^{31}P NMR analysis has demonstrated that the SCWP of *Ano. tepidamans* GS5-97^T is linked to muramic acid by a common phosphodiester linkage.[181]

Third group: In this group, the linkage-region disaccharide (compare with the *first group*) of teichoic acids is much further extended. In addition to linear chains, branched repeating units containing neutral sugars and/or amino sugars have also been observed. The charge-neutral SCWP isolated from *Ane. thermoaerophilus* DSM 10155 represents a unique bacterial glycan structure containing repeats of the sequence →3)-α-D-Glc*p*NAc-(1→3)-β-D-Man*p*NAc-(1→4)-β-D-Gal*p*NAc-(1→ (Fig. 24).[87] The interpretation of the observed NMR data supporting a single branched GalNAc unit, however, should be treated with caution, since the ^{13}C NMR chemical shifts of the carbon atoms in the branched residue are very similar to those located in the linear chain.

A branched structure was found in the SCWP from *G. stearothermophilus* PV72/p2, which, however, had undergone treatment with HF for cleavage of the linkage to

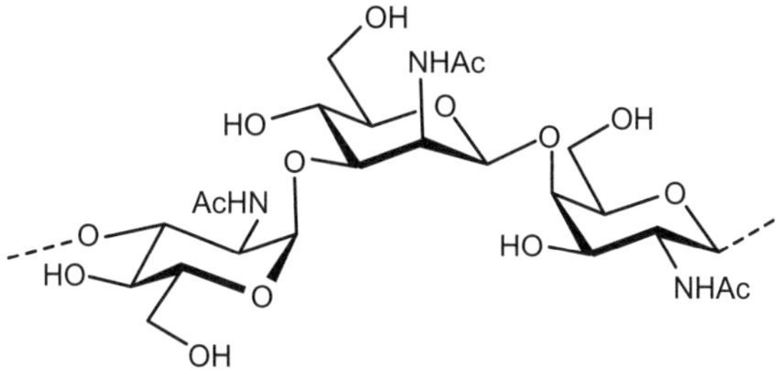

FIG. 24. Structure of the repeating unit of the SCWP from *Ane. thermoaerophilus* DSM 10155.[87]

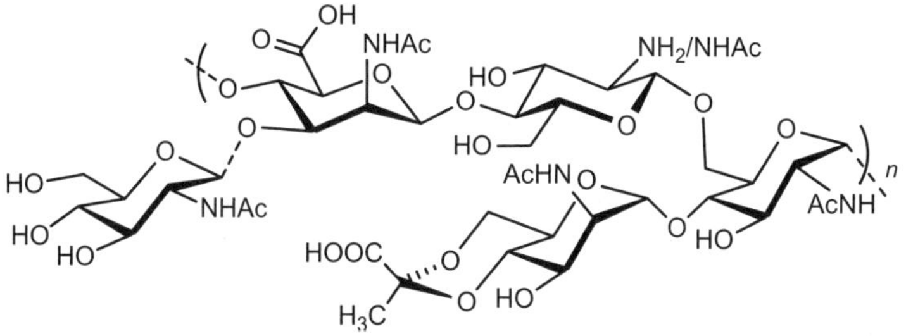

FIG. 25. Structure of the repeating unit of the acid-degraded SCWP from *G. stearothermophilus* PV72/p2.[182] Reprinted from Ref. 182. Copyright (2008), with permission from Elsevier.

peptidoglycan. Hence information on the linkage region as well as the extent of pyruvate substitution and proof for *N*-deacetylated glycoses is compromised (Fig. 25).[182]

The SCWP contains a pentasaccharide repeating unit of the sequence →4)-[β-D-Glc$p$NAc-(1→3)]$_{\sim0.3}$-β-D-Man$p$NAcA-(1→4)-β-D-Glc$p$N/NAc-(1→6)-[4,6-(*S*)-Pyr-α-D-Man$p$NAc-(1→4)]-α-D-Glc$p$NAc-(1→. Similar to the SCWPs of the *first group*, the (*S*)-configured pyruvate residues are present on the D-Man$p$NAc residues, albeit not within the linear backbone chain, but extending from the branching 4,6-disubstituted α-D-Glc$p$NAc residue.

Secondary cell-wall polymers of *Bacillus anthracis* and *Bacillus cereus* may also be classified into the third group. *B. anthracis*, the causative agent of anthrax disease, harbors antigenic cell-surface carbohydrate determinants which may be utilized for diagnostic purposes and vaccine development, respectively. Nonclassical SCWPs of *B. anthracis* having approximate molecular masses in the range of 12–22 kDa mediate binding to the surface-layer homology (SLH) domain of the paracrystalline surface-array protein Sap and the extractable antigen 1, respectively.[183,184] The SCWP of several *B. anthracis* strains was released from its covalent linkage to peptidoglycan by treatment with HF, and the resulting purified fractions were analyzed by high-field NMR spectroscopy and by mass spectrometry techniques.[185] The repeating unit of the SCWP comprises the backbone sequence →6)-α-D-Glc*p*NAc-(1→4)-β-D-Man*p*NAc-(1→4)-β-D-Glc*p*NAc-(1→ with α-galactopyranosyl sidechain units linked to O-3 of the α-D-Glc*p*NAc and β-D-Glc*p*NAc units, respectively, and a β-D-Gal*p* residue linked to O-4 of the α-D-Glc*p*NAc residue (Fig. 26). The trisaccharide backbone forms a consensus motif shared by many *B. anthracis* and *B. cereus* strains, respectively. Cross-reactive epitopes are present in these SCWPs as well as epitopes being specific for *B. anthracis* and also *B. cereus*, respectively.[185,186]

The β-D-galactosyl moiety attached to the nonterminal GlcNAc residue constitutes a unique binding epitope to endolysins from *B. anthracis* bacteriophages.[187,188] Galactosylation, however, is completely absent in the SCWP of the avirulent strain *B. anthracis* CDC 648 (Ba684).[189] Additional modifications observed in that SCWP pertain to acetylation at O-3 and unsubstituted amino groups of the β- and α-configured glucosamine residues, respectively, at the terminal, nonrepeating sequence.

A structure similar to the *B. anthracis* SCWP has been reported to constitute the SCWP of pathogenic *B. cereus* strains G9242, 03BB87, and 03BB102, which features an additional substitution at O-3 of the β-ManNAc unit by an α-galactosyl residue in the former two strains (Fig. 27).[190]

Unique structural features have been found for *B. cereus* strains ATCC 10987 and ATCC 14579.[191] The trisaccharide backbone is composed of the repeating unit →6)-α-D-Gal*p*NAc-(1→4)-β-D-Man*p*NAc-(1→4)-β-D-Glc*p*NAc-(1→ substituted by a single β-D-Gal*p* residue at O-3 of the Gal*p*NAc unit and an *O*-acetyl group located at O-3 of the central β-D-ManNAc moiety (Fig. 28).

FIG. 26. Structure of HF-treated *B. anthracis* SCWP.[185]

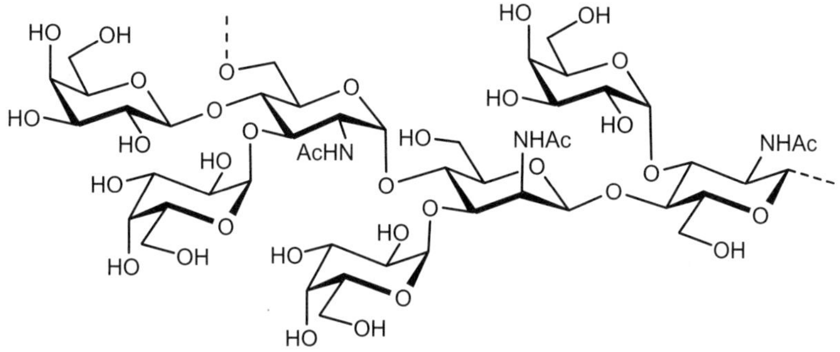

FIG. 27. Structure of the HF-treated SCWP from *B. cereus* strains G92141 and 03BB87.[190]

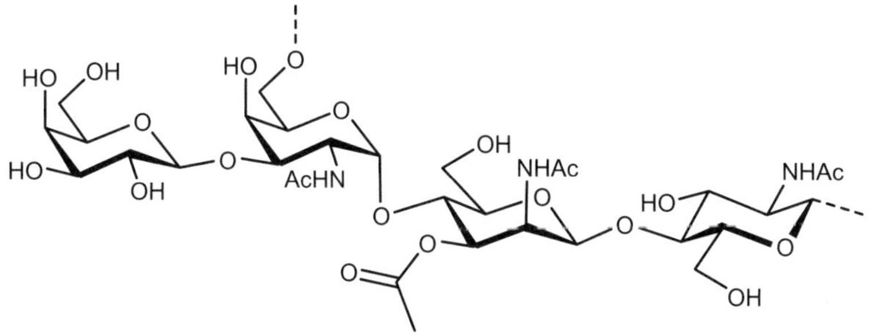

FIG. 28. Structure of the HF-treated SCWP from *B. cereus* ATCC 10987.[191] Adapted from Ref. 191. © The American Society for Biochemistry and Molecular Biology.

B. cereus strain ATCC 14579 produces two structurally different SCWPs, depending on environmental conditions.[192] The structure of the first SCWP, which is constantly being expressed, is closely related to the canonical sequence motif comprising the sequence of β-D-Glc*p*NAc-α-D-Hex*p*NAc-β-D-Man*p*NAc (as in *B. cereus* strain 10987), but it contains a highly branched α-D-Gal*p*NAc residue (Fig. 29).

The second, completely unrelated, SCWP was identified as being present in a phase-specific manner, both in the planktonic phase and in late biofilm formation. The presence of both SCWPs, as well as the enrichment of the latter form in biofilm, was established by high-resolution (HR)-MAS NMR spectroscopy. The charged second SCWP could be separated from the neutral SCWP by anion-exchange chromatography as well as by mild acid hydrolysis. The SCWP contains a 4-amino-4,6-dideoxy-D-Gal*p*NAc residue extended by an amide-linked α-2-(*R*)-hydroxyglutaric acid residue (Fig. 30).

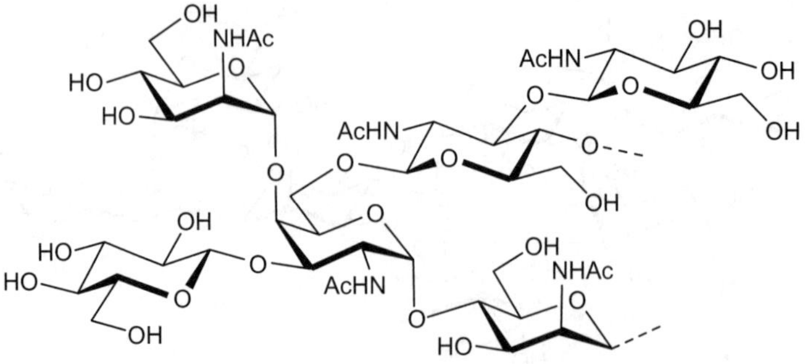

FIG. 29. Structure of the HF-treated SCWP from *B. cereus* ATCC 14579.[192]

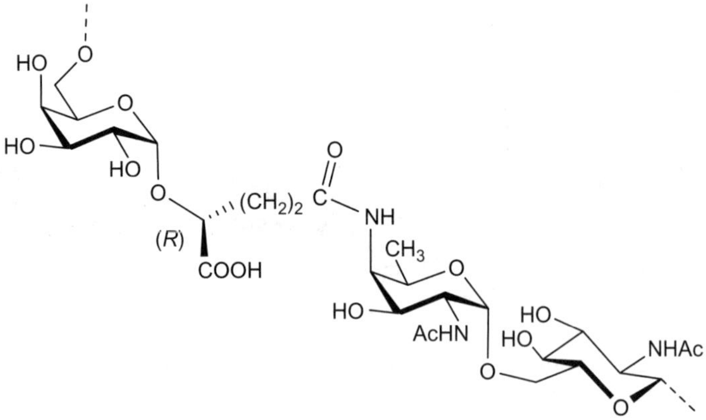

FIG. 30. Structure of the charged SCWP from *B. cereus* ATCC 14579.[192]

Pyruvate substitution has been detected in the HF-treated SCWPs from *B. anthracis* (Sterne 34F$_2$, 7702) and *B. cereus* strain G9241, 03BB87, 03BB102.[190] The extent of pyruvate substitution remains ambiguous, since the cleavage of the SCWP from peptidoglycan with 48% HF also leads to the liberation of the acid-sensitive ketal groups, thus preventing any quantification and even detection of pyruvate. For the *B. anthracis* SCWP CDC 684, the location of a terminal (*S*)-pyruvate substituent at O-4 and O-6 of the β-D-Man*p*NAc residue has been identified by NMR spectroscopy.[189]

b. General Considerations About Nonclassical SCWP Structure.—Comparison of the overall composition of all nonclassical SCWPs analyzed so far reveals that the more-complex glycans of the *second group* exhibit the same alternating order of *gluco* and *manno* sugars as glycans of the *first group*. The glycan chain starts with a GlcNAc residue at the reducing end and ends with Man*p*-2,3-diNAcA. It is possible that glycan structures of the *second group* have evolved from the simpler *first-group* structures by the introduction of such residues as glucose or Man*p*-2,3-diNAcA through the action of strain-specific enzymes. The same scenario could be imagined for glycans of the *third group*. However, biosynthesis pathways of greater complexity have to be envisaged and, so far, no experimental evidence is available to support this notion.

The linkages between SCWPs and PG are currently being investigated in more detail. Interestingly, for all SCWPs investigated, the glycose residue linked to the bridging phosphate residue is in the α configuration. Whether this has any impact on the biosynthesis of the SCWPs remains to be established.

   c. Interactions of SCWPs and S-Layers from *Bacillaceae*.—In addition to the general features previously mentioned, novel properties for SCWPs have emerged from research on S-layers.[67,69] S-Layers are two-dimensional crystalline protein lattices on the outermost surface layer of bacteria from almost all phylogenetic branches.[67,69] The S-layer proteins are of interest because of their involvement in important physiological processes, which include, in the case of pathogenic bacteria, the infection mechanism.[193] If present, S-layer (glyco)proteins are the most abundant cellular proteins. Provision of any kind of selection advantage by the S-layer to a bacterium in its natural environment requires the stable attachment of the S-layer to the cell surface *in vivo*. During evolution, Gram-positive bacteria have developed various strategies for displaying proteins on their surface. These strategies include predominantly covalent binding of LPXTG-carrying proteins to PG, but a number of noncovalent binding strategies have also been developed.[193] Reattachment experiments of isolated S-layer glycoproteins from thermophilic clostridia revealed the noncovalent character of the interaction between S-layer and PG.[194] For *Lb. buchneri*, it was shown very early that hydroxyl groups of a neutral cell-wall polysaccharide are responsible for the attachment of the S-layer protein to the cell wall.[195]

   d. Possible Cell-Wall-Targeting Mechanisms.—Recently, the cell-wall-targeting mechanism of S-layer proteins has been investigated in more detail.[164,165,167,174] These studies suggest that, in general, S-layer proteins have two functional regions: a cell-wall-targeting domain, which in most of the organisms thus far investigated is located at the N-terminus, and a C-terminal self-assembly domain. The existence of a cell-wall-targeting domain in S-layer proteins was substantiated by Fujino *et al.*[196] and Lupas *et al.*,[197] who identified motifs of approximately 55 amino acids, containing 10–15 conserved residues, which were designated SLH (S-layer homology) domains. These SLH domains, usually composed of one to three modules, are the means frequently used for targeting proteins to the cell surface. They are found not only in various S-layer proteins but also in many other surface-associated proteins, such as the cellulosomes[198] or other surface-associated enzymes.[199] In the case of *B. anthracis*, the etiological agent of anthrax, the molecular basis of the interaction between SLH domains and SCWPs was established by elucidating their binding properties.[164] Both S-layer proteins of *B. anthracis* (EA1 and Sap) possess SLH domains that bind the S-layer proteins directly to PG.[200,201] While the overall sequence similarity of SLH domains is rather low, the highly conserved four amino acid motif TRAE has been found to play a key role for the binding function to SCWP.[167,202,203] The crystal structure recently elucidated of one S-layer protein from *B. anthracis* shows the SLH domains arranged in threefold pseudo-symmetry with the TRAE (or similar) motifs arranged so that they would be accessible to the SCWP.[203]

A different binding mechanism was proposed for the S-layer proteins of *G. stearothermophilus* PV72/p2[204,205] and *P. alvei* CCM 2051[T].[167] In these organisms, two different binding domains were identified in the N-terminal region of the S-layer proteins, one for SCWP and another for PG. In *Thermus thermophilus*, a similar binding mechanism was identified between an S-layer–outer membrane complex and the cell wall. There is a strong interaction of the SLH domain of the S-layer protein with a pyruvated component of a highly immunogenic SCWP.[174]

Originally, the involvement of pyruvate groups in S-layer binding was inferred from observations in *B. anthracis*.[164] It was demonstrated that the pyruvate transferase CsaB is involved in the addition of pyruvate to a PG-associated polysaccharide fraction and that this modification is necessary for the binding of the S-layers via SLH domains. Interestingly, the *csaAB* operon was found to be present in several bacterial species.[164] Pyruvate was also identified in the SCWP repeating units of *L. sphaericus* CCM 2177[177] and *P. alvei* CCM 2051,[176] which may be taken as an indication that pyruvate, or more generally speaking, negative charges of SCWPs, constitutes a widespread mechanism for the anchoring of S-layer proteins containing SLH domains to the bacterial cell wall. Besides the anchoring mechanism involving SLH domains, another mechanism that possibly utilizes basic amino acids, present in the cell-wall-targeting region and known for their direct interaction with carbohydrates, may apply for S-layer proteins devoid of SLH domains. In the genus *Geobacillus*, SLH domains have only been identified on the S-layer protein SbsB of *G. stearothermophilus* PV72/p2.[206] All other investigated *G. stearothermophilus* strains possess S-layer proteins devoid of SLH domains.[74]

Interestingly, none of the SCWPs of the organisms that possess S-layer proteins lacking SLH domains are modified with pyruvate groups, and some of them have a net-neutral charge.[166] These observations support the notion that, in addition to the previously discussed involvement of pyruvate, other mechanisms can be involved in the binding of S-layer proteins to the PG.[164,165,167,174] In addition, the nonconserved character of S-layer-binding mechanisms is shown by the observation that the cell-wall-targeting domain is not necessarily located in the N-terminal region of the S-layer protein. Well-documented examples of C-terminal anchoring are the S-layer proteins of *Lactobacillus acidophilus* ATCC 4556[207] and *Lactobacillus crispatus*.[208]

The diversity observed among different SCWP structures of *Bacillaceae* follows a general theme that is well known from other cell-surface structures, such as the serotypes of lipopolysaccharides[123] and capsular polysaccharides.[209] Presumably, this diversity is responsible for creating microenvironments in which different organisms can survive under unfavorable conditions. Current data indicate that nonclassical SCWPs function as mediators for anchoring S-layer (glyco)proteins from *Bacillaceae* to the bacterial cell wall. The complete elucidation of both the structure and biosynthesis of several nonclassical SCWPs would contribute to our general understanding of the various mechanisms underlying the tethering of S-layer (glyco)proteins to the cell surface of Gram-positive bacteria.

IV. Structural Analysis

1. Isolation of Cell-Envelope Polysaccharides and Glycopeptides

Cell-envelope polysaccharides and glycopeptides are characterized by a high degree of structural variability regarding their carbohydrate constituents (resembling the broad spectrum of sugars found in bacterial lipopolysaccharides), interunit linkages, and glycosylation sites of the protein, which are distributed over the whole S-layer protein but are mainly present within the proposed self-assembly domains.[74,117] While O-linked polysaccharides seem to be limited only to S-layer glycoprotein glycans, N-glycans have also frequently been found in archaea. Usually, the polysaccharides harbor approximately 15–50 repeating units as linear or branched homo- and heteroglycans, respectively. Structural analysis has also to take into account the microheterogeneity of both the glycan and the peptide domain. In general, the isolation of inherently water-insoluble S-layer glycoproteins has been achieved by extraction from purified cell-wall preparations, employing such chaotropic agents, as 5 M LiCl, 5 M guanidine hydrochloride, or 6 M urea, followed by pronase or tryptic digestion and removal of the peptide fragments by passage over a cation-exchange resin. Subsequent final purification is usually accomplished by size-exclusion chromatography and reverse-HPLC separation, resulting in glycopeptide fractions amenable to structural analysis. Release of the O-glycans can eventually be induced by β-elimination. Hydrolysis to give the individual monosaccharide constituents has typically been performed by treatment with 4 M HCl or 2 M trifluoroacetic acid at 110 °C for several hours. The quantitation of the glycan components was achieved by high-performance anion-exchange chromatography with pulsed amperometric detection (HPAEC PAD) or by gas chromatography–mass-spectrometric analysis of the trimethylsilylated methyl glycosides. The absolute configuration of the monosaccharides was based on GC–MS data of the TMS 2-butyl derivatives (for a review, see Ref. 7).

SCWPs usually exhibit an overall molecular mass in the range of 4–6 kDa.[175] The polymers were mostly released from the murein-containing sacculus by treatment with 48% hydrofluoric acid at low temperature, which cleaves the phosphodiester link to C-6 of the peptidoglycan.[210] The acidic conditions, however, also lead to cleavage of the acetal linkages of pyruvate substituents (see sections III.2.a. and IV.2.a.). Alternatively, the phosphodiester bridge to the peptidoglycan may be preserved, allowing the determination of the linkage and substitution pattern of the SCWP to the murein backbone.[87] Separation of SCWPs from S-layer glycoprotein glycans can be effected by size-exclusion chromatography with 0.1 M NaCl as eluent or by RP-4 HPLC chromatography. Following degradation of proteins by trypsin digestion and re-N-acetylation, the material is subjected to digestion by lysozyme, dialysis, size-exclusion chromatography on Sephadex G-50, and final HPLC purification on RP-18 to give SCWP samples amenable to NMR analysis.[180]

2. Degradation Reactions

a. HF Treatment of SCWPs.—Cleavage of the phosphate linkages of SCWPs from the murein by the action of HF ($pK_a = 3.2$) leads to hydrolytic degradation of the polymer under the acidic conditions. Degradation may affect the glycosidic linkages, O- and N-acetyl groups, and also pyruvate groups. The latter group has been suggested to act as a noncovalent anchor of the SCWP to confer binding to the S-layer homology (SLH) domain.[164–166,174] Thus, in several structural reports on SCWPs, the actual amount given for the pyruvate substituent may be too low, or the native pyruvate groups may have not been detected at all in the acid-degraded material. The difference in pyruvate contents depending on the isolation protocol used may be demonstrated by comparing two SCWPs from *L. sphaericus* CCM 2177, isolated by HF treatment,[177] and *P. alvei* CCM 2051, treated with lysozyme, respectively.[176] In the latter case, full substitution of the β-D-Man$p$NAc residues by 4,6-pyruvate acetal groups in the intact SCWP was found (Fig. 20), but this was decreased by up to 50% upon mild acid treatment. An approximately 50% degree of substitution was also reported for the related SCWP from *P. alvei* CCM 2051 as extracted by HF treatment.[177]

b. Smith Degradation.—Smith degradation[211,212] has frequently been employed as a common tool to detect protein glycosylation using the periodic acid–Schiff reagent, but has also been proved to be of considerable value in establishing the structure of several SCWPs and S-layer glycoprotein glycans. Smith degradation, employing periodate oxidation followed by sodium borohydride reduction and acid hydrolysis of the acetal linkages, was applied for the removal of distal side-chain units in the HF-degraded SCWP from *G. stearothermophilus* PV72/p2[182] and S-layer glycoprotein glycans from *Ane. thermoaerophilus* L420-91,[133,137] *Thb. thermohydrosulfuricus* L77-66,[145] *P. alvei* CCM 2051,[138] and *G. stearothermophilus* NRS 2004/3a.[131] Smith degradation has also been used for in-chain cleavage to release oligosaccharide fragments, fragments originating from the core portion of S-layer glycopeptides,[136,139,178,213] and to confirm the 3-*O*-methyl substitution of a terminal rhamnopyranosyl unit in *Thb. thermohydrosulfuricus* L111-69.[136]

3. Structure Elucidation by Nuclear Magnetic Resonance Spectroscopy

The tools of modern NMR spectroscopy have been instrumental in the detailed structure elucidation of S-layer glycans and SCWPS. NMR-based findings include the identification of distal end-group modifications, substoichiometric substitution by *O*- and *N*-acetyl groups, the detailed structure of core oligosaccharides at the reducing end of the polysaccharides and the binding regions of the peptide and murein fragments, respectively. Information on the individual glycose constituents is now routinely derived from one- and two-dimensional NMR spectroscopic techniques employing ^1H–^1H correlation spectroscopy (COSY), total correlation spectroscopy (TOCSY), ^1H–^{13}C heteronuclear single quantum coherence spectroscopy (HSQC), and also ^1H–^{13}C HSQC-TOCSY experiments, which additionally help in identifying the glycosylation sites by virtue of the observed ^{13}C NMR glycosylation shifts.[214,215] Assignment of the anomeric configuration is based on the values of the homonuclear coupling constants $J_{H1,H2}$ as well as being confirmed by the value of the one-bond heteronuclear coupling constant $J_{C1,H1}$. Information on glycosyl sequences is mainly achieved from nuclear Overhauser effect spectroscopy (NOESY) and ^1H–^{13}C heteronuclear multiple-bond coherence spectroscopy (HMBC) data, which additionally also support the complete assignment of the individual glycosyl spin systems. Whereas

HMBC records three-bond connectivities, the H2BC technique monitors two-bond correlations only, thereby clarifying ambiguous assignments made from HMBC or HSQC-TOCSY data.[182,216] Comparison of the integral values arising from nondegenerate signals at the termini of the polysaccharide chain with the degenerate signals of the repeat units provides an estimate of the chain length of the cell-envelope polymers.

Specific structural features of SCWPs relate to the presence of pyruvate groups and the linkage region to the peptidoglycan. The presence of pyruvate acetals is readily identified via HMBC correlations from the high-field-shifted signal of the methyl group to carbon 2 at ~102–103 ppm and carbon 1 of the carboxyl group. The (S)-configuration is deduced from the low-field-shifted ^{13}C NMR signal of the equatorially arranged methyl group.[176,182,189,190,217] The linkage of SCWPs to muramic acid residues of PG has been determined for only a very limited number of examples because of the heterogeneity of the peptidoglycan backbone and substantial signal broadening. Both phosphodiester[180,181] and putative diphosphate linkages from a GlcpNAc residue have been suggested, based on ^{31}P NMR data.[176,181] A connectivity of the anomeric proton to the adjacent phosphorus atom could be established by ^{1}H{^{31}P} decoupling difference experiments as well as by using ^{1}H–^{31}P HMBC experiments.[180] Notably, in the case of the diphosphate unit that is suggested to occur in the SCWP of *P. alvei* CCM 2051, a reversible change in the integral values of two phosphorus signals was observed, depending on the presence or absence of micelle-breaking agents.[176] Low-field-shifted signals of CH_2 groups correlated to ^{31}P NMR signals agree with the existence of an attachment site at carbon 6 of the muramyl units in the PG backbone. Additional evidence is required to substantiate the presence of these diphosphate entities.

The structural elucidation of end groups occurring in S-layer glycoprotein glycans has to cover the capping components—assumed to be involved in chain termination of the polymer during biosynthesis—as well as the structurally diverse core units and their linkage to the peptide portion of the S-layer glycoprotein. Methylation at O-2 and O-3 of terminal α-L-rhamnopyranosyl residues has been found in the S-layer glycoprotein glycans from *G. stearothermophilus* NRS 2004/3a[131] and *Ane. thermoaerophilus* L420-91^T and GS4-97,[137] as well as *Thb. thermohydrosulfuricus* L111-69 and L110-69,[136] respectively. The location of the methylation site was established through HMBC correlation of the methyl carbon atom to the respective proton of the rhamnose residue.[120] A unique capping component was identified in *Ano. tepidamans* GS5-97^T harboring (R)-N-acetylmuramic acid and GlcpNAc as distal sugar units[132]; diffusion difference experiments were used to rule out any accompanying contamination by low-molecular-weight components.[139]

S-Layer glycoprotein glycans frequently contain the unusual O-glycosidic linkage to tyrosine residues, which was readily identified by the two doublet proton signals in the aromatic region (albeit occurring in low intensity). The tyrosine-connected sugars found thus far are glucopyranosyl and galactopyranosyl groups in a β-glycosidic linkage.[135,136,139,140,218] Other glycosyl amino acid units harbor O-linked threonine/serine as well as N-linked asparagine residues. Analysis of the glycopeptide part is complicated by the inherent microheterogeneity and low intensity of the respective NMR signals. Elucidation of these domains has been accomplished by NMR analysis of fragments obtained by acid hydrolysis as well as by measurements performed in 90% H_2O/10% D_2O solution with efficient suppression of the water signal using a WATERGATE pulse-sequence. In that way, correlations arising from the NH amide signals were utilized to elucidate the amino acid sequence of the glycopeptide domain. Because of heterogeneity present in a number of core structures and the peptide part, however, a full structural elucidation has usually to be based on a combination of NMR data and mass-spectrometric evidence.

4. Mass Spectrometry

Matrix-assisted laser-desorption ionization time-of-flight (MALDI) MS has proved to be a valuable tool for the study of S-layer glycans as well as SCWPs. MALDI-TOF mass spectra of glycopeptide pools have been recorded in both negative- and positive-ion mode, giving insight into structural heterogeneities of the sample, chain-length distribution, and also the molecular mass of the repeat units.[131,136,139] Electrospray-ionization time-of-flight (ESI Q-TOF) mass spectrometry provides enhanced sensitivity and, by analysis of multiply charged species and fragmentation patterns, allows sequencing of core units and glycopeptide fragments.[132,219–222]

For analyzing patterns of protein glycosylation at the nanoscale, LC-ESI-MS/MS can be employed, as has been recently exemplified with *T. forsythia*.[106] Here, the TfsA and TfsB S-layer glycoproteins, as well as four additional carbohydrate-positive protein bands, were excised from SDS–polyacrylamide gels and the O-glycans were isolated by applying in-gel reductive β-elimination with 1 M $NaBH_4$ in 0.5 M NaOH at 50 °C followed by removal of excess salt and subsequent LC-ESI-MS/MS analysis. In different bacteria,[181] such as in the *T. forsythia* S-layer glycan, mannosaminuron-amides are present which are hydrolyzed to mannosaminuronic acids under these conditions. This change is evidenced by a mass difference of 1 Da between released and bound O-glycan. Alternatively, ammonia-based nonreductive β-elimination can be applied,[223] followed by prolonged exposure to sodium hydroxide at 50 °C, which leads to a deamidation reaction that can be readily monitored by MS.

V. CELL-ENVELOPE GLYCAN BIOSYNTHESIS

1. Genetic Basis for S-Layer Glycoprotein Biosynthesis

If present, S-layer proteins are among the most abundant cellular proteins, with up to 20% of the total protein biosynthetic effort of a bacterium being devoted to production of S-layer protein. This large amount of S-layer protein is required for a complete coverage of the bacterium by a closed S-layer lattice during all stages of its growth cycle. At high growth rates of an average-sized cell, approximately 500 S-layer subunits must be synthesized per second, translocated to the cell surface, and incorporated into the preexisting S-layer lattice.[224] All of these steps have to be strictly coordinated with the glycosylation process to afford a glycosylated S-layer lattice. According to recent data, S-layer glycosylation lags behind biosynthesis of S-layer protein and is likely to proceed cosecretionally, with the underlying regulatory mechanism being currently unknown.

The provision of this extraordinary high amount of S-layer protein is ensured at the molecular level by a combination of strong S-layer gene promoters and high mRNA stability. In the context of gene regulation of S-layer protein biosynthesis, it should be mentioned that some bacteria (such as *C. fetus* or lactobacilli) can express different S-layer proteins.[103,105] S-Layer variation might have evolved as a pivotal strategy of bacteria to respond to changing environmental conditions.[101,102,104,105]

The genetic information for S-layer protein glycosylation is encoded by S-layer glycosylation (*slg*) gene clusters (polycistronic units) or *slg* gene loci (comprising several transcriptional units), working in concert with housekeeping genes. Housekeeping genes map outside the *slg* gene regions, and the encoded gene products are required for the biosynthesis of the reducing-end glycoses of the S-layer glycans. The gene that encodes the respective S-layer target protein is transcribed independently of the genes in the *slg* gene region, and its chromosomal location is not necessarily in close vicinity. Currently, most data are available from the Gram-positive bacteria *G. stearothermophilus* NRS 2004/3a (GenBank AF328862),[225] *Ano. tepidamans* GS5-97^T (GenBank AY883421),[226] and *Ane. thermoaerophilus* strains L420-91^T (GenBank AY442352) and DSM 10155/G$^+$ (GenBank AF324836)[227] as well as from the Gram-negative *T. forsythia* ATCC 43037.[106] Depending on the complexity of the encoded S-layer glycan, the *slg* gene regions are ~16 to ~25 kb in size. They include genes of the nucleotide-sugar pathway that are arranged consecutively, glycosyltransferase genes, glycan-processing genes, and transporter genes. The presence of insertion sequences and the decrease of the G + C content at these gene regions in comparison to the respective bacterial genome suggest that lateral gene transfer is the source of S-layer glycan variation (for review, see Ref. 116).

The comparison of the *slg* gene clusters encoding the extended tripartite S-layer glycans of *G. stearothermophilus* NRS 2004/3a and *Ano. tepidamans* GS5-97^{T} [115] revealed that the clusters are organized in a similar way.[226] The central, variable part of the *slg* gene clusters is responsible for the biosynthesis of the individual repeating units and terminating elements, whereas the region having higher homology codes for proteins involved in assembling the core region, transport of the glycan to the cell surface, and its ligation to the S-layer protein. This behavior resembles the organization of O-antigen gene clusters in Gram-negative organisms, where the variability of O-antigens is considered to be a result of recombination events in the central region of the O-antigen gene clusters.[228,229]

A special situation exists with *T. forsythia,* where theoretical analysis of the S-layer O-glycosylation indicated the involvement of a 6.8-kb gene locus that is conserved among different bacteria from the *Bacteroidetes* phylum.[106]

2. Nucleotide Sugar Biosynthesis

Based on the identification of the S-layer glycan-specific nucleotide-sugar genes in the *slg* gene regions, the encoded proteins were cloned and overexpressed in *E. coli.* This procedure, together with the establishment of functional assays for the recombinant enzymes, led to the characterization of the biosynthesis pathways for dTDP-β-L-Rha$p$,[230] dTDP-α-D-Fuc$p$3NAc,[231] dTDP-α-D-Qui$p$3NAc,[232] GDP-D-*glycero*-α-D-*manno*-heptose,[233] GDP-α-D-Rha$p$,[234] and dTDP-α-D-Fuc$p$[226] in Gram-positive bacteria. In all these pathways, dTDP-"4-dehydro-6-deoxyglucose" (6-deoxy-D-*xylo*-hexos-4-ulose) plays a central role (Fig. 31).

Consequently, pathways for S-layer protein glycosylation provide a spectrum of rare enzymes that may be used for glycoengineering purposes in heterologous hosts (see Section VI). Furthermore, some of these enzymes, for example, most of the Rml enzymes from *Ane. thermoaerophilus* DSM 10155/G^{+}, because of their thermophilic origin, exhibit significantly higher stability at 37 °C than the enzymes from the mesophilic strain *Salmonella enterica*. This advantage could lead to the development of improved high-throughput screening systems for specific sugars.[235]

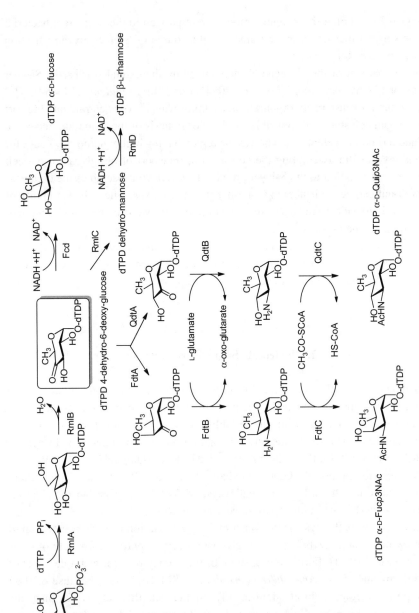

Fig. 31. Biosynthesis pathways for nucleotide sugars as required for S-layer protein glycosylation.[226,230–232]

3. Multispecific Glycosyltransferases

As already outlined, G. *stearothermophilus* NRS 2004/3a possesses S-layer O-glycans having the repeating-unit structure [→2)-α-L-Rha$p$-(1→3)-β-L-Rha$p$-(1→2)-α-L-Rha$p$-(1→]$_{n=13-18}$, including a 2-*O*-methyl group capping the terminal repeating unit at the nonreducing end and a →2)-α-L-Rha$p$-[(1→3)-α-L-Rha$p$]$_{n=1-2}$ (1→3)- adaptor linked via a β-D-Gal$p$ residue to distinct threonine and serine sites of the S-layer protein SgsE (for structure, see Fig. 7).[131] During *in vitro* analysis of the biosynthesis of this S-layer glycan, using purified enzyme preparations and synthetic substrates,[236] an unusual multifunctional methylrhamnosyltransferase (WsaE) was identified. The N-terminal portion of WsaE is responsible for the *S*-adenosylmethionine-dependent methylation reaction of the terminal α-(1→3)-linked L-rhamnose residue, and the central and C-terminal portions are involved in the transfer of L-rhamnose from dTDP-β-L-rhamnose to the adaptor saccharide to form the α-(1→2)- and α-(1→3)-linkages during elongation of the S-layer glycan chain, with the methylation and the glycosylation reactions occurring independently.

In *P. alvei* CCM 2051^T, the S-layer polysaccharide consisting of →3)-β-D-Gal$p$-(1[α-D-Glc$p$-(1→6)]→4)-β-D-Man$p$NAc-(1→ repeating units is O-glycosidically linked via an adaptor having the structure -[GroA-2→O)-P(=O)OH-(O→4)-β-D-Man$p$NAc-(1→4)]→3)-α-L-Rha$p$-(1→3)-α-L-Rha$p$-(1→3)-α-L-Rha$p$-(1→3)-β-D-Gal$p$-(1→ to specific tyrosine residues of the S-layer protein SpaA (for structure, see Fig. 4). A 24.3-kb S-layer *slg* gene locus comprising 18 open reading frames encodes the information necessary for the biosynthesis of this glycan. The protein encoded by *wsfC* is the largest found in this gene locus, coding for a tripartite transferase of 147.3 kDa. Two glycosyltransferase family 2 motifs are predicted in the central and C-terminal part, while a single CDP-glycerol:poly(glycerophosphate) glycerophosphotransferase motif is present at the N-terminal part of the protein. It is conceivable that the two former motifs synthesize the repeating-unit backbone →3)-β-D-Gal$p$-(1→4)-β-D-Man$p$NAc-(1→ of the S-layer glycan,[134] and we speculate that the latter motif might be involved in chain-length termination by catalyzing the transfer of a phosphoglyceric acid. This conclusion strongly suggests a multifunctional role of this enzyme in the biosynthesis of the S-layer O-glycan of *P. alvei*.[130]

The involvement of multifunctional enzymes in the biosynthesis of polysaccharides is not unusual. In polysaccharides of *E. coli* O8 and O9a, for instance, chain extension is performed by a multidomain serotype-specific mannosyltransferase WbdA.[238] In contrast to WsaE of *G. stearothermophilus* NRS 2004/3a, methylation or phosphomethylation of the nonreducing end occurs in separate steps by the action of WdbD. Although it is reported that methylation plays a critical role in chain-length determination, and that overexpression of WdbD reduces O-PS chain length in serotypes O8 and O9a, it is unclear how WdbD activity is regulated.[239]

4. Proposed Pathway for S-Layer Glycoprotein Biosynthesis

Based on sound structural data and knowledge of the underlying *slg* gene regions, a picture of how S-layer glycoprotein diversity is created by nature is unraveling. Biosynthesis of S-layer glycan glycoprotein can be viewed as a sequence of distinct stages: initiation, assembly, and transfer to the protein. Initial approaches for investigating the details concerning the machinery of S-layer protein glycosylation benefited from the molecular information available on the biosynthesis of LPS O-antigens, because due to the structural similarities (see Section II) it was reasonable to assume that both cell-surface glycoconjugates are assembled via similar biosynthetic routes, involving comparable enzyme activities.

a. Initiation.—Synthesis of the glycan chain is initiated by transfer of the first sugar residue from its nucleotide-activated form onto the membrane-bound undecaprenylphosphate (undP) lipid carrier. The initiation enzymes of S-layer-glycan biosynthesis in *G. stearothermophilus* NRS 2004/3a and *P. alvei* CCM 2051^T are the UDP-Gal:phosphoryl-polyprenol Gal-1-phosphate-transferases WsaP and WsfP, respectively. WsaP was shown to transfer Gal-1-P from UDP-Gal to phosphoryl-polyprenol *in vitro* and to reconstitute the function of WbaP in WbaP-deficient strains of *E. coli* and *S. enterica*.[221] Deletion of *wsfP* in *P. alvei* CCM 2051^T resulted in loss of S-layer glycosylation in this strain, while the original phenotype could be restored by expression of either WsfP or WsaP.[240] WsaP and WsfP share 60% identity and 75% similarity, and both enzymes show high sequence homology to members of the poly-isoprenyl-phosphate hexose-1-phosphate transferase (PHPT) family such as WbaP.[241] The involvement of WbaP homologues in the S-layer glycosylation pathway of *G. stearothermophilus* NRS 2004/3a and *P. alvei* CCM 2051^T is rather unexpected, since in the corresponding *slg* gene regions, the presence of the ABC-transporter constituents Wzm and Wzt is predicted. Consequently, by analogy to what is known from the well-investigated routes for biosynthesis of LPS O-antigen, the initiation of S-layer glycan assembly should be catalyzed by a WecA homologue (a GlcNAc-1-phosphate transferase), which is the usual initiation enzyme in the ABC-transporter-dependent pathway of O-polysaccharide biosynthesis.[123] Thus, despite many similarities between the biosynthesis of S-layer glycans and that of O-polysaccharides, the S-layer-specific pathways utilize a new combination of modules that has not been observed before.

b. Assembly of the S-Layer Glycan.—S-Layer glycans in *G. stearothermophilus* NRS 2004/3a and *P. alvei* CCM 2051^T are assembled stepwise by the consecutive transfer of glycose residues from the nucleotide-activated forms to the lipid-bound growing glycan chain, with the individual transfer reactions being catalyzed by specific glycosyltransferases.[130] This scheme implies the growth of the glycan chain from the nonreducing end and, thus, resembles the ABC-transporter-dependent pathway of O-polysaccharide biosynthesis.[123]

Assembly of the poly-L-rhamnan S-layer glycan of *G. stearothermophilus* NRS 2004/3a has been studied in different *in vitro* setups. The *slg* gene cluster of this bacterium encodes four rhamnosyltransferases, named WsaC through WsaF, which are involved in the biosynthesis of the S-layer glycan. The individual reactions catalyzed by each of the four enzymes have been elucidated in detail by their respective ability to transfer rhamnose from dTDP-β-L-Rha to different synthetic substrates, thus resembling the different stages of the lipid-bound growing glycan chain. Membrane-anchored WsaD was found to catalyze the initial rhamnosylation reaction, which is the transfer of activated β-L-Rha to the undP-bound galactose primer. Subsequently, WsaC adds one or two additional L-Rha residues, thereby completing the adaptor saccharide. The repeating units are synthesized in a concerted action of the enzyme proteins WsaE and WsaF. WsaF catalyzes the formation of the β-(1→2)-linkage,[242] while WsaE is a trifunctional enzyme with direct involvement in chain-length determination through its methyltransferase activity.

Insight into the assembly of the S-layer glycan of *P. alvei* CCM 2051^T was gained through a systematic gene-deletion approach of the individual genes encoded by the *slg* gene locus of this bacterium.[130] The biosynthesis pathway was proposed based on detailed analysis of the glycosylation pattern of the individual mutants. The first building block of the *P. alvei* CCM 2051^T S-layer glycan is the adaptor saccharide having the structure -[GroA-2→O)-P(=O)OH-(O→4)-β-D-Man*p*NAc-(1→4)]→ 3)-α-L-Rha*p*-(1→3)-α-L-Rha*p*-(1→3)-α-L-Rha*p*-(1→3)-β-D-Gal*p*-(1→ (for structure, see Fig. 4).[135] WsfG transfers the first L-Rha residue onto the initial lipid-linked Gal obtained upon catalysis of WsfP. Subsequently, WsfF adds two more L-Rha residues. The branching ManNAc residue is predicted to be transferred by WsfE, and the GroA-2-phosphate is likely to be added by the CDP-glycerol:poly (glycerophosphate) glycerophosphotransferase domain of WsfC. WsfC is furthermore involved in biosynthesis of the repeating-unit backbone. A branching glucose residue would be α-(1→6)-linked to the repeating-unit ManNAc residue via a separate mechanism that was shown to depend on the activity of WsfH and WsfD. According to the current interpretation, WsfH would transfer a single glucose residue from

UDP-Glc to a separate lipid carrier, from which it would be further transferred to the S-layer glycan by the activity of WsfD.

c. Transfer of the S-Layer Glycan onto the Protein.—The *slg* gene regions from both *G. stearothermophilus* NRS 2004/3a and *P. alvei* CCM 2051^T contain the integral membrane protein Wzm and the nucleotide-binding protein Wzt constituting an ABC transporter.[243–245] By analogy to the ABC-dependent pathway of O-polysaccharide synthesis, the ABC transporter would be responsible for the transport of complete lipid-bound S-layer glycans across the cytoplasmic membrane. Transfer of the elongated glycan chain onto the S-layer protein most likely occurs cosecretionally. This final step of transferring the complete glycan to the O-glycosylation sites of the S-layer protein requires the action of a dedicated oligosaccharyl:protein transferase,[246] which would act on the S-layer protein upon cleavage of the signal peptide and transfer across the cytoplasmic membrane. Eventually, the S-layer protein would undergo 2-D crystallization on the surface of the bacterial cell and form a closed S-layer lattice.

5. SCWP Glycosylation Gene Clusters

Upstream and downstream of the S-layer gene *spaA* of *P. alvei* CCM 2051^T several genes were identified, which are predicted to be involved in the biosynthesis of the bacterium's SCWP with the structure [(Pyr4,6)-β-D-Man*p*NAc-(1 → 4)-β-D-Glc*p*NAc-(1 → 3)$_{n\sim 11}$-(Pyr4,6)-β-D-Man*p*NAc-(1 → 4)-α-D-Glc*p*NAc-(1 → (for structure, see Fig. 21).[247] The prediction of transcription units for the complete ~14-kb DNA region resulted in a polycystronic RNA-containing *orf1*, *csaB*, and *tagA* in addition to four separate monocystronic RNAs for *tagO*, *slhA*, *spaA*, and *orf7*. The putative gene products of the SCWP biosynthesis locus were analyzed by database comparison.[247] Orf1 might act as an exporter for the SCWP-like CsaA in the biosynthesis of other pyruvated SCWPs.[164,174] Both *tagA* and *tagO* reveal high similarity to the glycosyl-transferases TagA and TagO of different *Bacillaceae*. In *B. anthracis*, these enzyme proteins mediate assembly of SCWP linkage units and tether pyruvated SCWP to the *B. anthracis* envelope. In *B. subtilis*, both enzymes are involved in the biosynthesis of teichoic acids, with TagO coupling GlcNAc to the membrane-embedded lipid undP and TagA catalyzing the addition of ManNAc to produce the undP-linked GlcNAc-ManNAc disaccharide.[248,249] Since a ManNAc-GlcNAc backbone disaccharide motif is found in the SCWP of *P. alvei* CCM 2051^T, it is conceivable that both enzymes are involved in the biosynthesis of this SCWP.[135,248]

In the context of the anchoring function of SCWP for the S-layer protein SpaA of *P. alvei* CCM 2051^T, CsaB is of special interest, because the *csaB* gene is highly homologous to genes coding for pyruvate transferases (CsaB) in various *Bacillus* strains. In *B. anthracis* and *Ths. thermophilus*, CsaB is involved in the addition of pyruvate groups to the respective SCWPs, a prerequisite for anchoring of cell-wall-associated proteins containing SLH domains.[164–166,174] The presence of a pyruvate-containing SCWP in *P. alvei* CCM 2051^T suggests that CsaB is responsible for SCWP pyruvation also in this organism.[176] To analyze directly the contribution of the pyruvate residues for the interaction with the SLH domains of the *P. alvei* CCM 2051^T S-layer protein, an attempt was made to inactivate *csaB*; this, however, generated a lethal phenotype.[167] This is in line with the impossibility of inactivating *spaA*,[167] which indicates that cell-envelope integrity is essential for the viability of *P. alvei* CCM 2051^T cells. In this context, it is noteworthy that also for *B. anthracis*, no null mutations of either *tagO* or *tagA* could be obtained.[250]

For *B. anthracis*, comparably little is known about SCWP biosynthesis, even though for this bacterium the important role of CsaB for proper SCWP function *in vivo* was reported for the first time.[164] Previous work described the SCWP of *B. anthracis* as a polysaccharide having the repeating structure [→4)-β-ManNAc-(1→4)-β-GlcNAc-(1→6)-α-GlcNAc-(1→]$_n$, where α-GlcNAc is substituted with α-Gal and β-Gal at O-3 and O-4, respectively, and the β-GlcNAc is substituted with α-Gal at O-3 (for structure, see Fig. 26).[185] A SCWP strand is tethered via a murein linkage unit (GlcNAc–ManNAc) and a phosphodiester bond to the C-6 hydroxyl group of *N*-acetylmuramic acid (MurNAc) within the peptidoglycan.[184,251] Mesnage and coworkers[164] identified the two-gene operon *csaAB* for cell-surface anchoring of the SLH domain-containing S-layer proteins EA1 and Sap in *B. anthracis*. Biochemical analysis of cell-wall components showed that CsaB was involved in the addition of a pyruvate group to a peptidoglycan-associated polysaccharide fraction and that this modification was necessary for binding of the SLH domain. The *csaAB* operon was found to be present in several bacterial species that synthesize SLH-containing proteins. This observation, and the presence of pyruvate in the cell wall of the corresponding bacteria, suggested that this mechanism is widespread among bacteria. In contrast to *P. alvei* CCM 2051^T, deletion of *csaAB* affected only attachment of the *B. anthracis* S-layer proteins EA1 and Sap, but did not lead to a lethal phenotype.[164]

Very recent studies revealed that the SCWP of *B. anthracis* is modified with acetalpyruvate and acetyl groups.[184,189] Acetylation of SCWP appears to be a universal feature of *B. anthracis* and *B. cereus*, and the sites of acetylation and pyruvation on the polysaccharide appear similar between *B. anthracis* CDC 684 and *B. anthracis* Sterne.[186,189,252] Nevertheless, the genetic basis for such modification is not yet known in detail. Recently, two putative acyl-transferase systems, designated PatA1 and PatA2, were identified in *B. anthracis* and were proposed to function as peptidoglycan acetyl-transferases.[253] Investigation of the contributions of PatA1 and PatA2 on SCWP acetylation and S-layer assembly revealed that mutations in *patA1* and *patA2* affect the chain lengths of *B. anthracis* vegetative forms and perturb the deposition of the BslO murein hydrolase at cell-division septa.[254] Furthermore, *patA1* and *patA2* mutants are defective for the assembly of EA1 in the envelope, but retain the ability of S-layer formation with Sap. The SCWP isolated from the *patA1/patA2* mutant lacked acetyl moieties identified in the wild-type SCWP and failed to associate with the SLH domains of EA1. However, it still remains to be investigated experimentally whether PatA1 and PatA2 also act to acetylate the peptidoglycan of *B. anthracis*.

VI. Glycan Engineering and Applications

1. The S-Layer Glycobiology Toolbox

Glycosylation of proteins accounts for much of the biological diversity generated at the proteome level. S-Layer glycoproteins as a special class of glycoproteins are present on a wide variety of prokaryotic cell surfaces. They offer, on account of their high degree of variability with regard to glycoses, glycosidic linkages, and structures involving many rare glycoses, an unsurpassed glycobiology toolbox (compare with Section II).[11,111] As outlined in detail in Section II, S-layer glycoproteins additionally possess an intrinsic self-assembly feature into 2-D crystalline structures, which cover entire bacterial cells as a closed monolayer with lattice-like appearance (for reviews, see Refs. 67,74). Accounting for this intrinsic nanometer-scale cell-surface display feature of bacterial S-layer glycans via the S-layer protein matrix, we have coined the neologism *S-layer nanoglycobiology*, which comprises structural, functional, and biosynthetic aspects of S-layers.[111,115,116] The awareness of the promising application potential of S-layer glycoproteins as a unique matrix with inherent self-assembly properties (for reviews, see Refs. 80,115,122) is the driving force behind the efforts for structural elucidation of novel S-layer glycans of different bacteria; recent focus is on medically relevant bacteria and the characterization of S-layer glycoprotein biosynthesis at the molecular level (compare with Section V).

Carbohydrate-active enzymes from pathways for S-layer protein glycosylation, especially those from nucleotide-sugar biosynthesis, offer interesting options for therapeutic intervention. For instance, the immediate source for L-rhamnose, a common component of the cell wall and the capsule of many pathogenic bacteria as well as of many S-layer glycans, is dTDP-L-Rha. Since to date neither rhamnose nor the RmlABCD genes responsible for dTDP-L-Rha biosynthesis have been identified in humans, these enzymes are ideal targets for inhibiting rhamnose-mediated bacterial pathogenicity.[255] Other enzymes from the S-layer glycome can be beneficial for synthetic purposes because of their unusual specificities or thermostability. Thermostable Rml enzymes from *Ane. thermoaerophilus* DSM 10155 allow higher productivity of dTDP-L-Rha (compare with Section V).[230] Other enzymes, such as an isomerase derived from *Ane. thermoaerophilus* DSM 420-91^T capable of synthesizing DTDP-6-deoxy-D-xylo-hexos-3-ulose ("dTDP-3-keto-6-deoxygalactose") from dTDP-6-deoxy-D-xylo-hexos-4-ulose ("dTDP-4-keto-6-deoxyglucose")[230] (Fig. 31), may be relevant for the synthesis of antibiotics that contain C-3-aminated deoxy sugars, such as erythromycin or tylosin.

Eventually, detailed insights into the structures of S-layer glycans and the molecular understanding of the governing biosynthetic machinery will increase the possibilities offered by the glycobiology toolbox aiming at the rational design of S-layer glycans and, consequently, the generation of artificial S-layer *neo*glycoconjugates having rationally designed and biofunctional glycosylation motifs.[247,256,257] Since the S-layer glycome is connected with a molecular self-assembly system, our approach has a strong link to the field of nanobiotechnology, because the means for organizing materials, such as biologically functional glycans, at the nanometer level are prime candidates for the fabrication of supramolecular structures and devices.[93]

2. Glycosylation Engineering

Considering the constantly increasing number of reports about prokaryotic glycoproteins that are important in pathogenesis,[8,9,57,60,126] the molecular understanding of S-layer protein glycosylation (see Sections V. and VI.1.) should eventually allow the rational design of glycosylation patterns for S-layer proteins to obtain bioactive S-layer neoglycoproteins.[11] The common trend of glycoengineering is reflected by many recent articles on that topic.[47,53,127,258–260] In general, the display of heterologous (glyco)proteins on the surface of bacteria, enabled by means of recombinant DNA technology, has become increasingly used as a strategy in various applications in microbiology, nanobiotechnology, and vaccinology.[247,261,262] Besides outer membrane proteins, lipoproteins, autotransporters, or subunits of surface appendages, the use of the glycosylated S-layer proteins as a surface-anchoring structure is a very attractive and promising alternative. It offers the unique advantage of providing a crystalline, regular "immobilization matrix" that should eventually allow the controlled and periodic surface display of functional glycosylation motifs.[122,247] Nanobiotechnology applications of such tailored S-layer neoglycoproteins may include the fields of receptor mimics, vaccine design, or drug delivery using carbohydrate recognition. The possibility for incorporating additional functional domains into the S-layer protein portion may permit additional tuning of structural and functional features of these S-layer neoglycoproteins.

Glycosylation engineering of S-layer glycoproteins constitutes a rather new area of research, and only a few examples have been published so far. Previous studies have demonstrated that several functional domains introduced into S-layer proteins at distinct positions by genetic engineering do not interfere with the intrinsic self-assembly property of the S-layer system, and simultaneously they fully retain their specific biological properties.[256] SgsE is a naturally *O*-glycosylated protein of *G. stearothermophilus* NRS 2004/3a which has been used in a combined carbohydrate/protein engineering approach for the production of two types of S-layer neoglycoproteins in *E. coli*. Based on the identification of a suitable periplasmic targeting system for the SgsE self-assembly protein as a cellular prerequisite for protein glycosylation, and on engineering of one of the natural protein *O*-glycosylation sites into a target for *N*-glycosylation, the heptasaccharide from the AcrA protein of *C. jejuni* and the O7 polysaccharide of *E. coli* were transferred to the S-layer protein by the action of the oligosaccharyltransferase PglB. The degree of glycosylation of the S-layer neoglycoproteins after purification from the periplasmic fraction reached completeness.[257]

Considering the manifold properties glycans provide to proteins, including stabilizing functions,[263] enhanced thermal stability,[264] and protection from proteases,[265] glycosylation engineering seems to be promising for the future generation of glycoproteins having improved characteristics. Recently, "cross-glycosylation" of nonnative proteins was successfully performed in the closely related organisms *T. forsythia* and *B. fragilis*.[127] Both organisms are known to possess general *O*-glycosylation systems sharing a conserved amino acid motif that is targeted for glycosylation. Since the glycan structure of *T. forsythia* is already known (Fig. 17),[106] we next partially characterized the native *B. fragilis* O-glycan in order to determine whether that glycan would be added to *T. forsythia* proteins expressed in *B. fragilis*. Mass-spectrometric analysis of native *B. fragilis* glycoproteins subjected to β-elimination revealed the glycan to be an oligosaccharide consisting of nine sugar residues (Fig. 18).[127] Notably, the proposed structure based on MS data resembles that of the *T. forsythia* O-glycan in several aspects (compare with Section II). As in the *T. forsythia* glycan, the deoxyhexose of the *B. fragilis* glycan most likely is a fucose residue, as glycoproteins of *B. fragilis* are readily detected with *A. aurantia* lectin and depend on GDP-fucose biosynthesis genes[126]; however, there was no indication for the presence of a sialic acid-like sugar as in the *T. forsythia* S-layer glycan.[106] Knowing the size and approximate composition of the *B. fragilis* glycan, the respective glycans were transferred onto nonnative proteins. The known S-layer proteins TfsA and TfsB from *T. forsythia* were chosen for heterologous glycosylation in *B. fragilis*.[84,125] Conversely, two model glycoproteins of *B. fragilis*, BF2494 and BF3567, both of unknown function,[126] were selected for heterologous expression in *T. forsythia*. The successful transfer of the O-glycans was confirmed by the appearance of novel glycoproteins with the respective O-glycans linked via serine/threonine residues to the *Bacteroides* glycosylation sequon D(S/T)(A/I/L/V/M/T).[126] Moreover, glycan transfer appeared to be highly efficient, as none of the peptides spanning a putative glycosylation site were observed as unmodified.[127] Considering the vast amount of glycoproteins synthesized by *B. fragilis*[260]—and most likely also by *T. forsythia*—it is likely that O-glycosylation of proteins has a more general function, for instance, in conferring protein stability. Analyzing the influence of (heterologous) glycosylation on protein stability in *T. forsythia* and *B. fragilis* will contribute to a better understanding of protein glycosylation in general, as well as trigger efforts specifically to improve protein stability through glycoengineering.

VII. Concluding Remarks

The remarkable diversity observed among the S-layer glycan and SCWP structures presented here reflects the high level of adaptation developed by bacteria and archaea in the course of evolution as a consequence of ongoing changes in their respective microenvironments and habitats. The important function of secondary cell-wall polysaccharides in anchoring the S-layer remains to be studied in more detail in order to unravel the mechanisms leading to the regularly patterned display of S-layers on the surface of the bacterial cell. In addition, elucidation of the underlying genetic basis and biosynthetic pathways has already opened a fertile playground in the forthcoming years for numerous applications in the fields of nanotechnology and biomedicine, using rationally designed glycoengineering tools.

References

1. A. Kobata, The carbohydrates of glycoproteins, In: V. Ginsburg and P. W. Robbins, (Eds.) *Biology of Carbohydrates*, Vol. 2, John Wiley & Sons, New York, 1984, pp. 87–161.
2. J. Montreuil, Spatial conformation of glycans and glycoproteins, *Biol. Cell*, 51 (1984) 115–131.

3. R. Kornfeld and S. Kornfeld, Assembly of asparagine-linked oligosaccharides, *Annu. Rev. Biochem.*, 54 (1985) 631–664.
4. A. Varki, R. D. Cummings, J. D. Esko, H. H. Freeze, P. Stanley, C. R. Bertozzi, G. W. Hart, and M. E. Etzler, (Eds.), In: Essentials of Glycobiology, 2nd ed. Cold Spring Harbor Laboratory Press, Cold Spring Harbor, NY, 2009, p. 784.
5. P. Messner and U. B. Sleytr, Bacterial surface layer glycoproteins, *Glycobiology*, 1 (1991) 545–551.
6. S. Moens and J. Vanderleyden, Glycoproteins in prokaryotes, *Arch. Microbiol.*, 168 (1997) 169–175.
7. C. Schäffer, M. Graninger, and P. Messner, Prokaryotic glycosylation, *Proteomics*, 1 (2001) 248–261.
8. M. A. Schmidt, L. W. Riley, and I. Benz, Sweet new world: Glycoproteins in bacterial pathogens, *Trends Microbiol.*, 11 (2003) 554–561.
9. M. Szymanski and B. W. Wren, Protein glycosylation in bacterial mucosal pathogens, *Nat. Rev. Microbiol.*, 3 (2005) 225–237.
10. J. D. Esko, T. L. Doering, and C. R. H. Raetz, Eubacteria and Archaea, In: A. Varki, R. D. Cummings, J. D. Esko, H. H. Freeze, P. Stanley, C. R. Bertozzi, G. W. Hart, and M. E. Etzler, (Eds.), *Essentials of Glycobiology*, 2nd ed. Cold Spring Harbor Laboratory Press, Cold Spring Harbor, NY, 2009, pp. 293–299.
11. P. Messner, Prokaryotic protein glycosylation is rapidly expanding from "curiosity" to "ubiquity", *ChemBioChem*, 13 (2009) 2151–2154.
12. I. Hug and M. F. Feldman, Analogies and homologies in lipopolysaccharide and glycoprotein biosynthesis in bacteria, *Glycobiology*, 21 (2011) 138–151.
13. P. T. McKenney, A. Driks, and P. Eichenberger, The *Bacillus subtilis* endospore: Assembly and functions of the multilayered coat, *Nat. Rev. Microbiol.*, 11 (2013) 33–44.
14. M. F. Mescher, Glycoproteins as cell surface structural components, *Trends Biochem. Sci.*, 6 (1981) 97–99.
15. J. Lechner and F. Wieland, Structure and biosynthesis of prokaryotic glycoproteins, *Annu. Rev. Biochem.*, 58 (1989) 173–194.
16. M. Sumper and F. T. Wieland, Bacterial glycoproteins, In: J. Montreuil, J. F. G. Vliegenthart, and H. Schachter, (Eds.), *Glycoproteins,* Elsevier, Amsterdam, 1995, pp. 455–473.
17. J. Eichler and M. W. W. Adams, Posttranslational protein modification in *Archaea, Microbiol. Mol. Biol. Rev.*, 69 (2005) 393–425.
18. P. Messner and C. Schäffer, Prokaryotic glycoproteins, In: W. Herz, H. Falk, and G. W. Kirby, (Eds.), *Progressin the Chemistry of Organic Natural Products,* Vol. 85, Springer, Wien, 2003, pp. 51–124.
19. A. Varki, Biological roles of oligosaccharides—All of the theories are correct, *Glycobiology*, 3 (1993) 97–130.
20. R. Spiro, Protein glycosylation: Nature, distribution, enzymatic formation, and disease implications of glycopeptide bonds, *Glycobiology*, 12 (2002) 43R–56R.
21. T. Endo, Structure, function and pathology of O-mannosyl glycans, *Glycoconj. J.*, 21 (2004) 3–7.
22. E. Weerapana and B. Imperiali, Asparagine-linked protein glycosylation: From eukaryotic to prokaryotic systems, *Glycobiology*, 16 (2006) 91R–101R.
23. M. F. Mescher and J. L. Strominger, Purification and characterization of a prokaryotic glycoprotein from the cell envelope of *Halobacterium salinarium, J. Biol. Chem.*, 251 (1976) 2005–2014.
24. U. B. Sleytr and K. J. I. Thorne, Chemical characterization of the regularly arrayed surface layers of *Clostridium thermosaccharolyticum* and *Clostridium thermohydrosulfuricum, J. Bacteriol.*, 126 (1976) 377–383.
25. O. Kandler and H. König, Cell envelopes of archaebacteria, In: C. R. Woese and R. S. Wolfe, (Eds.) *The Bacteria, Archaebacteria,* Vol. VIII, Academic Press, New York, 1985, pp. 413–457.
26. M. Sumper, Halobacterial glycoprotein biosynthesis, *Biochim. Biophys. Acta*, 906 (1987) 69–79.
27. L. E. Sandercock, A. M. MacLeod, E. Ong, and R. A. J. Warren, Non-S-layer glycoproteins in eubacteria, *FEMS Microbiol. Lett.*, 118 (1994) 1–8.

28. T. Kawamura and G. D. Shockman, Purification and some properties of the endogenous, autolytic N-acetylmuramoylhydrolase of *Streptococcus faecium*, a bacterial glycoenzyme, *J. Biol. Chem.*, 258 (1983) 9514–9521.
29. C. Lindenthal and E. A. Elsinghorst, Identification of a glycoprotein produced by enterotoxigenic *Escherichia coli*, *Infect. Immun.*, 67 (1999) 4084–4091.
30. S. M. Logan, Flagellar glycosylation—A new component of the motility repertoire? *Microbiology*, 152 (2006) 1249–1262.
31. F. Wieland, G. Paul, and M. Sumper, Halobacterial flagellins are sulfated glycoproteins, *J. Biol. Chem.*, 260 (1985) 15180–15185.
32. S. Moens, K. Michiels, and J. Vanderleyden, Glycosylation of the flagellin of the polar flagellum of *Azospirillum brasilense*, a Gram-negative nitrogen-fixing bacterium, *Microbiology*, 141 (1995) 2651–2657.
33. P. Thibault, S. M. Logan, J. F. Kelly, J.-R. Brisson, C. P. Ewing, T. J. Trust, and P. Guerry, Identification of the carbohydrate moieties and glycosylation motifs in *Campylobacter jejuni* flagellin, *J. Biol. Chem.*, 276 (2001) 34862–34870.
34. M. Schirm, E. C. Soo, A. J. Aubry, J. Austin, P. Thibault, and S. M. Logan, Structural, genetic and functional characterization of the flagellin glycosylation process in *Helicobacter pylori*, *Mol. Microbiol.*, 48 (2003) 1579–1592.
35. M. Schirm, I. C. Schoenhofen, S. M. Logan, K. C. Waldron, and P. Thibault, Identification of unusual bacterial glycosylation by tandem mass spectrometry analyses of intact proteins, *Anal. Chem.*, 77 (2005) 7774–7782.
36. C. Schoenhofen, D. J. McNally, E. Vinogradov, D. Whitfield, N. M. Young, S. Dick, W. W. Wakarchuk, J.-R. Brisson, and S. M. Logan, Functional characterization of dehydratase/aminotransferase pairs from *Helicobacter* and *Campylobacter*: Enzymes distinguishing the pseudaminic acid and bacillosamine biosynthetic pathways, *J. Biol. Chem.*, 281 (2006) 723–732.
37. C. Schoenhofen, E. Vinogradov, D. M. Whitfield, J.-R. Brisson, and S. M. Logan, The CMP-legionaminic acid pathway in *Campylobacter*: Biosynthesis involving novel GDP-linked precursors, *Glycobiology*, 19 (2009) 715–725.
38. P. Castric, F. J. Cassels, and R. W. Carlson, Structural characterization of the *Pseudomonas aeruginosa* 1244 pilin glycan, *J. Biol. Chem.*, 276 (2001) 26479–26485.
39. Y. A. Knirel, E. T. Rietschel, R. Marre, and U. Zähringer, The structure of the O-specific chain of *Legionella pneumophila* serogroup 1 lipopolysaccharide, *Eur. J. Biochem.*, 221 (1994) 239–245.
40. Y. A. Knirel, A. S. Shashkov, Y. E. Tsvetkov, P. E. Jansson, and U. Zähringer, 5,7-Diamino-3,5,7,9-tetradeoxynon-2-ulosonic acids in bacterial glycopolymers: Chemistry and biochemistry, *Adv. Carbohydr. Chem. Biochem.*, 58 (2003) 371–417.
41. E. Kiss, A. Kereszt, F. Barta, S. Stephens, B. L. Reuhs, Á. Kondorosi, and P. Putnoky, The *rkp-3* gene region of *Sinorhizobium meliloti* RM41 contains strain-specific genes that determine K antigen structure, *Mol. Plant-Microbe Interact.*, 14 (2001) 1395–1403.
42. M. Szymanski, R. Yao, C. P. Ewing, T. J. Trust, and P. Guerry, Evidence for a system of general protein glycosylation in *Campylobacter jejuni*, *Mol. Microbiol.*, 32 (1999) 1022–1030.
43. S. M. Logan, J. F. Kelly, P. Thibault, C. P. Ewing, and P. Guerry, Structural heterogeneity of carbohydrate modifications affects serospecificity of *Campylobacter* flagellins, *Mol. Microbiol.*, 46 (2002) 587–597.
44. R. M. Macnab, How bacteria assemble flagella, *Annu. Rev. Microbiol.*, 57 (2003) 77–100.
45. N. M. Young, J.-R. Brisson, J. Kelly, D. C. Watson, L. Tessier, P. H. Lanthier, H. C. Jarrell, N. Cadotte, F. St. Michael, E. Aberg, and C. M. Szymanski, Structure of the *N*-linked glycan present on multiple glycoproteins in the Gram-negative bacterium, *Campylobacter jejuni*, *J. Biol. Chem.*, 277 (2002) 42530–42539.
46. Q. Yan and W. J. Lennarz, Studies on the function of oligosaccharyl transferase subunits. Stt3p is directly involved in the glycosylation process, *J. Biol. Chem.*, 277 (2002) 47692–47700.

47. M. Wacker, D. Linton, P. G. Hitchen, M. Nita-Lazar, S. M. Haslam, S. J. North, M. Panico, H. R. Morris, A. Dell, B. W. Wren, and M. Aebi, N-linked glycosylation in *Campylobacter jejuni* and its functional transfer into *E. coli, Science*, 298 (2002) 1790–1793.
48. D. Linton, N. Dorrell, P. G. Hitchen, S. Amber, A. V. Karlyshev, H. R. Morris, A. Dell, M. A. Valvano, M. Aebi, and B. W. Wren, Functional analysis of the *Campylobacter jejuni* N-linked protein glycosylation pathway, *Mol. Microbiol.*, 55 (2005) 1695–1703.
49. M. F. Feldman, M. Wacker, M. Hernandez, P. G. Hitchen, C. L. Marolda, M. Kowarik, H. R. Morris, A. Dell, M. A. Valvano, and M. Aebi, Engineering N-linked protein glycosylation with diverse O antigen lipopolysaccharide structures in *Escherichia coli, Proc. Natl. Acad. Sci. U.S.A.*, 102 (2005) 3016–3021.
50. M. Kowarik, S. Numao, M. F. Feldman, B. L. Schulz, N. Callewaert, E. Kiermaier, I. Catrein, and M. Aebi, N-linked glycosylation of folded proteins by the bacterial oligosaccharyltransferase, *Science*, 314 (2006) 1148–1150.
51. K. Ilg, E. Yavuz, C. Maffioli, B. Priem, and M. Aebi, Glycomimicry: Display of the GM3 sugar epitope on *Escherichia coli* and *Salmonella enterica* sv Typhimurium, *Glycobiology*, 20 (2010) 1289–1297.
52. C. Lizak, S. Gerber, S. Numao, M. Aebi, and K. P. Locher, X-ray structure of a bacterial oligosaccharyltransferase, *Nature*, 474 (2011) 350–355.
53. J. D. Valderrama-Rincon, A. C. Fisher, J. H. Merritt, Y. Y. Fan, C. A. Reading, K. Chhiba, C. Heiss, P. Azadi, M. Aebi, and M. P. DeLisa, An engineered eukaryotic protein glycosylation pathway in *Escherichia coli, Nat. Chem. Biol.*, 8 (2012) 434–436.
54. M. P. Jennings, M. Virji, D. Evans, V. Foster, Y. N. Srikhanta, L. Steeghs, P. van der Ley, and E. R. Moxon, Identification of a novel gene involved in pilin glycosylation in *Neisseria meningitidis, Mol. Microbiol.*, 29 (1998) 975–984.
55. A. Faridmoayer, M. A. Fentabil, D. C. Mills, J. S. Klassen, and M. F. Feldman, Functional characterization of bacterial oligosaccharyltransferases involved in O-linked protein glycosylation, *J. Bacteriol.*, 189 (2007) 8088–8098.
56. M. D. Hartley, M. J. Morrison, F. E. Aas, B. Børud, M. Koomey, and B. Imperiali, Biochemical characterization of the O-linked glycosylation pathway in *Neisseria gonorrhoeae* responsible for biosynthesis of protein glycans containing N,N'-diacetylbacillosamine, *Biochemistry*, 50 (2011) 4936–4948.
57. R. K. Upreti, M. Kumar, and V. Shankar, Bacterial glycoproteins: Functions, biosynthesis and applications, *Proteomics*, 3 (2003) 363–379.
58. P. Messner, Prokaryotic glycoproteins: Unexplored but important, *J. Bacteriol.*, 186 (2004) 2517–2519.
59. P. Guerry and C. M. Szymanski, *Campylobacter* sugars sticking out, *Trends Microbiol.*, 16 (2008) 428–435.
60. H. Nothaft and C. M. Szymanski, Protein glycosylation in bacteria: Sweeter than ever, *Nat. Rev. Microbiol.*, 8 (2010) 765–778.
61. B. Børud, F. E. Aas, Å. Vik, H. C. Winther-Larsen, W. Egge-Jacobsen, and M. Koomey, Genetic, structural, and antigenic analyses of glycan diversity in the O-linked protein glycosylation systems of human *Neisseria* species, *J. Bacteriol.*, 192 (2010) 2816–2829.
62. I. Hug, B. Zheng, B. Reiz, R. M. Whittal, A. Fentabil, J. S. Klassen, and M. F. Feldman, Exploiting bacterial glycosylation machineries for the synthesis of a Lewis antigen-containing glycoprotein, *J. Biol. Chem.*, 286 (2011) 37887–37894.
63. M. T. Madigan, J. M. Martinko, D. Stahl, and D. P. Clark, *Brock Biology of Microorganisms*, 13th ed. (2010) Benjamin Cummings, San Francisco, CA, 1152.
64. R. G. E. Murray, On the cell wall structure of *Spirillum serpens, Can. J. Microbiol.*, 9 (1963) 381–392.
65. M. Glauert, Moiré patterns in electron micrographs of a bacterial membrane, *J. Cell Sci.*, 1 (1966) 425–428.

66. M. J. Thornley, Cell envelopes with regularly arranged surface subunits in *Acinetobacter* and related bacteria, *CRC Crit. Rev. Microbiol.*, 4 (1975) 65–100.
67. U. B. Sleytr, Regular arrays of macromolecules on bacterial cell walls: Structure, chemistry, assembly and function, *Int. Rev. Cytol.*, 53 (1978) 1–64.
68. T. J. Beveridge, Ultrastructure, chemistry, and function of the bacterial wall, *Int. Rev. Cytol.*, 72 (1981) 229–317.
69. U. B. Sleytr, P. Messner, D. Pum, and M. Sára, Crystalline surface layers on eubacteria and archaeobacteria, In: U. B. Sleytr, P. Messner, D. Pum, and M. Sára, (Eds.), *Crystalline Bacterial Cell Surface Proteins,* R.G. Landes/Academic Press, Austin, TX, 1996, pp. 211–225.
70. H. Engelhardt and J. Peters, Structural research on surface layers: A focus on stability, surface layer homology domains, and surface layer-cell wall interactions, *J. Struct. Biol.*, 124 (1998) 276–302.
71. S. Åvall-Jääskeläinen and A. Palva, *Lactobacillus* surface layers and their applications, *FEMS Microbiol. Rev.*, 29 (2005) 511–529.
72. H. Claus, E. Akça, T. Debaerdemaeker, C. Evrard, J.-P. Declercq, J. R. Harris, B. Schlott, and H. König, Molecular organization of selected prokaryotic S-layer proteins, *Can. J. Microbiol.*, 51 (2005) 731–743.
73. U. B. Sleytr and P. Messner, Crystalline cell surface layers (S layers), In: M. Schaechter, (Ed.), 3rd ed. *Encyclopedia of Microbiology*, Vol. 1, Academic Press/Elsevier, San Diego, CA, 2009, pp. 89–98.
74. P. Messner, C. Schäffer, E.-M. Egelseer, and U. B. Sleytr, Occurrence, structure, chemistry, genetics, morphogenesis, and functions of S-layers, In: H. König, H. Claus, and A. Varma, (Eds.), *Prokaryotic Cell Wall Compounds—Structure and Biochemistry,* Springer, Berlin, 2010, pp. 53–109.
75. O. Kandler and H. König, Cell envelopes of archaea: Structure and chemistry, In: M. Kates, D. J. Kushner, and A. T. Matheson, (Eds.), *The Biochemistry of Archaea (Archaebacteria),* Elsevier, Amsterdam, 1993, pp. 223–259.
76. H. Claus, E. Akça, T. Debaerdemaeker, C. Evrard, J.-P. Declercq, and H. König, Primary structure of selected archaeal mesophilic and extremely thermophilic outer surface layer proteins, *System. Appl. Microbiol.*, 25 (2002) 3–12.
77. A. Veith, A. Klingl, B. Zolghadr, K. Lauber, R. Mentele, F. Lottspeich, R. Rachel, S.-V. Albers, and A. Kletzin, *Acidianus, Sulfolobus* and *Metallosphaera* surface layers: structure, composition and gene expression, *Mol. Microbiol.*, 73 (2009) 58–72.
78. W. Baumeister, I. Wildhaber, and B. M. Phipps, Principles of organization in eubacterial and archaebacterial surface proteins, *Can. J. Microbiol.*, 35 (1989) 215–227.
79. U. B. Sleytr and P. Messner, Self-assembly of crystalline bacterial cell surface layers, In: H. Plattner, (Ed.), *Electron Microscopy of Subcellular Dynamics,* CRC Press, Boca Raton, FL, 1989, pp. 13–31.
80. U. B. Sleytr, P. Messner, D. Pum, and M. Sára, Crystalline bacterial cell surface layers (S-layers): From supramolecular cell structure to biomimetics and nanotechnology, *Angew. Chem. Int. Ed. Engl.*, 38 (1999) 1034–1054.
81. J. Müller, W. Baumeister, and A. Engel, Conformational change of the hexagonally packed intermediate layer of *Deinococcus radiodurans* monitored by atomic force microscopy, *J. Bacteriol.*, 178 (1996) 3025–3030.
82. Y. F. Dufrêne, Application of atomic force microscopy to microbial surfaces: From reconstituted cell surface layers to living cells, *Micron,* 32 (2001) 153–165.
83. D. Pum, J. Tang, P. Hinterdorfer, J.-L. Toca-Herrera, and U. B. Sleytr, S-layer protein lattices studied by scanning force microscopy, In: C. S. S. R. Kumar, (Ed.), *Nanomaterial for the Life Sciences, Biomimetic and Bioinspired Nanomaterials,* Vol. 7, Wiley-VCH, Weinheim, Germany, 2010, pp. 459–510.
84. G. Sekot, G. Posch, Y. J. Oh, S. Zayni, H. F. Mayer, D. Pum, P. Messner, P. Hinterdorfer, and C. Schäffer, Analysis of the cell surface layer ultrastructure of the oral pathogen *Tannerella forsythia, Arch. Microbiol.*, 194 (2012) 525–539.

85. U. B. Sleytr, M. Sára, D. Pum, and B. Schuster, Crystalline bacterial cell surface layers (S-layers): A versatile self-assembly system, In: A. Ciferri, (Ed.), *Supramolecular Polymers,* 2nd ed. Taylor & Francis, Boca Raton, FL, 2005, pp. 583–616.
86. T. Pavkov-Keller, S. Howorka, and W. Keller, The structure of bacterial S-layer proteins, *Prog. Mol. Biol. Transl. Sci.*, 103 (2011) 73–130.
87. C. Steindl, C. Schäffer, T. Wugeditsch, M. Graninger, I. Matecko, N. Müller, and P. Messner, The first biantennary bacterial secondary cell wall polymer from bacteria and its influence on S-layer glycoprotein assembly, *Biochem. J.*, 368 (2002) 483–494.
88. M. Jarosch, E. M. Egelseer, C. Huber, D. Moll, D. Mattanovich, U. B. Sleytr, and M. Sára, Analysis of the structure-function relationship of the S-layer protein SbsC of *Bacillus stearothermophilus* ATCC 12980 by producing truncated forms, *Microbiology*, 147 (2001) 1353–1363.
89. D. Rünzler, C. Huber, D. Moll, G. Köhler, and M. Sára, Biophysical characterization of the entire bacterial surface layer protein SbsB and its two distinct functional domains, *J. Biol. Chem.*, 279 (2004) 5207–5215.
90. T. Pavkov, E. M. Egelseer, M. Tesarz, D. I. Svergun, U. B. Sleytr, and W. Keller, The structure and binding behavior of the bacterial cell surface layer protein SbsC, *Structure*, 16 (2008) 1226–1237.
91. A. Đorđić, E. M. Egelseer, M. Tesarz, U. B. Sleytr, W. Keller, and T. Pavkov-Keller, Crystallization of domains involved in self-assembly of the S-layer protein SbsC, *Acta Cryst.*, F 68 (2012) 1511–1514.
92. E. Baranova, R. Fronzes, A. Garcia-Pino, N. Van Gerven, D. Papapostolou, G. Péhau-Arnaudet, E. Pardon, J. Steyaert, S. Howorka, and H. Remaut, SbsB structure and lattice reconstruction unveil Ca^{2+} triggered S-layer assembly, *Nature*, 487 (2012) 119–122.
93. U. B. Sleytr, E.-M. Egelseer, N. Ilk, P. Messner, C. Schäffer, D. Pum, and B. Schuster, Nanobiotechnological applications of S-layers, In: H. König, H. Claus, and A. Varma, (Eds.), *Prokaryotic Cell Wall Compounds—Structure and Biochemistry,* Springer, Berlin, 2010, pp. 459–481.
94. C. Horejs, D. Pum, U. B. Sleytr, and R. Tscheliessnig, Structure prediction of an S-layer protein by the mean force method, *J. Chem. Phys.*, 128 (2008) 065106.
95. C. Horejs, M. K. Mitra, D. Pum, U. B. Sleytr, and M. Muthukumar, Monte Carlo study of the molecular mechanisms of surface-layer protein self-assembly, *J. Chem. Phys.*, 134 (2011) 125103.
96. U. B. Sleytr, B. Schuster, E. M. Egelseer, D. Pum, C. M. Horejs, R. Tscheliessnig, and N. Ilk, Nanobiotechnology with S-layer proteins as building blocks, *Prog. Mol. Biol. Transl. Sci.*, 103 (2011) 277–352.
97. U. B. Sleytr, Basic and applied S-layer research: An overview, *FEMS Microbiol. Rev.*, 20 (1997) 5–12.
98. U. B. Sleytr, M. Sára, D. Pum, B. Schuster, P. Messner, and C. Schäffer, Self-assembly protein systems: Microbial S-layers, In: A. Steinbüchel and S. R. Fahnestock, (Eds.) *Biopolymers, Polyamides and Complex Proteinaceous Matrices I*, Vol. 7, Wiley-VCH, Weinheim, Germany, 2002, pp. 285–338.
99. H. Engelhardt, Are S-layers exoskeletons? The basic function of protein surface layers revisited, *J. Struct. Biol.*, 160 (2007) 115–124.
100. G. Posch, G. Sekot, V. Friedrich, Z. A. Megson, A. Koerdt, P. Messner, and C. Schäffer, Glycobiology aspects of the periodontal pathogen *Tannerella forsythia*, *Biomolecules*, 2 (2012) 467–482.
101. M. Sára, B. Kuen, H. F. Mayer, F. Mandl, K. C. Schuster, and U. B. Sleytr, Dynamics in oxygen-induced changes in S-layer protein synthesis from *Bacillus stearothermophilus* PV72 and the S-layer-deficient variant T5 in continuous culture and studies of the cell wall composition, *J. Bacteriol.*, 178 (1996) 2108–2117.
102. J. Dworkin and M. J. Blaser, Molecular mechanisms of *Campylobacter fetus* surface layer protein expression, *Mol. Microbiol.*, 26 (1997) 433–440.
103. M. J. Blaser, *Campylobacter fetus*—Emerging infection and model system for bacterial pathogenesis at mucosal surfaces, *Clin. Infect. Dis.*, 27 (1998) 256–258.

104. H. C. Scholz, E. Riedmann, A. Witte, W. Lubitz, and B. Kuen, S-layer variation in *Bacillus stearothermophilus* PV72 is based on DNA rearrangements between the chromosome and the naturally occurring megaplasmids, *J. Bacteriol.*, 183 (2001) 1672–1679.
105. M. Jakava-Viljanen, S. Åvall-Jääskeläinen, P. Messner, U. B. Sleytr, and A. Palva, Isolation of three new surface layer protein genes (*slp*) from *Lactobacillus brevis* ATCC 14869 and characterization of the change in their expression under aerated and anaerobic conditions, *J. Bacteriol.*, 184 (2002) 6786–6795.
106. G. Posch, M. Pabst, L. Brecker, F. Altmann, P. Messner, and C. Schäffer, Characterization and scope of S-layer protein *O*-glycosylation in *Tannerella forsythia*, *J. Biol. Chem.*, 286 (2011) 38714–38724.
107. K.-L. Hsu, K. T. Pilobello, and L. K. Mahal, Analyzing the dynamic bacterial glycome with a lectin microarray approach, *Nat. Chem. Biol.*, 2 (2006) 153–157.
108. J. Sakakibara, K. Nagano, Y. Murakami, N. Higuchi, H. Nakamura, K. Shimozato, and F. Yoshimura, Loss of adherence ability to human gingival epithelial cells in S-layer protein-deficient mutants of *Tannerella forsythensis*, *Microbiology*, 153 (2007) 866–876.
109. R. P. Settem, K. Honma, T. Nakajima, C. Phansopa, S. Roy, G. P. Stafford, and A. Sharma, A bacterial glycan core linked to surface (S)-layer proteins modulates host immunity through Th17 suppression, *Mucosal Immunol.*, 6 (2013) 415–426.
110. C. Schäffer and P. Messner, Glycobiology of surface layer proteins, *Biochimie*, 83 (2001) 591–599.
111. R. Ristl, K. Steiner, K. Zarschler, S. Zayni, P. Messner, and C. Schäffer, The S-layer glycome—Adding to the sugar coat of bacteria, *Int. J. Microbiol.*, 2011 (2011) 127870.
112. F. Jarrell, G. M. Jones, L. Kandiba, D. B. Nair, and J. Eichler, S-layer glycoproteins and flagellins: Reporters of archaeal posttranslational modifications, *Archaea*, 2010 (2010) 612948.
113. S.-V. Albers and B. H. Meyer, The archaeal cell envelope, *Nat. Rev. Microbiol.*, 9 (2011) 414–426.
114. Z. Guan, S. Naparstek, D. Calo, and J. Eichler, Protein glycosylation as an adaptive response in Archaea: Growth at different salt concentrations leads to alterations in *Haloferax volcanii* S-layer glycoprotein N-glycosylation, *Environ. Microbiol.*, 14 (2012) 743–753.
115. C. Schäffer and P. Messner, Surface layer glycoproteins: An example for the diversity of bacterial glycosylation with promising impacts on nanobiotechnology, *Glycobiology*, 14 (2004) 31R–42R.
116. P. Messner, K. Steiner, K. Zarschler, and C. Schäffer, S-layer nanoglycobiology of bacteria, *Carbohydr. Res.*, 343 (2008) 1934–1951.
117. P. Messner, F. Hollaus, and U. B. Sleytr, Paracrystalline cell wall surface layers of different *Bacillus stearothermophilus* strains, *Int. J. Syst. Bacteriol.*, 34 (1984) 202–210.
118. K. Meier-Stauffer, H.-J. Busse, F. A. Rainey, J. Burghardt, A. Scheberl, F. Hollaus, B. Kuen, A. Makristathis, U. B. Sleytr, and P. Messner, Description of *Bacillus thermoaerophilus* sp. nov., to include sugar beet isolates and *Bacillus brevis* ATCC 12990, *Int. J. Syst. Bacteriol.*, 46 (1996) 532–541.
119. C. Schäffer, W. L. Franck, A. Scheberl, P. Kosma, T. R. McDermott, and P. Messner, Classification of two isolates from locations in Austria and Yellowstone National Park as *Geobacillus tepidamans* sp. nov., *Int. J. Syst. Evol. Microbiol.*, 54 (2004) 2361–2368.
119a. A. Coorevits, A. E. Dinsdale, G. Halket, L. Lebbe, P. De Vos, A. van Landschoot, and N. A. Logan. Taxonomic revision of the genus *Geobacillus*: emendation of *Geobacillus*, *G. stearothermophilus, G. jurassicus, G. toebii, G. thermodenitrificans* and *G. thermoglucosidans* (nom. corrig., formerly '*thermoglucosidasius*'); transfer of *Bacillus thermantarcticus* to the genus as *G. thermantarcticus* comb. nov.; proposal of *Caldibacillus debilis* gen. nov., comb. nov.; transfer of *G. tepidamans* to *Anoxybacillus* as *A. tepidamans* comb. nov.; and proposal of *Anoxybacillus caldiproteolyticus* sp. nov., *Int. J. Syst. Evol. Microbiol.*, 62 (2012) 1470–1485.
120. A. Möschl, C. Schäffer, U. B. Sleytr, P. Messner, R. Christian, and G. Schulz, Characterization of the S-layer glycoproteins of two lactobacilli, In: T. J. Beveridge and S. F. Koval, (Eds.) *Advances in Paracrystalline Bacterial Surface Layers*, Plenum Press, New York, 1993, pp. 281–284.

121. S. Heinl, D. Wibberg, F. Eikmeyer, R. Szczepanowski, J. Blom, B. Linke, A. Goesmann, R. Grabherr, H. Schwab, A. Pühler, and A. Schlüter, Insights into the completely annotated genome of *Lactobacillus buchneri* CD034, a strain isolated from stable grass silage, *J. Biotechnol.*, 161 (2012) 153–166.
122. E.-M. Egelseer, N. Ilk, D. Pum, P. Messner, C. Schäffer, B. Schuster, and U. B. Sleytr, S-layers, microbial, biotechnological applications, In: M. C. Flickinger, (Ed.), *Encyclopedia of Industrial Biotechnology: Bioprocess, Bioseparation, and Cell Technology*, Vol.7, John Wiley & Sons, Hoboken, NJ, 2010, pp. 4424–4448.
123. R. H. Raetz and C. Whitfield, Lipopolysaccharide endotoxins, *Annu. Rev. Biochem.*, 71 (2002) 635–700.
124. K. Greenfield and C. Whitfield, Synthesis of lipopolysaccharide O-antigens by ABC transporter-dependent pathways, *Carbohydr. Res.*, 356 (2012) 12–24.
125. S.-W. Lee, M. Sabet, H.-S. Um, J. Yang, H. C. Kim, and W. Zhu, Identification and characterization of the genes encoding a unique surface (S-) layer of *Tannerella forsythia*, *Gene*, 371 (2006) 102–111.
126. M. Fletcher, M. J. Coyne, O. F. Villa, M. Chatzidaki-Livanis, and L. E. Comstock, A general O-glycosylation system important to the physiology of a major human intestinal symbiont, *Cell*, 137 (2009) 321–331.
127. G. Posch, M. Pabst, L. Neumann, M. J. Coyne, F. Altmann, P. Messner, L. E. Comstock, and C. Schäffer, "Cross-glycosylation" of proteins in *Bacteriodales* species, *Glycobiology*, 23 (2013) 568–577.
128. P. Messner and C. Schäffer, Surface layer glycoproteins of Bacteria and Archaea, In: R. J. Doyle, (Ed.), *Glycomicrobiology*, Kluwer Academic/Plenum Publishers, New York, 2000, pp. 93–125.
130. K. Zarschler, B. Janesch, M. Pabst, F. Altmann, P. Messner, and C. Schäffer, Protein tyrosine O-glycosylation—Insights into a rather unexplored prokaryotic glycosylation system, *Glycobiology*, 20 (2010) 787–798.
131. C. Schäffer, T. Wugeditsch, H. Kählig, A. Scheberl, S. Zayni, and P. Messner, The surface layer (S-layer) glycoprotein of *Geobacillus stearothermophilus* NRS 2004/3a. Analysis of its glycosylation, *J. Biol. Chem.*, 277 (2002) 6230–6239.
132. H. Kählig, D. Kolarich, S. Zayni, A. Scheberl, P. Kosma, C. Schäffer, and P. Messner, N-Acetylmuramic acid as capping element of α-D-fucose-containing S-layer glycoprotein glycans from *Geobacillus tepidamans* GS5-97^T, *J. Biol. Chem.*, 280 (2005) 20292–20299.
133. P. Kosma, T. Wugeditsch, R. Christian, S. Zayni, and P. Messner, Glycan structure of a heptose-containing S-layer glycoprotein of *Bacillus thermoaerophilus*, *Glycobiology*, 5 (1995) 791–796.
134. E. Altman, J.-R. Brisson, P. Messner, and U. B. Sleytr, Chemical characterization of the regularly arranged surface layer glycoprotein of *Bacillus alvei* CCM 2051, *Biochem. Cell Biol.*, 69 (1991) 72–78.
135. P. Messner, R. Christian, C. Neuninger, and G. Schulz, Similarity of "core" structures in two different glycans of tyrosine-linked eubacterial S-layer glycoproteins, *J. Bacteriol.*, 177 (1995) 2188–2193.
136. K. Bock, J. Schuster-Kolbe, E. Altman, G. Allmaier, B. Stahl, R. Christian, U. B. Sleytr, and P. Messner, Primary structure of the O-glycosidically linked glycan chain of the crystalline surface layer glycoprotein of *Thermoanaerobacter thermohydrosulfuricus* L111-69. Galactosyl tyrosine as a novel linkage unit, *J. Biol. Chem.*, 269 (1994) 7137–7144.
137. C. Schäffer, N. Müller, R. Christian, M. Graninger, T. Wugeditsch, A. Scheberl, and P. Messner, Complete glycan structure of the S-layer glycoprotein of *Aneurinibacillus thermoaerophilus* GS4-97, *Glycobiology*, 9 (1999) 407–414.
138. E. Altman, J.-R. Brisson, P. Messner, and U. B. Sleytr, Chemical characterization of the regularly arranged surface layer glycoprotein of *Clostridium thermosaccharolyticum* D120-70, *Eur. J. Biochem.*, 188 (1990) 73–82.
139. C. Schäffer, K. Dietrich, B. Unger, A. Scheberl, F. A. Rainey, H. Kählig, and P. Messner, A novel type of carbohydrate-protein linkage region in the tyrosine-bound S-layer glycan of *Thermoanaerobacterium thermosaccharolyticum* D120-70, *Eur. J. Biochem.*, 267 (2000) 5482–5492.

140. R. Christian, G. Schulz, J. Schuster-Kolbe, G. Allmaier, E. R. Schmid, U. B. Sleytr, and P. Messner, Complete structure of the tyrosine-linked saccharide moiety from the surface layer glycoprotein of *Clostridium thermohydrosulfuricum* S102-70, *J. Bacteriol.*, 175 (1993) 1250–1256.
141. J. Anzengruber, M. Pabst, L. Neumann, G. Sekot, S. Heinl, R. Grabherr, F. Altmann, P. Messner, and C. Schäffer, Protein *O*-glucosylation in *Lactobacillus buchneri*. *Glycoconj. J.*, http://dx.doi.org/10.1007/s10719-013-9505-7
142. B. Lindberg, Components of bacterial polysaccharides, *Adv. Carbohydr. Chem. Biochem.*, 48 (1990) 279–318.
143. R. Apweiler, H. Hermjakob, and N. Sharon, On the frequency of protein glycosylation, as deduced from analysis of the SWISS-PROT database, *Biochim. Biophys. Acta*, 1473 (1999) 4–8.
144. P. Kosma, C. Neuninger, R. Christian, G. Schulz, and P. Messner, Glycan structure of *Bacillus* sp. L420-91, *Glycoconj. J.*, 12 (1995) 99–107.
145. E. Altman, J.-R. Brisson, S. M. Gagné, J. Kolbe, P. Messner, and U. B. Sleytr, Structure of the glycan chain from the surface layer glycoprotein of *Clostridium thermohydrosulfuricus* L77-66, *Biochim. Biophys. Acta*, 1117 (1992) 71–77.
146. E. Altman, C. Schäffer, J.-R. Brisson, and P. Messner, Characterization of the glycan structure of a major glycopeptide from the surface layer glycoprotein of *Clostridium thermosaccharolyticum* E207-71, *Eur. J. Biochem.*, 229 (1995) 308–315.
147. K. Larue, M. S. Kimber, R. Ford, and C. Whitfield, Biochemical and structural analysis of bacterial *O*-antigen chain length regulator proteins reveals a conserved quaternary structure, *J. Biol. Chem.*, 284 (2009) 7395–7403.
148. H. Claus and H. König, Cell envelopes of methanogens, In: H. König, H. Claus, and A. Varma, (Eds.), *Prokaryotic Cell Wall Compounds—Structure and Biochemistry*, Springer, Berlin, 2010, pp. 231–251.
149. J. Eichler, M. Abu-Qarn, Z. Konrad, H. Magidovich, N. Plavner, and S. Yurist-Doutsch, The cell envelopes of haloarchaea: Staying in shape in a world of salt, In: H. König, H. Claus, and A. Varma, (Eds.), *Prokaryotic Cell Wall Compounds—Structure and Biochemistry*, Springer, Berlin, 2010, pp. 253–270.
149a. J. Eichler, Extreme sweetness: Protein glycosylation in archaea, *Nat. Rev. Microbiol.*, 11 (2013) 151–156.
150. L. Kaminski, M. Abu-Qarn, Z. Guan, S. Naparstek, V. V. Ventura, C. R. H. Raetz, P. G. Hitchen, A. Dell, and J. Eichler, AglJ adds the first sugar of the N-linked pentasaccharide decorating the *Haloferax volcanii* S-layer glycoprotein, *J. Bacteriol.*, 192 (2010) 5572–5579.
151. L. Kaminski, Z. Guan, M. Abu-Qarn, Z. Konrad, and J. Eichler, AglR is required for addition of the final mannose residue of the N-linked glycan decorating the *Haloferax volcanii* S-layer glycoprotein, *Biochim. Biophys. Acta*, 1820 (2012) 1664–1670.
152. C. Cohen-Rosenzweig, S. Yurist-Doutsch, and J. Eichler, AglS, a novel component of the *Haloferax volcanii* N-glycosylation pathway, is a dolichol phosphate-mannose mannosyltransferase, *J. Bacteriol.*, 194 (2012) 6909–6916.
153. L. Kandiba, Z. Guan, and J. Eichler, Lipid modification gives rise to two distinct *Haloferax volcanii* S-layer glycoprotein populations, *Biochim. Biophys. Acta*, 1828 (2013) 938–943.
154. D. Calo, Z. Guan, and J. Eichler, Glyco-engineering in *Archaea*: Differential N-glycosylation of the S-layer glycoprotein in a transformed *Haloferax volcanii* strain, *Microb. Biotechnol.*, 4 (2011) 461–470.
155. S. Voisin, R. S. Houliston, J. Kelly, J.-R. Brisson, D. Watson, S. L. Bardy, K. F. Jarrell, and S. M. Logan, Identification and characterization of the unique *N*-linked glycan common to the flagellins and S-layer glycoprotein of *Methanococcus voltae*, *J. Biol. Chem.*, 280 (2005) 16586–16593.
156. B. Chaban, S. M. Logan, J. F. Kelly, and K. F. Jarrell, AglC and AglK are involved in biosynthesis and attachment of diacetylated glucuronic acid to the N-glycan in *Methanococcus voltae*, *J. Bacteriol.*, 191 (2009) 187–195.

157. K. F. Jarrell, G. M. Jones, and D. B. Nair, Biosynthesis and role of N-linked glycosylation in cell surface structures of Archaea with a focus on flagella and S layers, *Int. J. Microbiol.*, 2010 (2010) 470138.
158. D. J. VanDyke, J. Wu, S. M. Logan, J. F. Kelly, S. Mizuno, S.-I. Aizawa, and K. F. Jarrell, Identification of genes involved in the assembly and attachment of a novel flagellin N-linked tetrasaccharide important for motility in the archaeon *Methanococcus maripaludis*, *Mol. Microbiol.*, 72 (2009) 633–644.
159. G. M. Jones, J. Wu, Y. Ding, K. Uchida, S.-I. Aizawa, A. Robotham, S. M. Logan, J. Kelly, and K. F. Jarrell, Identification of genes involved in the acetamidino group modification of the flagellin N-linked glycan of *Methanococcus maripaludis*, *J. Bacteriol.*, 194 (2012) 2693–2702.
160. E. Peyfoon, B. Meyer, P. G. Hitchen, M. Panico, H. R. Morris, S. M. Haslam, S.-V. Albers, and A. Dell, The S-layer glycoprotein of the crenarchaeote *Sulfolobus acidocaldarius* is glycosylated at multiple sites with chitobiose-linked N-glycans, *Archaea*, 2010 (2010) 754101.
161. U. Zähringer, H. Moll, T. Hettmann, Y. A. Knirel, and G. Schäfer, Cytochrome $b_{558/566}$ from the archaeon *Sulfolobus acidocaldarius* has a unique Asn-linked highly branched hexasaccharide chain containing 6-sulfoquinovose, *Eur. J. Biochem.*, 267 (2000) 4144–4149.
162. B. H. Meyer, B. Zolghadr, E. Peyfoon, M. Pabst, M. Panico, H. R. Morris, S. M. Haslam, P. Messner, C. Schäffer, A. Dell, and S.-V. Albers, Sulfoquinovose synthase—An important enzyme in the N-glycosylation pathway of *Sulfolobus acidocaldarius*, *Mol. Microbiol.*, 82 (2011) 1150–1163.
163. Z. Guan, B. H. Meyer, S.-V. Albers, and J. Eichler, The thermoacidophilic archaeon *Sulfolobus acidocaldarius* contains an unusually short, highly reduced dolichyl phosphate, *Biochim. Biophys. Acta*, 1811 (2011) 607–616.
164. S. Mesnage, T. Fontaine, T. Mignot, M. Delepierre, M. Mock, and A. Fouet, Bacterial SLH domain proteins are non-covalently anchored to the cell surface via a conserved mechanism involving wall polysaccharide pyruvylation, *EMBO J.*, 19 (2000) 4473–4484.
165. M. Sára, Conserved anchoring mechanisms between crystalline cell surface S-layer proteins and secondary cell wall polymers in Gram-positive bacteria, *Trends Microbiol.*, 9 (2001) 47–49.
166. C. Schäffer and P. Messner, The structure of secondary cell wall polymers: How Gram-positive bacteria stick their cell walls together, *Microbiology*, 151 (2005) 643–651.
167. B. Janesch, P. Messner, and C. Schäffer, Are the SLH-domains essential for cell surface display and glycosylation of the S-layer protein from *Paenibacillus alvei* CCM 2051^T? *J. Bacteriol.*, 195 (2013) 565–575.
168. S. G. Griffiths and W. H. Lynch, Characterization of *Aeromonas salmonicida* variants with altered cell surfaces and their use in studying surface protein assembly, *Arch. Microbiol.*, 154 (1990) 308–312.
169. A. R. Archibald, I. C. Hancock, and C. R. Harwood, Cell wall structure, synthesis and turnover, In: A. Sonenshein, J. A. Hoch, and R. Losick, (Eds.), *Bacillus subtilis and Other Gram-Positive Bacteria*, American Society for Microbiology, Washington, DC, 1993, pp. 381–410.
170. I. C. Hancock and J. Baddiley, Biosynthesis of the bacterial envelope polymers teichoic acid and teichuronic acid, In: A. N. Martonosi, (Ed.), 2nd ed. *The Enzymes of Biological Membranes*, Vol. 2, Plenum Press, New York, NY, 1985, pp. 279–307.
171. Y. Araki and E. Ito, Linkage units in cell walls of Gram-positive bacteria, *CRC Crit. Rev. Microbiol.*, 17 (1989) 121–135.
172. I. B. Naumova and A. S. Shashkov, Anionic polymers in cell walls of Gram-positive bacteria, *Biochemistry (English translation of Biokhimiya)*, 62 (1997) 809–840.
173. H. M. Pooley and D. Karamata, Teichoic acid synthesis in *Bacillus subtilis*: Genetic organisation and biological roles, In: J.-M. Ghuysen and R. Hakenbeck, (Eds.) *Bacterial Cell Wall*, Elsevier, Amsterdam, 1994, pp. 187–198.
174. F. Cava, M. A. de Pedro, H. Schwarz, A. Henne, and J. Berenguer, Binding to pyruvylated compounds as an ancestral mechanism to anchor the outer envelope in primitive bacteria, *Mol. Microbiol.*, 52 (2004) 677–690.

175. P. Messner, E. M. Egelseer, U. B. Sleytr, and C. Schäffer, Surface layer glycoproteins and secondary cell wall polymers, In: A. P. Moran, P. J. Brennan, O. Holst, and M. von Itzstein, (Eds.), *Microbial Glycobiology: Structures, Relevance and Applications,* Academic Press–Elsevier, San Diego, CA, 2009, pp. 109–128.
176. C. Schäffer, N. Müller, P. K. Mandal, R. Christian, S. Zayni, and P. Messner, A pyrophosphate bridge links the pyruvate-containing secondary cell wall polymer of *Paenibacillus alvei* CCM 2051 to muramic acid, *Glycoconj. J.*, 17 (2000) 681–690.
177. N. Ilk, P. Kosma, M. Puchberger, E. M. Egelseer, H. F. Mayer, U. B. Sleytr, and M. Sára, Structural and functional analyses of the secondary cell wall polymer of *Bacillus sphaericus* CCM 2177 that serves as an S-layer-specific anchor, *J. Bacteriol.*, 181 (1999) 7643–7646.
178. E. Altman, C. Schäffer, J.-R. Brisson, and P. Messner, Isolation and characterization of an amino sugar-rich glycopeptide from the surface layer glycoprotein of *Thermoanaerobacterium thermosaccharolyticum* E207-71, *Carbohydr. Res.*, 295 (1996) 245–253.
179. P. Messner, U. B. Sleytr, R. Christian, G. Schulz, and F. M. Unger, Isolation and structure determination of a diacetamidodideoxyuronic acid-containing glycan chain from the S-layer glycoprotein of *Bacillus stearothermophilus* NRS 2004/3a, *Carbohydr. Res.*, 168 (1987) 211–218.
180. C. Schäffer, H. Kählig, R. Christian, G. Schulz, S. Zayni, and P. Messner, The diacetamidodideoxyuronic-acid-containing glycan chain of *Bacillus stearothermophilus* NRS 2004/3a represents the secondary cell wall polymer of wild-type *B. stearothermophilus* strains, *Microbiology*, 145 (1999) 1575–1583.
181. C. Steindl, C. Schäffer, V. Smrečki, P. Messner, and N. Müller, The secondary cell wall polymer of *Geobacillus tepidamans* strain GS5-97^T: Structure of different glycoforms, *Carbohydr. Res.*, 340 (2005) 2290–2296.
182. B. O. Petersen, M. Sára, C. Mader, H. F. Mayer, U. B. Sleytr, M. Pabst, M. Puchberger, A. Hofinger, J.Ø. Duus, and P. Kosma, Structural characterization of the acid-degraded secondary cell wall polysaccharide of *Geobacillus stearothermophilus* PV72/p2, *Carbohydr. Res.*, 343 (2008) 1346–1358. Corrigendum: *Carbohydr. Res.*, 373 (2013) 52.
183. V. J. Kern, J. W. Kern, J. A. Theriot, O. Schneewind, and D. Missiakis, Surface-layer (S-layer) proteins Sap and EA1 govern the binding of the S-layer-associated protein BslO at the cell septa of *Bacillus anthracis*, *J. Bacteriol.*, 194 (2012) 3833–3840.
184. J. Kern, C. Ryan, K. Faull, and O. Schneewind, *Bacillus anthracis* surface-layer proteins assemble by binding to the secondary cell wall polysaccharide in a manner that requires *csaB* and *tagO*, *J. Mol. Biol.*, 401 (2010) 757–775.
185. B. Choudhury, C. Leoff, E. Saile, P. Wilkins, C. P. Quinn, E. L. Kannenberg, and R. W. Carlson, The structure of the major cell wall polysaccharide of *Bacillus anthracis* is species-specific, *J. Biol. Chem.*, 281 (2006) 27932–27941.
186. C. Leoff, E. Saile, J. Rauvolfova, C. P. Quinn, A. Hoffmaster, W. Zhong, A. S. Mehta, G.-J. Boons, R. W. Carlson, and E. L. Kannenberg, Secondary cell wall polysaccharides of *Bacillus anthracis* are antigens that contain specific epitopes which cross-react with three pathogenic *Bacillus cereus* strains that caused severe disease, and other epitopes common to all the *Bacillus cereus* strains tested, *Glycobiology*, 19 (2009) 665–673.
187. R. Schuch, D. Nelson, and V. A. Fischetti, A bacteriolytic agent that detects and kills *Bacillus anthracis*, *Nature*, 418 (2002) 884–889.
188. K.-F. Mo, X. Li, H. Li, L. Y. Low, C. P. Quinn, and G.-J. Boons, Endolysins of *Bacillus anthracis* bacteriophages recognize unique carbohydrate epitopes of vegetative cell wall polysaccharides with high affinity and selectivity, *J. Am. Chem. Soc.*, 134 (2012) 15556–15562.
189. L. S. Forsberg, T. G. Abshire, A. Friedlander, C. P. Quinn, E. L. Kannenberg, and R. W. Carlson, Localization and structural analysis of a conserved pyruvylated epitope in *Bacillus anthracis* secondary cell wall polysaccharides and characterization of the galactose-deficient wall polysaccharide from avirulent *B. anthracis* CDC 684, *Glycobiology*, 22 (2012) 1103–1117.

190. L. S. Forsberg, B. Choudhury, C. Leoff, C. K. Marston, A. R. Hoffmaster, E. Saile, C. P. Quinn, E. L. Kannenberg, and R. W. Carlson, Secondary cell wall polysaccharides from *Bacillus cereus* strains G9241, 03BB87 and 03BB102 causing fatal pneumonia share similar glycosyl structures with polysaccharides from *Bacillus anthracis, Glycobiology*, 21 (2011) 934–948.
191. C. Leoff, B. Choudhury, E. Saile, C. P. Quinn, R. W. Carlson, and E. L. Kannenberg, Structural elucidation of the nonclassical secondary cell wall polysaccharide from *Bacillus cereus* ATCC 10987, *J. Biol. Chem.*, 283 (2008) 29812–29821.
192. T. Candela, E. Maes, E. Garénaux, Y. Rombouts, F. Krzewinski, M. Gohar, and Y. Guérardel, Environmental and bio-film dependent changes in a *Bacillus cereus* secondary cell wall polysaccharide, *J. Biol. Chem.*, 286 (2011) 31250–31262.
193. W. W. Navarre and O. Schneewind, Surface proteins of gram-positive bacteria and mechanisms of their targeting to the cell wall envelope, *Microbiol. Mol. Biol. Rev.*, 63 (1999) 174–229.
194. U. B. Sleytr, Self-assembly of the hexagonally and tetragonally arranged subunits of bacterial surface layers and their reattachment to cell walls, *J. Ultrastruct. Res.*, 55 (1976) 360–377.
195. K. Masuda and T. Kawata, Reassembly of a regularly arranged protein in the cell wall of *Lactobacillus buchneri* and its reattachment to cell walls: Chemical modification studies, *Microbiol. Immunol.*, 29 (1985) 927–938.
196. T. Fujino, P. Béguin, and J. P. Aubert, Organization of a *Clostridium thermocellum* gene cluster encoding the cellulosomal scaffolding protein CipA and a protein possibly involved in attachment of the cellulosome to the cell surface, *J. Bacteriol.*, 175 (1993) 1891–1899.
197. A. Lupas, H. Engelhardt, J. Peters, U. Santarius, S. Volker, and W. Baumeister, Domain structure of the *Acetogenium kivui* surface layer revealed by electron crystallography and sequence analysis, *J. Bacteriol.*, 176 (1994) 1224–1233.
198. E. A. Bayer, J.-P. Belaich, Y. Shoham, and R. Lamed, The cellulosomes: Multienzyme machines for degradation of plant cell wall polysaccharides, *Annu. Rev. Microbiol.*, 58 (2004) 521–554.
199. S.-Y. Liu, F. C. Gherardini, M. Matuschek, H. Bahl, and J. Wiegel, Cloning, sequencing, and expression of the gene encoding a large S-layer-associated endoxylanase from *Thermoanaerobacterium* sp. strain JW/SL-YS 485 in *Escherichia coli, J. Bacteriol.*, 178 (1996) 1539–1547.
200. S. Mesnage, E. Tosi-Couture, M. Mock, P. Gounon, and A. Fouet, Molecular characterization of the *Bacillus anthracis* S-layer component: Evidence that it is the major cell-associated antigen, *Mol. Microbiol.*, 23 (1997) 1147–1155.
201. S. Mesnage, E. Tosi-Couture, M. Mock, and A. Fouet, The S-layer homology domain as a means for anchoring heterologous proteins on the cell surface of *Bacillus anthracis, J. Appl. Microbiol.*, 87 (1999) 256–260.
202. A. May, T. Pusztahelyi, N. Hoffmann, R. J. Fischer, and H. Bahl, Mutagenesis of conserved charged amino acids in SLH domains of *Thermoanaerobacterium thermosulfurigenes* EM1 affects attachment to cell wall sacculi, *Arch. Microbiol.*, 185 (2006) 263–269.
203. J. Kern, R. Wilton, R. Zhang, T. A. Binkowski, A. Joachimiak, and O. Schneewind, Structure of surface layer homology (SLH) domains from *Bacillus anthracis* surface array protein, *J. Biol. Chem.*, 286 (2011) 26042–26049.
204. W. Ries, C. Hotzy, I. Schocher, U. B. Sleytr, and M. Sára, Evidence for the N-terminal part of the S-layer protein from *Bacillus stearothermophilus* PV72/p2 recognizes a secondary cell wall polymer, *J. Bacteriol.*, 179 (1997) 3892–3898.
205. M. Sára, E. M. Egelseer, C. Dekitsch, and U. B. Sleytr, Identification of two binding domains, one for peptidoglycan and another for a secondary cell wall polymer on the N-terminal part of the S-layer protein SbsB from *Bacillus stearothermophilus* PV72/p2, *J. Bacteriol.*, 180 (1998) 6780–6783.
206. B. Kuen, A. Koch, E. Asenbauer, M. Sára, and W. Lubitz, Molecular characterization of the *Bacillus stearothermophilus* PV72 S-layer gene *sbsB* induced by oxidative stress, *J. Bacteriol.*, 179 (1997) 1664–1670.

207. E. Smit, F. Oling, R. Demel, B. Martinez, and P. H. Pouwels, The S-layer protein of *Lactobacillus acidophilus* ATCC 4356: Identification and characterisation of domains responsible for S-protein assembly and cell wall binding, *J. Mol. Biol.*, 305 (2001) 245–257.
208. J. Antikainen, L. Anton, L. J. Sillanpää, and T. K. Korhonen, Domains in the S-layer protein CbsA of *Lactobacillus crispatus* involved in adherence to collagens, laminin and lipoteichoic acids and in self-assembly, *Mol. Microbiol.*, 46 (2002) 381–394.
209. I. W. Sutherland, Microbial polysaccharide products, *Biotechnol. Genet. Eng. Rev.*, 16 (1999) 217–229.
210. E. Egelseer, K. Leitner, M. Jarosch, C. Hotzy, S. Zayni, U. B. Sleytr, and M. Sára, The S-layer proteins of two *Bacillus stearothermophilus* wild-type strains are bound via their N-terminal region to a secondary cell wall polymer of identical chemical composition, *J. Bacteriol.*, 180 (1998) 1488–1495.
211. J. M. Bobbitt, Periodate oxidation of carbohydrates, *Adv. Carbohydr. Chem. Biochem.*, 48 (1956) 1–41.
212. A. S. Perlin, Glycol-cleavage oxidation, *Adv. Carbohydr. Chem. Biochem.*, 60 (2006) 183–250.
213. T. Wugeditsch, N. E. Zachara, M. Puchberger, P. Kosma, A. A. Gooley, and P. Messner, Structural heterogeneity in the core oligosaccharide of the S-layer glycoprotein from *Aneurinibacillus thermoaerophilus* DSM 10155, *Glycobiology*, 9 (1999) 787–795.
214. K. Bock, C. Pedersen, and H. Pedersen, Carbon-13 nuclear magnetic resonance data for oligosaccharides, *Adv. Carbohydr. Chem. Biochem.*, 42 (1984) 193–225.
215. A. S. Shashkov, G. M. Lipkind, Y. A. Knirel, and N. K. Kochetkov, Stereochemical factors determining the effects of glycosylation on the ^{13}C chemical shifts in carbohydrates, *Magn. Res. Chem.*, 26 (1988) 735–747.
216. B. O. Petersen, Y. Vinogradov, W. Kay, P. Würtz, N. T. Nyberg, J.Ø. Duus, and O. W. Sørensen, H2BC: A new technique for NMR analysis of complex carbohydrates, *Carbohydr. Res.*, 341 (2006) 550–556.
217. P. J. Garegg, P.-E. Jansson, B. Lindberg, F. Lindh, J. Lönngren, I. Kvarnström, and W. Nimmich, Configuration of the acetal carbon atom of pyruvic acid acetals in some bacterial polysaccharides, *Carbohydr. Res.*, 78 (1980) 127–132.
218. P. Messner, R. Christian, J. Kolbe, G. Schulz, and U. B. Sleytr, Analysis of a novel linkage unit of O-linked carbohydrates from the crystalline surface layer glycoprotein of *Clostridium thermohydrosulfuricum* S102-70, *J. Bacteriol.*, 174 (1992) 2236–2240.
219. G. Allmaier, C. Schäffer, P. Messner, U. Rapp, and F. J. Mayer-Posner, Accurate determination of the molecular weight of the major surface layer protein isolated from *Clostridium thermosaccharolyticum* by time-of-flight mass spectrometry, *J. Bacteriol.*, 177 (1995) 1402–1404.
220. L. Bindila, K. Steiner, C. Schäffer, P. Messner, M. Mormann, and J. Peter-Katalinić, Sequencing of O-glycopeptides derived from a S-layer glycoprotein of *Geobacillus stearothermophilus* NRS 2004/3a containing up to 51 monosaccharide residues at a single glycosylation site by Fourier transform ion cyclotron resonance infrared multiphoton dissociation mass spectrometry, *Anal. Chem.*, 79 (2007) 3271–3279.
221. K. Steiner, R. Novotny, K. Patel, E. Vinogradov, C. Whitfield, M. A. Valvano, P. Messner, and C. Schäffer, Functional characterization of the initiation enzyme of S-layer glycoprotein glycan biosynthesis in *Geobacillus stearothermophilus* NRS 2004/3a, *J. Bacteriol.*, 189 (2007) 2590–2598.
222. K. Steiner, R. Novotny, D. B. Werz, K. Zarschler, P. H. Seeberger, A. Hofinger, P. Kosma, C. Schäffer, and P. Messner, Molecular basis of S-layer glycoprotein glycan biosynthesis in *Geobacillus stearothermophilus*, *J. Biol. Chem.*, 283 (2008) 21120–21133.
223. Y. Huang, Y. Mechref, and M. V. Novotny, Microscale nonreductive release of O-linked glycans for subsequent analysis through MALDI mass spectrometry and capillary electrophoresis, *Anal. Chem.*, 73 (2001) 6063–6069.
224. U. B. Sleytr and P. Messner, Crystalline surface layers on bacteria, *Annu. Rev. Microbiol.*, 37 (1983) 311–339.

225. R. Novotny, C. Schäffer, J. Strauss, and P. Messner, S-Layer glycan-specific loci on the chromosome of *Geobacillus stearothermophilus* NRS 2004/3a and dTDP-L-rhamnose biosynthesis potential of *Geobacillus stearothermophilus* strains, *Microbiology*, 150 (2004) 953–965.
226. S. Zayni, K. Steiner, A. Pföstl, A. Hofinger, P. Kosma, C. Schäffer, and P. Messner, The dTDP-4-dehydro-6-deoxyglucose reductase encoding *fcd* gene is part of the surface layer glycoprotein glycosylation gene cluster of *Geobacillus tepidamans* GS5-97^T, *Glycobiology*, 17 (2007) 433–443.
227. R. Novotny, A. Pföstl, P. Messner, and C. Schäffer, Genetic organization of chromosomal S-layer glycan biosynthesis loci of *Bacillaceae*, *Glycoconj. J.*, 20 (2004) 435–447.
228. L. Wang, L. K. Romana, and P. R. Reeves, Molecular analysis of a *Salmonella enterica* group E1 *rfb* gene cluster: O antigen and the genetic basis of the major polymorphism, *Genetics*, 130 (1992) 429–443.
229. J. B. Kaplan, M. B. Perry, L. L. MacLean, D. Furgang, M. E. Wilson, and D. H. Fine, Structural and genetic analyses of O polysaccharide from *Actinobacillus actinomycetemcomitans* serotype f, *Infect. Immun.*, 69 (2001) 5375–5384.
230. M. Graninger, B. Kneidinger, K. Bruno, A. Scheberl, and P. Messner, Homologs of the Rml enzymes from *Salmonella enterica* are responsible for dTDP-β-L-rhamnose biosynthesis in the Gram-positive thermophile *Aneurinibacillus thermoaerophilus* DSM 10155, *Appl. Environ. Microbiol.*, 68 (2002) 3708–3715.
231. A. Pföstl, A. Hofinger, P. Kosma, and P. Messner, Biosynthesis of dTDP-3-acetamido-3,6-dideoxy-α-D-galactose in *Aneurinibacillus thermoaerophilus* L420-91^T, *J. Biol. Chem.*, 278 (2003) 26410–26417.
232. A. Pföstl, S. Zayni, A. Hofinger, P. Kosma, C. Schäffer, and P. Messner, Biosynthesis of dTDP-3-acetamido-3,6-dideoxy-α-D-glucose, *Biochem. J.*, 410 (2008) 187–194.
233. B. Kneidinger, M. Graninger, M. Puchberger, P. Kosma, and P. Messner, Biosynthesis of nucleotide-activated D-*glycero*-D-*manno*-heptose, *J. Biol. Chem.*, 276 (2001) 20935–20944.
234. B. Kneidinger, M. Graninger, G. Adam, M. Puchberger, P. Kosma, S. Zayni, and P. Messner, Identification of two GDP-6-deoxy-D-*lyxo*-4-hexulose reductases synthesizing GDP-D-rhamnose in *Aneurinibacillus thermoaerophilus* L420-91^T, *J. Biol. Chem.*, 276 (2001) 5577–5583.
235. Y. Ma, R. J. Stern, M. S. Scherman, V. D. Vissa, W. Yan, V. Cox Jones, F. Zhang, S. G. Franzblau, W. H. Lewis, and M. R. McNeil, Drug targeting *Mycobacterium tuberculosis* cell wall synthesis: Genetics of dTDP-rhamnose synthetic enzymes and development of a microtiter plate-based screen for inhibitors of conversion of dTDP-glucose to dTDP-rhamnose, *Antimicrob. Agents Chemother.*, 45 (2001) 1407–1416.
236. K. Steiner, G. Pohlentz, K. Dreisewerd, S. Berkenkamp, P. Messner, J. Peter-Katalinić, and C. Schäffer, New insights into the glycosylation of the surface layer protein SgsE from *Geobacillus stearothermophilus* NRS 2004/3a, *J. Bacteriol.*, 188 (2006) 7914–7921.
238. N. Kido, T. Sugiyama, T. Yokochi, H. Kobayashi, and Y. Okawa, Synthesis of *Escherichia coli* O9a polysaccharide requires the participation of two domains of WbdA, a mannosyltransferase encoded within the *wb** gene cluster, *Mol. Microbiol.*, 27 (1998) 1213–1221.
239. B. R. Clarke, L. Cuthbertson, and C. Whitfield, Nonreducing terminal modifications determine the chain length of polymannose O antigens of *Escherichia coli* and couple chain termination to polymer export via an ATP-binding cassette transporter, *J. Biol. Chem.*, 279 (2004) 35709–35718.
240. K. Zarschler, B. Janesch, S. Zayni, C. Schäffer, and P. Messner, Construction of a gene knockout system for application in *Paenibacillus alvei* CCM 2051^T, exemplified by the S-layer glycan biosynthesis initiation enzyme WsfP, *Appl. Environ. Microbiol.*, 75 (2009) 3077–3085.
241. L. Wang, D. Liu, and P. R. Reeves, C-terminal half of *Salmonella enterica* WbaP (RfbP) is the galactosyl-1-phosphate transferase domain catalyzing the first step of O-antigen synthesis, *J. Bacteriol.*, 178 (1996) 2598–2604.
242. K. Steiner, G. Hagelueken, P. Messner, C. Schäffer, and J. H. Naismith, Structural basis of substrate binding in WsaF, a rhamnosyltransferase from *Geobacillus stearothermophilus*, *J. Mol. Biol.*, 397 (2010) 436–447.

243. D. Bronner, B. R. Clarke, and C. Whitfield, Identification of an ATP-binding cassette transport system required for translocation of lipopolysaccharide O-antigen side-chains across the cytoplasmic membrane of *Klebsiella pneumoniae* serotype O1, *Mol. Microbiol.*, 14 (1994) 505–519.
244. L. Cuthbertson, J. Powers, and C. Whitfield, The C-terminal domain of the nucleotide-binding domain protein Wzt determines substrate specificity in the ATP-binding cassette transporter for the lipopolysaccharide O-antigens in *Escherichia coli* serotypes O8 and O9a, *J. Biol. Chem.*, 280 (2005) 30310–30319.
245. L. Cuthbertson, M. S. Kimber, and C. Whitfield, Substrate binding by a bacterial ABC transporter involved in polysaccharide export, *Proc. Natl. Acad. Sci. U.S.A.*, 104 (2007) 19529–19534.
246. R. Ristl, B. Janesch, J. Anzengruber, A. Forsthuber, J. Blaha, P. Messner, and C. Schäffer, Description of a putative oligosaccharyl:S-layer protein transferase from the tyrosine O-glycosylation system of *Paenibacillus alvei* CCM 2051^T, *Adv. Microbiol.*, 2 (2012) 537–546.
247. K. Zarschler, B. Janesch, B. Kainz, R. Ristl, P. Messner, and C. Schäffer, Cell surface display of chimeric glycoproteins via the S-layer protein of *Paenibacillus alvei*, *Carbohydr. Res.*, 345 (2010) 1422–1431.
248. A. P. Bhavsar, R. Truant, and E. D. Brown, The TagB protein in *Bacillus subtilis* 168 is an intracellular peripheral membrane protein that can incorporate glycerol phosphate onto a membrane-bound acceptor *in vitro*, *J. Biol. Chem.*, 280 (2005) 36691–36700.
249. A. Formstone, R. Carballido-López, P. Noirot, J. Errington, and D. J. Scheffers, Localization and interactions of teichoic acid synthetic enzymes in *Bacillus subtilis*, *J. Bacteriol.*, 190 (2008) 1812–1821.
250. M. Desvaux, E. Dumas, I. Chafsey, and M. Hébraud, Protein cell surface display in Gram-positive bacteria: From single protein to macromolecular protein structure, *FEMS Microbiol. Lett.*, 256 (2006) 1–15.
251. N. Kojima, Y. Arakai, and E. Ito, Structure of the linkage units between ribitol teichoic acids and peptidoglycan, *J. Bacteriol.*, 161 (1985) 299–306.
252. C. Leoff, E. Saile, D. Sue, P. P. Wilkins, C. P. Quinn, R. W. Carlson, and E. L. Kannenberg, Cell wall carbohydrate compositions of strains from *Bacillus cereus* group of species correlate with phylogenetic relatedness, *J. Bacteriol.*, 190 (2008) 112–121.
253. M.-H. Laaberki, J. Pfeffer, A. J. Clarke, and J. Dworkin, O-Acetylation of peptidoglycan is required for proper cell separation and S-layer anchoring in *Bacillus anthracis*, *J. Biol. Chem.*, 286 (2010) 5278–5288.
254. J. M. Lunderberg, S.-M. Nguyen-Mau, G. S. Richter, Y.-T. Wang, J. Dworkin, D. M. Missiakas, and O. Schneewind, *Bacillus anthracis* acetyl-transferases PatA1 and PatA2 modify the secondary cell wall polysaccharide and affect the assembly of S-layer proteins, *J. Bacteriol.*, 195 (2013) 977–989.
255. M.-F. Giraud and J. H. Naismith, The rhamnose pathway, *Curr. Opin. Struct. Biol.*, 10 (2000) 687–696.
256. C. Schäffer, R. Novotny, S. Küpcü, S. Zayni, A. Scheberl, J. Friedmann, U. B. Sleytr, and P. Messner, Novel biocatalysts based on S-layer self-assembly of *Geobacillus stearothermophilus* NRS 2004/3a: A nanobiotechnological approach, *Small*, 3 (2007) 1549–1559.
257. K. Steiner, A. Hanreich, B. Kainz, P. G. Hitchen, A. Dell, P. Messner, and C. Schäffer, Recombinant glycans on an S-layer self-assembly protein: A new dimension for nanopatterned biomaterials, *Small*, 4 (2008) 1728–1740.
258. E. Saxon and C. R. Bertozzi, Chemical and biological strategies for engineering cell surface glycosylation, *Annu. Rev. Cell Dev. Biol.*, 17 (2001) 1–23.
259. F. Schwarz, W. Huang, C. Li, B. L. Schulz, C. Lizak, A. Palumbo, S. Numao, D. Neri, M. Aebi, and L.-X. Wang, A combined method for producing homogeneous glycoproteins with eukaryotic N-glycosylation, *Nat. Chem. Biol.*, 6 (2010) 264–266.
260. C. M. Fletcher, M. J. Coyne, and L. E. Comstock, Theoretical and experimental characterization of the scope of protein O-glycosylation in *Bacteroides fragilis*, *J. Biol. Chem.*, 286 (2011) 3219–3226.

261. P. Samuelson, E. Gunneriusson, P. A. Nygren, and S. Ståhl, Display of proteins on bacteria, *J. Biotechnol.*, 96 (2002) 129–154.
262. C. Dürr, H. Nothaft, C. Lizak, R. Glockshuber, and M. Aebi, The *Escherichia coli* glycophage display system, *Glycobiology*, 20 (2010) 1366–1372.
263. S. Krapp, Y. Mimura, R. Jefferis, R. Huber, and P. Sondermann, Structural analysis of human IgG-Fc glycoforms reveals a correlation between glycosylation and structural integrity, *J. Mol. Biol.*, 325 (2003) 979–989.
264. Y. Mimura, S. Church, R. Ghirlando, P. R. Ashton, S. Dong, M. Goodall, J. Lund, and R. Jefferis, The influence of glycosylation on the thermal stability and effector function expression of human IgG1-Fc: Properties of a series of truncated glycoforms, *Mol. Immunol.*, 37 (2000) 697–706.
265. M. L. Langsford, N. R. Gilkes, B. Singh, B. Moser, R. C. Miller, Jr., R. A. J. Warren, and D. G. Kilburn, Glycosylation of bacterial cellulases prevents proteolytic cleavage between functional domains, *FEBS Lett.*, 225 (1987) 163–167.

AUTHOR INDEX

Page numbers in roman type indicate that the listed author is cited on that page of an article in this volume; numbers in italic denote the reference number, in the list of references for that article, where the literature citation is given.

A

Abbas, S. (Z.), 165, *208*, 182, *208*, 182, *278*
Abbruzzese, J.L., 142, *96*
Abdelfattah, M.S., 107, *149*
Aberg, E., 213, *45*
Abrams, J.N., 91, *111*, 93, *113*
Abshire, T.G., 236, *189*, 238, *189*, 244, *189*, 253, *189*
Abu-Qarn, M., 230, *149*, 230, *150*, 230, *151*
Acharya, K.R., 130, *34*, 156, *182*, 159, *190*
Ackerman, S.J., 130, *34*, 159, *190*
Adachi, K., 91, *86*
Adam, G., 247, *234*
Adamo, R., 111, *165*, 111, *166*, 111, *167*
Adamo, V., 151, *170*
Adams, M.W.W., 211, *17*, 214, *17*, 217, *17*, 230, *17*
Adams, W.J., 173, *247*
Adibekian, A., 74, *32*, 74, *33*, 74, *34*, 79, *42*, 79, *46*, 111, *170*
Adlard, M.W., 91, *87*
Adluri, S., 140, *72*
Aebi, M., 213, *47*, 213, *48*, 213, *49*, 213, *50*, 213, *51*, 213, *52*, 213, *53*, 255, *47*, 255, *53*, 255, *259*, 255, *262*
Aggarwal, B.B., 101, *125*
Ahern, T.J., 137, *58*
Ahmad, N., 133, *44*, 134, *44*, 152, *44*, 165, *213*, 165, *215*
Ahmed, H., 130, *34*, 148, *148*, 161, *194*

Ahmed, Md. M., 83, *51*, 83, *54*, 85, *67*, 86, *68*, 86, *70*, 86, *73*, 86, *74*, 87, *75*
Aizawa, S.-I., 230, *158*, 230, *159*
Akahani, S., 130, *27*, 130, *33*, 143, *104*, 144, *27*, 144, *116*, 147, *27*
Akça, E., 214, *72*, 214, *76*, 217, *72*
Akhmedov, N.G., 103, *138*, 109, *157*
Akhtar, A., 149, *157*, 151, *157*
Akita, K., 142, *95*
Alban, S., 180, *274*
Albers, S.-V., 214, *77*, 217, *113*, 230, *113*, 231, *113*, 231, *160*, 231, *162*, 231, *163*, 232, *163*
Allbutt, A.D., 101, *134*
Allen, H.J., 148, *148*, 161, *194*
Allen, T.M., 185, *291*
Allmaier, G., 219, *136*, 220, *136*, 220, *140*, 223, *136*, 227, *140*, 229, *136*, 243, *136*, 245, *219*, 245, *136*, 245, *140*
Almkvist, J., 158, *185*
Alphey, M.S., 128, *17*, 128, *18*, 129, *17*
Al-Rifai, I., 141, *79*
Altevogt, P., 140, *73*
Altheide, T.K., 142, *95*
Altman, E., 219, *134*, 219, *136*, 220, *134*, 220, *136*, 220, *138*, 223, *136*, 224, *145*, 224, *134*, 225, *134*, 226, *145*, 226, *146*, 229, *136*, 233, *138*, 233, *178*, 234, *178*, 243, *136*, 243, *138*, 243, *145*, 243, *178*, 245, *136*, 249, *134*

273

AUTHOR INDEX

Altmann, F., 217, *106*, 218, *106*, 218, *127*, 219, *130*, 221, *141*, 222, *106*, 222, *127*, 227, *106*, 227, *141*, 228, *106*, 228, *127*, 245, *106*, 246, *106*, 247, *106*, 249, *130*, 251, *130*, 255, *127*, 256, *106*, 256, *127*, 257, *106*, 257, *127*
Alvarez, M., 141, *91*, 142, *91*, 144, *91*, 147, *91*
Amber, S., 213, *48*
Ananthaswamy, H.N., 167, *224*, 169, *224*
Anderson, A., 130, *23*
Anderson, A.C., 144, *113*
Anderson, B., 46, *47*, 46, *49*, 47, *49*
Anderson, K.C., 149, *159*, 150, *159*
André, S., 127, *10*, 130, *23*, 133, *44*, 134, *44*, 144, *117*, 152, *44*, 165, *209*, 165, *210*, 165, *213*, 165, *214*, 165, *216*, 166, *209*, 166, *210*, 166, *216*, 166, *218*, 167, *218*, 167, *221*, 167, *223*, 169, *221*, 169, *228*, 170, *228*
Andresen, M.S., 180, *271*
Angata, T., 141, *93*, 142, *95*
Angeletti, C.A., 144, *110*
Anthony, D.C., 150, *163*, 169, *163*, 169, *226*
Antikainen, J., 241, *208*
Anton, L., 241, *208*
Antonopoulos, A., 140, *75*
Anzengruber, J., 221, *141*, 227, *141*, 252, *246*
Aoki, H., 91, *86*
Apgar, J.R., 147, *137*
Apweiler, R., 222, *143*
Arakai, Y., 253, *251*
Araki, Y., 231, *171*, 232, *171*, 233, *171*
Arata, Y., 134, *46*
Archibald, A.R., 231, *169*, 232, *169*
Ariel, A., 133, *37*, 143, *37*
Arikawa, T., 144, *113*
Arimoto, Y., 141, *85*
Arnoux, P., 160, *192*
Arnusch, C.J., 167, *221*, 169, *221*, 169, *228*, 170, *228*
Aruffo, A., 141, *87*, 171, *238*, 173, *87*
Aryan, M., 180, *271*
Asa, D., 165, *208*, 182, *208*
Asenbauer, E., 240, *206*
Ashton, P.R., 256, *264*
Astorgues-Xerri, L., 170, *236*, 171, *236*

Astronomo, R.D., 126, *6*
Attrill, H., 128, *18*
Au, C.W.G., 91, *107*
Aub, J.C., 126, *1*
Aubert, J.P., 239, *196*
Aubert, M., 175, *257*
Aubry, A.J., 212, *34*
Auden, A.L., 84, *62*
Austin, C.A., 101, *127*
Austin, J., 212, *34*
Austin, M.B., 90, *84*
Åvall-Jääskeläinen, S., 214, *71*, 217, *105*, 246, *105*
Ayafor, J.P., 108, *154*
Ayuso, M.J., 101, *127*
Azad, N., 102, *136*
Azadi, P., 213, *53*, 255, *53*
Azrak, S.S., 101, *127*
Azuma, M., 144, *113*

B

Ba, X.Q., 178, *269*, 180, *275*
Babu, R.S., 88, *77*, 89, *82*, 91, *111*, 93, *113*, 94, *114*, 99, *122*, 115, *172*
Baccarini, S., 149, *153*, 151, *153*, 151, *166*
Bachhawat-Sikder, K., 165, *215*
Bacia, K., 49, *62*
Backvall, J.K., 73, *30*
Baddiley, J., 231, *170*, 232, *170*
Baenziger, J.U., 165, *207*
Baeuerle, P.A., 144, *112*
Bahl, H., 240, *199*, 240, *202*
Baig, I., 107, *149*
Baiga, T.J., 90, *84*
Baileul, F., 108, *155*
Bailey, D., 91, *94*
Baker, S.R., 79, *44*
Bakkers, J., 170, *232*
Balachari, D., 84, *63*, 84, *64*, 84, *65*
Balaji, P.V., 133, *38*
Balan, V., 148, *147*, 150, *147*
Baldini, L., 166, *217*
Ballan, E., 144, *117*
Bamford, M.J., 182, *279*
Banaszek, A., 58, *6*
Banda, K., 142, *95*
Banh, A., 149, *150*

Banteli, R., 182, *279*
Baranova, E., 215, *92*
Barboni, E.A.M., 152, *172*
Barclay, A.N., 129, *19*
Bardy, S.L., 230, *155*
Barondes, S.H., 130, *26*, 130, *34*, 130, *36*, 133, *41*, 133, *42*, 141, *41*, 152, *173*, 156, *173*, 165, *173*
Barone, K.M., 135, *55*, 137, *55*
Barta, F., 212, *41*, 217, *41*
Barthel, S.R., 148, *149*
Bast, R.C., 142, *98*
Basu, M., 43, *25*, 43, *26*, 43, *27*, 44, *27*, 44, *28*, 45, *27*, 45, *28*, 45, *33*, 45, *34*, 45, *35*, 45, *36*, 45, *37*, 45, *38*, 45, *39*, 45, *40*, 46, *27*, 46, *39*, 47, *55*,
Basu, S., 41, *8*, 41, *9*, 41, *10*, 41, *11*, 41, *12*, 41, *14*, 41, *15*, 41, *16*, 42, *8*, 42, *9*, 42, *10*, 42, *11*, 42, *12*, 42, *14*, 42, *15*, 42, *23*, 43, *8*, 43, *9*, 43, *10*, 43, *11*, 43, *12*, 43, *16*, 43, *25*, 43, *26*, 43, *27*, 44, *9*, 44, *12*, 44, *16*, 44, *27*, 44, *28*, 44, *29*, 45, *16*, 45, *27*, 45, *28*, 45, *29*, 45, *33*, 45, *34*, 45, *35*, 45, *36*, 45, *37*, 45, *38*, 45, *39*, 45, *40*, 46, *27*, 46, *29*, 46, *39*, 47, *53*, 47, *55*
Baum, L.G., 130, *26*, 130, *31*, 143, *31*, 143, *103*, 144, *115*, 148, *148*, 165, *206*, 170, *232*
Baumeister, W., 214, *78*, 214, *81*, 239, *197*
Baumhueter, S., 137, *58*
Bawumia, S., 152, *172*
Bayer, E.A., 240, *198*
Bayley, D., 175, *258*
Beal, T.L., 149, *158*
Beatson, R., 140, *75*
Beauharnois, M.E., 182, *278*
Becchetti, E., 127, *9*
Béguin, P., 239, *196*
Behalf, C.I., 151, *170*
Behrens, D., 171, *237*, 185, *237*, 185, *237*
Behrens, N., 45, *42*, 45, *43*
Beierbeck, H., 127, *8*
Belaich, J.-P., 240, *198*
Belitsky, J.M., 165, *206*
Bell, A.A., 91, *95*
Bellefleur, M.A., 154, *179*, 155, *179*, 164, *179*
Bellnier, D.A., 166, *217*

Bendas, G., 180, *274*
Bendt, K.M., 174, *254*, 175, *254*
Benoliel, A.M., 174, *255*, 175, *255*
Benz, I., 211, *8*, 214, *8*, 255, *8*
Berchuck, A., 142, *98*, 144, *114*
Berenguer, J., 232, *174*, 239, *174*, 240, *174*, 241, *174*, 242, *174*, 252, *174*, 253, *174*
Berg, E.L., 141, *90*
Bergh, A., 130, *35*, 155, *35*
Berkenkamp, S., 249, *236*
Berndt, M.C., 135, *53*
Bernfield, M., 176, *262*
Bernimoulin, M., 177, *265*
Berry, B.P., 83, *54*, 85, *67*
Bertolasi, V., 76, *39*
Bertozzi, C.R., 145, *128*, 149, *151*, 211, *4*, 219, *4*, 222, *4*, 255, *258*
Beuning, P., 109, *157*
Beuten, J., 142, *95*
Beuth, J., 146, *136*, 147, *136*, 151, *136*, 151, *169*
Bevan, M.J., 126, *5*
Beveridge, T.J., 214, *68*
Bevilacqua, M. (P.), 141, *87*, 173, *87*, 173, *250*
Bhatia, S.K., 178, *270*
Bhavsar, A.P., 252, *248*
Bianco, G.A., 162, *195*, 164, *195*
Bicknell, R., 143, *107*
Bielenberg, D., 167, *224*, 169, *224*
Bigbee, W.L., 137, *62*
Biliran, H., 144, *118*
Bindila, L., 245, *220*
Bindschädler, P., 74, *33*
Binkowski, T.A., 240, *203*
Biot, C., 158, *186*
Bird, K., 180, *272*
Bird, M., 182, *279*
Bischoff, J., 135, *49*, 176, *262*
Blaha, J., 252, *246*
Blanchard, H., 130, *20*, 133, *40*, 147, *141*, 156, *40*, 157, *40*, 158, *20*, 158, *40*, 163, *201*, 163, *202*, 164, *141*
Blanco, J., 145, *129*, 172, *129*
Blaser, M.J., 217, *102*, 217, *103*, 246, *102*, 246, *103*,
Blazar, B.R., 144, *113*

Blechert, S., 91, *104*
Blixt, O., 127, *12*, 128, *12*, 129, *12*
Blom, J., 218, *121*
Bloom, E.J., 137, *61*, 140, *61*
Blystone, S.D., 180, *272*
Bobbitt, J.M., 243, *211*
Bock, C., 165, *210*, 166, *210*
Bock, K., 219, *136*, 220, *136*, 223, *136*, 229, *136*, 243, *214*, 243, *136*, 245, *136*
Bodkin, N.L., 158, *187*
Bodo, M., 127, *9*
Boehncke, W.H., 180, *274*
Bolgiano, B., 130, *34*
Bongaerts, R.J.M., 149, *160*
Bongrand, P., 174, *255*, 175, *255*
Bonin, M.-A., 160, *193*, 161, *193*
Boons, G.-J., 111, *168*, 236, *186*, 236, *188*, 253, *186*
Borges, E., 185, *292*
Borisova, S.A., 101, *123*
Borsig, L., 144, *120*, 144, *121*, 145, *120*, 145, *121*, 177, *120*, 177, *266*, 177, *267*, 180, *266*, 181, *276*, 184, *289*, 185, *289*
Bouché, L., 74, *35*
Boudker, O., 49, *61*
Bouley, D.M., 149, *150*
Bourne, Y., 130, *34*
Bozzaro, S., 49, *57*, 49, *58*
Bradfield, P., 128, *14*
Braekman, J.C., 162, *196*
Branalt, J., 63, *15*
Brandley, B.K., 165, *208*, 182, *208*
Brandwijk, R.J.M.G.E., 170, *232*
Brasholz, M., 80, *48*
Bravo, A., 141, *91*, 142, *91*, 144, *91*, 147, *91*
Bravo, D.T., 147, *140*
Bray, P.F., 145, *125*, 180, *125*
Brecker, L., 217, *106*, 218, *106*, 222, *106*, 227, *106*, 228, *106*, 245, *106*, 246, *106*, 247, *106*, 256, *106*, 257, *106*
Breitman, M.L., 140, *67*, 140, *77*
Bresalier, R. (S.), 130, *32*, 142, *32*, 142, *96*, 143, *100*, 144, *119*, 149, *153*, 151, *153*, 172, *119*
Brewer, C.F., 133, *44*, 134, *44*, 152, *44*, 165, *213*, 165, *215*
Briese, V., 172, *241*, 185, *241*

Brisson, J.-R., 182, *279*, 212, *33*, 212, *36*, 212, 37, 213, *45*, 217, *33*, 217, *36*, 219, *134*, 220, *134*, 224, *134*, 224, *145*, 226, *145*, 225, *134*, 226, *146*, 230, *155*, 243, *145*, 249, *134*
Brockhausen, I., 140, *75*, 145, *128*
Brodt, P., 144, *119*, 172, *119*, 172, *245*
Bronner, D., 252, *243*
Brossmer, R., 127, *11*
Brown, E.D., 252, *248*
Brown, J.R., 173, *252*, 174, *253*
Brown, N.S., 143, *107*
Bruno, K., 247, *230*, 248, *230*, 255, *230*
Bryce, D.M., 140, *77*
Bucana, C.D., 126, *5*
Bugarcic, A., 163, *202*
Buicu, C., 142, *98*
Bundle, D.R., 127, *7*
Burchell, J., 140, *75*
Burchell, J.M., 141, *86*
Burford, B., 141, *86*
Burger, M.M., 126, *5*
Burghardt, J., 218, *118*, 219, *118*
Burns, G.F., 135, *53*
Burton, D.R., 126, *6*
Buschmann, N., 91, *104*
Bushey, M.L., 83, *55*
Buskas, T., 111, *168*
Busse, H.-J., 218, *118*, 219, *118*
Butcher, E.C., 141, *90*
Buttitta, F., 144, *110*
Byrd, J.C., 130, *32*, 142, *32*, 143, *100*
Bystrom, S.E., 73, *30*

C

Cadotte, N., 213, *45*
Califice, S., 144, *114*
Calin, O., 79, *45*
Callewaert, N., 213, *50*
Calo, D., 217, *114*, 230, *114*, 230, *154*
Cambien, B., 145, *126*
Cambillau, C., 130, *34*
Camby, I., 130, *25*, 147, *25*
Camphausen, R.T., 135, *55*, 135, *57*, 137, *55*, 182, *57*, 145, *126*
Canales, A., 182, *280*
Cancelas, J.A., 145, *129*, 172, *129*

Candela, T., 237, *192*, 238, *192*
Cantau, P., 130, *34*
Cantz, M., 165, *214*
Cao, H., 149, *150*
Carballido-López, R., 252, *249*
Carinci, P., 127, *9*
Carlson, D.M., 41, *7*, 42, *7*, 43, *7*
Carlson, R. (W.), 111, *168*, 212, *38*, 217, *38*, 236, *185*, 236, *186*, 236, *189*, 236, *190*, 236, *191*, 237, *190*, 237, *191*, 238, *189*, 238, *190*, 244, *189*, 244, *190*, 253, *185*, 253, *186*, 253, *189*, 253, *252*
Carlsson, M.C., 158, *185*
Carlsson, S., 130, *28*, 156, *184*, 158, *185*
Carminatti, H., 45, *43*
Carothers, A.M., 164, *204*
Carter, G.T., 97, *120*
Carter, L., 173, *250*
Carver, B., 149, *159*, 150, *159*
Carver, J.P., 127, *8*, 140, *76*
Casnati, A., 166, *217*, 166, *218*, 167, *218*
Cassels, F.J., 212, *38*, 217, *38*
Castric, P., 212, *38*, 217, *38*
Castronovo, V., 130, *36*, 142, *98*, 144, *114*
Casu, B., 177, *266*, 180, *266*
Catrein, I., 213, *50*
Cauet, E., 158, *186*
Cava, F., 232, *174*, 239, *174*, 240, *174*, 241, *174*, 242, *174*, 252, *174*, 253, *174*
Ceccon, J., 91, *106*
Cedeno-Laurent, F., 148, *149*
Cederfur, C., 130, *30*, 133, *30*
Celis, J.E., 130, *34*
Černý, M., 58, *5*
Cha, D.Y., 85, *66*, 101, *66*
Chaban, B., 230, *156*
Chafsey, I., 253, *250*
Chai, W.G., 137, *58*
Chambers, A.F., 142, *97*
Chaney, W.G., 140, *77*
Chang, Y., 149, *159*, 150, *159*
Chatzidaki-Livanis, M., 218, *126*, 228, *126*, 255, *126*, 257, *126*
Chauhan, D., 149, *159*, 150, *159*
Cheingsongpopov, R., 140, *73*
Chella, A., 144, *110*
Chen, A., 135, *47*

Chen, G.Q., 143, *108*
Chen, H.Y., 147, *137*
Chen, M.S., 58, *11*
Chen, Q., 106, *148*, 115, *172*
Chen, T.T., 143, *108*
Chen, X., 170, *235*, 170, *236*, 171, *236*
Chen, Y.-L., 147, *139*
Chen, Z.H., 172, *242*, 178, *242*
Cheng, X., 91, *105*, 182, *280*
Chernajovsky, A., 143, *106*
Chesterman, C.N., 135, *53*
Cheung, N.K.V., 146, *134*
Chhiba, K., 213, *53*, 255, *53*
Chiang, W.-F., 147, *139*
Chien, J.L., 45, *34*
Chiron, C., 162, *196*
Cho, M., 175, *256*
Choi, S.H., 135, *52*, 177, *52*, 178, *52*, 179, *52*, 181, *52*
Choi, W.B., 70, *23*
Choudhury, B., 236, *185*, 236, *190*, 236, *191*, 237, *190*, 237, *191*, 238, *190*, 244, *190*, 253, *185*
Christensen, C., 154, *178*
Christian, R., 218, *120*, 219, *133*, 219, *135*, 219, *136*, 220, *133*, 220, *135*, 220, *136*, 220, *137*, 220, *140*, 221, *120*, 221, *135*, 221, *137*, 222, *120*, 223, *133*, 223, *136*, 224, *133*, 224, *137*, 224, *144*, 225, *133*, 225, *137*, 225, *144*, 227, *120*, 227, *140*, 229, *136*, 229, *137*, 233, *176*, 233, *179*, 233, *180*, 234, *179*, 234, *180*, 240, *176*, 242, *176*, 242, *180*, 243, *133*, 243, *136*, 243, *137*, 244, *120*, 244, *137*, 244, *176*, 244, *180*, 245, *135*, 245, *136*, 245, *140*, 245, *218*, 251, *135*, 252, *135*, 253, *176*
Christian, R.A., 175, *258*
Chu, Y.H., 172, *242*, 178, *242*
Chung, C.D., 143, *106*
Chung, Y.S., 141, *85*
Church, S., 256, *264*
Ciavardelli, D., 162, *195*, 164, *195*
Cinato, E., 185, *292*
Cisneros, A., 101, *125*
Ciufolini, M., 65, *18*
Clark, D.P., 214, *63*
Clarke, A.J., 254, *253*

Clarke, B.R., 250, *239*, 252, *243*
Clarke, D.E., 94, *116*
Claudy, A., 141, *79*
Claus, H., 214, *72*, 214, *76*, 217, *72*, 230, *148*
Clausen, H., 45, *41*
Clausse, N., 142, *98*
Clayman, G.L., 142, *96*
Clement, G., 147, *140*
Clissold, D.W., 79, *44*
Cloutier, P., 160, *193*, 161, *193*
Cohen-Rosenzweig, C., 230, *152*
Cohn, M., 126, *5*
Colegate, S.M., 91, *88*
Collins, B.E., 127, *12*, 128, *12*, 129, *12*
Collins, P.M., 130, *20*, 147, *141*, 147, *141*, 158, *20*, 164, *141*
Comb, D.G., 40, *4*, 40, *5*, 42, *18*
Comely, A.C., 88, *78*
Comess, K.M., 135, *55*, 137, *55*
Comstock, L.E., 218, *126*, 218, *127*, 222, *127*, 228, *126*, 228, *127*, 255, *126*, 255, *127*, 255, *260*, 256, *127*, 257, *126*, 257, *127*, 257, *260*
Connolly, J.D., 108, *154*
Cooks, T., 144, *117*
Cooper, D.N.W., 130, *21*, 130, *26*, 142, *98*
Coral, J.A., 91, *110*
Cordon-Cardo, C., 140, *69*, 140, *71*, 140, *72*, 141, *71*
Corless, C., 173, *250*
Cortes, F., 101, *127*
Cossy, J., 91, *108*
Couture-Tosi, E., 111, *163*
Cox Jones, V., 247, *235*
Coyne, M.J., 218, *126*, 218, *127*, 222, *127*, 228, *126*, 228, *127*, 255, *126*, 255, *127*, 255, *260*, 256, *127*, 257, *126*, 257, *127*, 257, *260*
Crabtree, R.H., 58, *10*
Craig, S.J., 175, *258*
Cramer, F., 141, *80*
Crich, D., 111, *169*
Crocker, P.R., 127, *11*, 127, *12*, 127, *13*, 128, *12*, 128, *14*, 128, *16*, 128, *17*, 128, *18*, 129, *12*, 129, *16*, 129, *17*, 129, *19*, 142, *94*
Crotte, C., 175, *257*
Cuccarese, M.F., 103, *142*

Cui, H.X., 172, *244*
Cui, J., 170, *230*
Cumashi, A., 151, *170*, 162, *195*, 164, *195*
Cumming, D.A., 135, *55*, 137, *55*
Cummings, R.D., 211, *4*, 219, *4*, 222, *4*, 130, *36*, 135, *54*, 135, *56*, 137, *56*, 148, *148*, 158, *185*, 175, *256*
Cumpstey, I., 155, *181*, 156, *184*, 163, *198*, 163, *199*, 164, *203*
Cunto-Amesty, G., 167, *220*
Cusack, J.C., 174, *254*, 175, *254*
Cuthbertson, L., 250, *239*, 252, *244*, 252, *245*
Cutler, S., 163, *201*

D

Dabelic, S., 147, *142*
Dagher, S.F., 130, *33*
Daly, G., 143, *106*
Danguy, A., 141, *79*
Danishefsky, S.J., 140, *69*, 63, *14*, 63, *16*, 63, *17*, 65, *18*
Darro, F., 162, *196*
Das, K., 45, *40*
Das, P.C., 91, *93*
Dasgupta, F., 165, *208*, 182, *208*
Dastgheib, S., 45, *39*, 46, *39*
Davis, B., 91, *95*
Davis, D., 91, *90*
de Pedro, M.A., 232, *174*, 239, *174*, 240, *174*, 241, *174*, 242, *174*, 252, *174*, 253, *174*
Deakin, J.A., 145, *123*, 180, *123*
Dean, B., 151, *65*
Debaerdemaeker, T., 214, *72*, 214, *76*, 217, *72*
DeBellard, M.E., 128, *14*, 129, *19*
Dechamps, I., 91, *108*
Declercq, J.-P., 214, *72*, 214, *76*, 217, *72*
D'Egidio, M., 151, *170*, 162, *195*, 164, *195*
Dekitsch, C., 240, *205*
Delaine, T., 164, *203*, 164, *204*
Delbaere, L.T.J., 127, *8*
Delepierre, M., 231, *164*, 233, *164*, 239, *164*, 240, *164*, 240, *164*, 241, *164*, 242, *164*, 252, *164*, 253, *164*
DeLisa, M.P., 213, *53*, 255, *53*
Dell, A., 140, *75*, 140, *76*, 213, *47*, 213, *48*, 213, *49*, 230, *150*, 231, *160*, 231, *162*, 255, *47*, 255, *257*, 256, *257*

Demel, R., 241, *207*
Demetriou, M., 133, *43*
Den, H., 47, *54*
Deng, G.R., 126, *2*, 144, *2*
DeNinno, M.P., 63, *16*
Dennis, J.W., 91, *94*, 126, *5*, 133, *43*, 140, *67*, 140, *73*, 140, *76*, 140, *77*, 141, *66*
Denton, R.W., 182, *280*, 183, *285*
Deprez, M., 142, *98*
Desai, P.R., 141, *79*
Descheny, L., 173, *251*
Desvaux, M., 253, *250*
Deutscher, S.L., 140, *78*, 148, *78*, 167, *222*, 169, *78*, 169, *225*
Dewelle, J., 162, *196*
Dharmesh, S.M., 130, *22*, 142, *22*, 151, *22*
Dick, S., 212, *36*, 217, *36*
Dielie, G., 162, *196*
Dietrich, K., 220, *139*, 223, *139*, 224, *139*, 243, *139*, 244, *139*, 245, *139*
Dilhas, A., 182, *280*
Dillin, A., 90, *84*
Dimitroff, C.J., 148, *149*, 173, *251*
Ding, Y., 230, *159*
Dings, R.P.M., 170, *231*, 170, *232*, 170, *233*, 170, *234*, 170, *235*, 170, *236*, 171, *236*
Dippold, W., 172, *244*, 173, *249*, 185, *293*
Distler, J.J., 42, *19*, 42, *22*
Doering, T.L., 211, *10*
Dohertey, L., 40, *2*
Dohi, T., 141, *85*, 175, *259*, 175, *260*, 176, *259*
Donat, T.L., 149, *157*, 151, *157*
Dondoni, A., 75, *37*, 75, *38*, 76, *39*
Dong, L.-F., 163, *201*
Dong, S., 256, *264*
Dong, X., 127, *10*
Donner, D.B., 164, *204*
Đorđić, A., 215, *91*
Dorfman, A., 40, *3*, 41, *3*
Dorling, P.R., 91, *88*, 91, *92*
Dorrell, N., 213, *48*
Dougherty, D.A., 153, *177*
Dougherty, T.J., 166, *217*
Downey, T.E., 107, *151*
Dragsten, P., 126, *5*

Dreaden, T.M., 149, *158*
Dreisewerd, K., 249, *236*
Dreja, H., 143, *106*
Drickamer, K., 126, *3*, 130, *36*
Driks, A., 211, *13*
Drings, P., 127, *10*
Dube, D.H., 145, *128*
Duesler, E.N., 91, *100*
Dufrêne, Y.F., 214, *82*
Dumas, E., 253, *250*
Dumic, J., 147, *142*
Dumy, P., 166, *217*, 101, *129*
Dunn, R., 44, *31*
Dürr, C., 255, *262*
Dwek, M.V., 141, *92*
Dworkin, J., 217, *102*, 246, *102*, 254, *253*, 254, *254*
Dziadek, S., 137, *63*, 140, *63*

E

Eaton, S.F., 135, *56*, 137, *56*
Echten-Deckert, G.V., 44, *30*
Edovitsky, E., 185, *290*
Eelkema, R., 88, *78*
Egelseer, E. (M.), 214, *74*, 215, *88*, 215, *90*, 215, *91*, 216, *74*, 216, *93*, 216, *96*, 218, *74*, 218, *122*, 222, *74*, 232, *175*, 233, *177*, 240, *74*, 240, *205*, 241, *74*, 242, *175*, 242, *177*, 242, *210*, 243, *177*, 254, *74*, 254, *122*, 255, *93*, 256, *122*
Eggers, J.P., 142, *94*
Ehemann, V., 165, *214*
Eichenberger, P., 211, *13*
Eichler, J., 211, *17*, 214, *17*, 217, *17*, 217, *112*, 217, *114*, 219, *112*, 230, *17*, 230, *112*, 230, *114*, 230, *149*, 230, *149a*, 230, *150*, 230, *151*, 230, *152*, 230, *153*, 230, *154*, 231, *163*, 232, *163*
Eid, M., 149, *158*
Eikmeyer, F., 218, *121*
El Nemr, A., 91, *98*
Elad-Sfadia, G., 144, *117*
El-Battari, A., 174, *255*, 175, *255*
Elbaz, H., 102, *136*
Elbein, A.D., 91, *90*, 91, *92*
Elbert, B.L., 130, *34*

Eldridge, G.R., 108, *156*
Elkin, M., 185, *290*
Ellerhorst, J., 150, *162*
El-Naggar, A.K., 142, *96*
Elsamman, E., 150, *162*
Elsinghorst, E.A., 212, *29*
Endo, K., 141, *89*
Endo, T., 211, *21*
Engel, A., 214, *81*
Engel, P., 135, *47*
Engelhardt, H., 214, *70*, 216, *99*, 239, *197*
Engelhardt, R., 141, *80*
Entwistle, D.A., 58, *8*
Erickson, H.P., 184, *288*
Ernst, B., 182, *279*, 182, *282*
Errington, J., 252, *249*
Escriou, V., 183, *286*
Esko, J.D., 211, *4*, 211, *10*, 219, *4*, 222, *4*, 172, *240*, 173, *240*, 173, *252*, 174, *253*
Etzler, M.E., 211, *4*, 219, *4*, 222, *4*
Evans, D., 213, *54*
Evans, P.A., 88, *81*
Evrard, C., 214, *72*, 214, *76*, 217, *72*
Ewing, C.P., 212, *33*, 212, *42*, 213, *42*, 213, *43*, 217, *33*

F

Fainboim, L., 141, *91*, 142, *91*, 144, *91*, 147, *91*
Fajka-Boja, R., 169, *229*, 170, *229*
Fallavollita, L., 144, *119*, 172, *119*, 172, *245*
Fan, Y.Y., 213, *53*, 255, *53*
Fantin, G., 75, *38*
Faridmoayer, A., 213, *55*
Farnworth, S.L., 130, *23*
Faull, K., 236, *184*, 253, *184*
Faure, M., 141, *79*
Feather, M.S., 167, *224*, 169, *224*
Feizi, T., 130, *34*, 130, *36*, 137, *58*
Feldman, M.F., 211, *12*, 213, *49*, 213, *50*, 213, *55*, 213, *62*
Fentabil, M. (A.), 213, *55*, 214, *62*
Fenteany, G., 170, *230*
Feramisco, J., 144, *120*, 145, *120*, 177, *120*
Feringa, B.L., 88, *78*
Fernandez, P.L., 144, *114*

Fichtner, I., 171, *237*, 185, *237*, 185, *237*
Fidler, I.J., 126, *5*, 141, *84*
Filbin, M.T., 128, *14*, 129, *19*
Filser, C., 182, *281*, 185, *281*
Fine, D.H., 247, *229*
Finley, R.L., 150, *162*
Fischer, E., 57, *1*, 57, *2*, 58, *1*, 58, *2*
Fischer, M.J.E., 169, *228*, 170, *228*
Fischer, R.J., 240, *202*
Fischetti, V.A., 236, *187*
Fish, W.R., 180, *272*
Fisher, A.C., 213, *53*, 255, *53*
Fitz, W., 58, *7*
Flader, C., 170, *233*
Fleet, G.W.J., 91, *95*, 91, *97*
Fletcher, C.M., 218, *126*, 228, *126*, 255, *126*, 255, *260*, 257, *126*, 257, *260*
Flogel, M., 147, *142*
Foday, A.D., 149, *154*, 150, *154*
Fogagnolo, M., 75, *38*
Fogel, M., 140, *73*
Fokin, V.V., 154, *178*
Folkman, J., 144, *109*
Fontaine, T., 231, *164*, 233, *164*, 239, *164*, 240, *164*, 240, *164*, 241, *164*, 242, *164*, 252, *164*, 253, *164*
Ford, R., 229, *147*
Formstone, A., 252, *249*
Forsberg, L.S., 236, *189*, 236, *190*, 237, *190*, 238, *189*, 238, *190*, 244, *189*, 244, *190*, 253, *189*
Forsthuber, A., 252, *246*
Fort, S., 133, *39*, 156, *39*
Foster, V., 213, *54*
Fouet, A., 111, *162*, *164*, *200*, *201*, 231, 233, 239, 240, 241, 242, 252, 253
Fourmarier, M., 141, *79*
Foxall, C., 165, *208*, 182, *208*
Fradin, C., 159, *190*
Franck, W.L., 218, *119*
Franco, S., 76, *39*
Frank, M.H., 148, *149*
Franks, L.M., 140, *73*
Franzblau, S.G., 247, *235*
Fredman, P., 171, *238*
Freeze, H.H., 211, *4*, 219, *4*, 222, *4*
Frejd, T., 164, *205*, 165, *212*

Friedlander, A., 236, *189*, 238, *189*, 244, *189*, 253, *189*
Friedmann, J., 255, *256*, 256, *256*
Friedrich, V., 216, *100*, 217, *100*, 219, *129*, 228, *100*, 231, *129*
Friese, K., 148, *144*, 165, *144*, 167, *144*, 172, *241*, 185, *241*
Fritz, G., 172, *244*, 173, *249*
Fritz, T.A., 172, *240*, 173, *240*
Fritzsche, J., 180, *274*
Fronzes, R., 215, *92*
Fryer, A., 145, *122*
Fu, W.L., 178, *270*
Fujino, T., 239, *196*
Fujita, S., 182, *283*
Fukami, A., 184, *287*
Fukuda, M., 140, *76*, 145, *131*, 145, *132*, 145, *133*, 187, *294*
Fukuda, M.N., 187, *294*
Fukui, S., 140, *70*
Fukumori, T., 148, *146*, 150, *162*
Fukunaga, R., 175, *259*, 176, *259*
Fulton, D.A., 166, *217*
Funahashi, H., 141, *88*
Funahashi, T., 158, *187*
Funakoshi, I., 140, *70*
Furgang, D., 247, *229*
Furuike, T., 165, *216*, 166, *216*
Furukawa, H., 141, *81*, 145, *81*
Fuster, M.M., 173, *252*
Futai, M., 134, *46*

G

Gabius, H. (J. or -J.), 127, *10*, 130, *23*, 133, *44*, 134, *44*, 141, *80*, 144, *117*, 152, *44*, 165, *210*, 165, *209*, 165, *214*, 165, *213*, 165, *215*, 165, *216*, 166, *209*, 166, *210*, 166, *216*, 166, *218*, 167, *218*, 167, *221*, 167, *223*, 169, *221*, 169, *228*, 170, *228*
Gabriel, A., 141, *90*
Gabutero, E., 147, *141*, 164, *141*
Gagné, S.M., 224, *145*, 226, *145*, 243, *145*
Gainers, M.E., 173, *251*
Galili, U., 126, *4*, 127, *4*, 135, *4*, 142, *4*, 147, *4*, 175, *4*
Gallagher, J.T., 145, *123*, 180, *123*

Gallop, M.A., 169, *227*
Galustian, C., 137, *58*
Gandhi, S.S., 140, *70*
Gao, D., 86, *71*, 86, *72*
Gao, M.Y., 149, *158*
Gao, X., 149, *154*, 150, *154*
Gao, Y., 180, *273*
Gao, Y.G., 172, *239*, 176, *239*, 177, *239*, 178, *269*, 179, *239*, 180, *239*, 180, *275*
García Fernández, J.M., 166, *219*
Garcia-Aparico, V., 101, *130*
Garcia-Pino, A., 215, *92*
Garegg, P.J., 244, *217*
Garénaux, E., 237, *192*, 238, *192*
Garner, P.P., 170, *230*
Garnier, F., 144, *114*
Garo, E., 108, *156*
Geary, R., 175, *258*
George, C.M., 140, *69*
Gerber, S., 213, *52*
Gerolami, R., 174, *255*, 175, *255*
Ghazizadeh, M., 141, *79*
Gherardini, F.C., 240, *199*
Ghirlando, R., 256, *264*
Ghosh, S., 41, *17*, 42, *17*, 45, *37*, 46, *17*, 46, *46*
Giaccia, A.J., 149, *150*
Gibier, P., 175, *257*
Gibson, D. (T.), 71, *26*, 71, *27*, 71, *28*
Giguere, D., 154, *179*, 155, *179*, 156, *183*, 160, *193*, 161, *183*, 161, *193*, 162, *183*, 164, *179*, 170, *183*
Gijsen, H.J.M., 58, *7*
Gilkes, N.R., 256, *265*
Gillen, A.C., 151, *167*
Gillenwater, A., 142, *96*
Gillet, C., 142, *98*
Gillett, C., 141, *86*
Giraud, M.-F., 255, *255*
Gires, O., 144, *112*
Gitt, M.A., 130, *26*, 130, *34*, 130, *36*
Glauert, M., 214, *65*
Glinskii, A.B., 150, *164*, 162, *164*, 169, *164*
Glinskii, O.V., 140, *78*, 148, *78*, 150, *163*, 150, *164*, 162, *164*, 169, *78*, 169, *163*, 169, *164*, 169, *225*, 169, *226*

Glinsky, G.V., 140, *78*, 148, *78*, 150, *163*,
150, *164*, 162, *164*, 162, *197*, 167, *197*,
167, *224*, 169, *78*, 169, *163*, 169, *164*,
169, *224*, 169, *225*, 169, *226*
Glinsky, V.V., 140, *78*, 148, *78*, 150, *163*,
150, *164*, 162, *164*, 162, *197*, 167, *197*,
167, *222*, 167, *224*, 169, *78*, 169, *163*,
169, *164*, 169, *224*, 169, *225*, 169, *226*
Glockshuber, R., 255, *262*
Goering, M.G., 108, *156*
Goesmann, A., 218, *121*
Gohar, M., 237, *192*, 238, *192*
Goldstein, I.J., 169, *227*
Golebiowski, A., 71, *29*
Goletz, S., 172, *241*, 185, *241*
Gonatas, J.O., 185, *292*
Gonatas, N.K., 185, *292*
Goodall, M., 256, *264*
Goodman, N.A., 69, *22*
Gooley, A.A., 243, *213*
Gore, P.M., 182, *279*
Gorelik, E., 126, *4*, 127, *4*, 135, *4*, 142, *4*, 147, *4*, 175, *4*
Gorter, A., 185, *291*
Goss, P.E., 91, *94*
Gouge-Ibert, V., 183, *286*
Gough, M.J., 91, *97*
Gounon, P., 240, *200*
G.Posch, 214, *84*, 216, *100*, 217, *84*, 217, *100*, 217, *106*, 218, *106*, 218, *127*, 219, *129*, 222, *127*, 222, *106*, 227, *106*, 228, *100*, 228, *106*, 228, *127*, 231, *129*, 245, *106*, 246, *106*, 247, *106*, 256, *106*, 257, *84*, 257, *106*, 255, *127*, 256, *127*, 257, *127*
Grabherr, R., 218, *121*, 221, *141*, 227, *141*
Graf, L., 40, *2*
Graninger, M., 211, *7*, 215, *87*, 220, *137*, 221, *137*, 222, *7*, 224, *137*, 225, *137*, 229, *137*, 234, *87*, 235, *87*, 242, *7*, 242, *87*, 243, *137*, 244, *137*, 247, *230*, 247, *233*, 247, *234*, 248, *230*, 255, *230*
Grazaini, E.I., 97, *120*
Green, L.G., 154, *178*
Greene, A.E., 91, *106*
Greenfield, K., 218, *124*
Grice, I.D., 147, *141*, 164, *141*
Griffioen, A.W., 170, *232*, 170, *233*, 170, *235*

Griffiths, R.C., 91, *95*
Griffiths, S.G., 231, *168*
Grigoriadis, A., 141, *86*
Grigorian, A., 133, *43*
Groom, A.C., 142, *97*
Gross, H.J., 128, *16*, 129, *16*
Gruner, J., 84, *62*
Guan, Z., 217, *114*, 230, *114*, 230, *150*, 230, *151*, 230, *153*, 230, *154*, 231, *163*, 232, *164*
Guérardel, Y., 237, *192*, 238, *192*
Guerry, P., 212, *33*, 212, *42*, 213, *42*, 213, *43*, 214, *59*, 217, *33*
Gum, J., 145, *128*
Gunneriusson, E., 255, *261*
Gunning, A.P., 149, *160*
Guo, H., 90, *83*, 90, *84*, 91, *109*, 91, *110*, 91, *111*, 93, *112*, 93, *113*, 111, *159*, 111, *160*, 111, *161*, 113, *171*
Guo, M., 143, *108*
Guppi, S.R., 91, *85*, 98, *121*, 99, *122*, 101, *123*
Guzei, I.A., 101, *128*
Gynkiewicz, G., 58, *6*

H

H. Lis, 126, *3*
Hafezi-Moghadam, A., 145, *126*
Hagelueken, G., 251, *242*
Hagmar, B., 151, *168*
Haklai, R., 144, *117*
Hakomori, S. (I.), 45, *41*, 137, *61*, 137, *64*, 140, *61*, 141, *82*, 141, *83*, 145, *64*, 145, *83*, 151, *65*
Halim, A., 130, *30*, 133, *30*
Halkes, K.M., 167, *223*
Halloran, M.M., 135, *50*
Hamilton, W.B., 140, *72*
Hammer, D.A., 178, *270*
Han, Y.L., 172, *242*, 178, *242*
Hanahan, D., 144, *109*
Hancock, I.C., 231, *169*, 231, *170*, 232, *169*, 232, *170*
Hang, H.C., 149, *151*
Hanreich, A., 255, *257*, 256, *257*
Hans, S., 91, *105*
Hansen, B.C., 158, *187*
Hao, S., 172, *239*, 176, *239*, 177, *239*, 178, *269*, 179, *239*, 180, *239*, 180, *273*

Haraguchi, Y., 145, *132*
Hargittai, B., 170, *233*
Harris, J.M., 83, *49*, 83, *50*, 84, *60*, 84, *61*
Harris, J.R., 214, *72*, 217, *72*
Hart, G.W., 140, *74*, 211, *4*, 219, *4*, 222, *4*
Hart, I.R., 126, *5*
Hartnell, A., 128, *14*
Harwood, C.R., 231, *169*, 232, *169*
Hasebe, O., 145, *131*
Hasegawa, A., 141, *87*, 165, *208*, 173, *87*, 182, *208*, 183, *284*
Haseman, J., 170, *233*
Hashidate, T., 134, *46*
Hashimoto, Y., 128, *17*, 129, *17*
Haskell, C.J., 135, *50*
Haslam, S., 140, *75*
Haslam, S.M., 213, *47*, 231, *160*, 231, *162*, 255, *47*
Haslett, C., 130, *23*
Haukaas, M.H., 83, *55*, 83, *56*, 83, *59*, 88, *76*, 103, *143*
Haux, J., 101, *124*, 101, *126*
Hawes, J.W., 45, *37*, 45, *39*, 46, *39*
Hayashi, A., 151, *167*
Hayashi, M., 184, *287*
Hays, D., 148, *149*
He, B.A., 147, *140*
He, D.L., 149, *159*, 150, *159*
He, H., 97, *120*
He, X., 148, *149*
Hébraud, M., 253, *250*
Hecht, S.M., 94, *116*
Hedlund, M., 130, *28*
Heinl, S., 218, *121*, 221, *141*, 227, *141*
Heiss, C., 213, *53*, 255, *53*
Hellebrekers, D.M.E.I., 170, *235*
Hemmerich, S., 137, *58*
Hemsley, M., 71, *27*
Henderson, B.W., 166, *217*
Henderson, I., 73, *31*
Henderson, N.C., 130, *23*
Henne, A., 232, *174*, 239, *174*, 240, *174*, 241, *174*, 242, *174*, 252, *174*, 253, *174*
Henrick, K., 152, *172*
Henry, C.J., 150, *163*, 169, *163*
Henzel, W., 137, *58*
Hermjakob, H., 222, *143*

Hernandez, M., 213, *49*
Hernandez, T., 91, *90*
Hernandez-Gay, J.J., 183, *285*
Hershkovitz, R., 133, *37*, 143, *37*
Herzberg, O., 130, *34*
Herzner, H., 185, *293*
Hetenyi, A., 169, *229*, 170, *229*
Hettche, F., 70, *24*
Hettmann, T., 231, *161*
Hidaka, A., 141, *89*
Hideshima, T., 149, *159*, 150, *159*
Higashi, H., 43, *26*
Higuchi, N., 217, *108*
Hilinski, M.K., 94, *116*
Hill, R.l., 48, *63*, 49, *63*
Hiller, K.M., 174, *254*, 175, *254*
Hinds, J.W., 103, *138*
Hindsgaul, O., 101, *135*, 133, *39*, 156, *39*
Hino, M., 91, *86*
Hinterdorfer, P., 214, *83*, 214, *84*, 217, *84*, 257, *84*
Hirabayashi, J., 130, *36*, 133, *37*, 134, *46*, 143, *37*, 143, *106*, 144, *113*
Hirai, K., 175, *259*, 176, *259*
Hirashima, M., 130, *23*, 134, *46*, 144, *113*
Hitchen, P.G., 213, *47*, 213, *48*, 213, *49*, 230, *150*, 231, *160*, 255, *47*, 255, *257*, 256, *257*
Hitomi, S., 141, *83*, 145, *83*
Hitron, J.A., 107, *149*
Ho, S.B., 142, *96*
Hoeflich, A., 130, *23*
Hoffman, F.M., 101, *128*
Hoffmann, N., 240, *202*
Hoffmaster, A. (R.), 236, *186*, 236, *190*, 237, *190*, 238, *190*, 244, *190*, 253, *186*
Hofinger, A., 235, *182*, 243, *182*, 244, *182*, 245, *222*, 246, *225*, 247, *225*, 247, *231*, 247, *232*, 248, *225*, 248, *231*, 248, *232*
Hogan, C., 143, *101*
Hogan, V., 130, *27*, 130, *33*, 144, *27*, 147, *27*, 148, *146*, 148, *147*, 49, *153*, 149, *157*, 150, *147*, 151, *153*, 151, *157*
Hoidal, J., 145, *122*
Hollaus, F., 218, *117*, 218, *118*, 219, *117*, 219, *118*, 241, *117*

Hollingsworth, M.A., 142, *94*
Holmes, D.S., 182, *279*
Hong, H.-C., 147, *139*
Hong, T.-M., 147, *139*
Honjo, Y., 130, *27*, 144, *27*, 147, *27*, 149, *153*, 151, *153*
Honma, K., 217, *109*
Horejs, C., 216, *94*, 216, *95*
Horejs, C.M., 216, *96*
Horie, H., 163, *202*
Hostettler, N., 177, *266*, 180, *266*
Hotta, K., 158, *187*
Hotzy, C., 240, *204*, 242, *210*
Hough, G.W., 108, *156*
Hough, L., 91, *96*
Houliston, R.S., 230, *155*
Howard, S.C., 182, *278*
Howorka, S., 214, *86*, 215, *92*
Hoye, T.R., 170, *233*, 170, *235*, 170, *236*, 171, *236*
Hoyer, L.C., 126, *5*
Hrachovinova, I., 145, *126*
Hsu, D.K., 130, *23*, 143, *104*, 147, *137*
Hsu, K.-L., 217, *107*
Hu, J.-F., 108, *156*
Huang, B.Q., 172, *239*, 176, *239*, 177, *239*, 179, *239*, 180, *239*
Huang, J.F., 181, *277*
Huang, W., 255, *259*
Huang, Y., 245, *223*
Huang, Y.C., 145, *122*
Hubener, S.F.C.F., 40, *2*
Huber, C., 215, *88*, 215, *89*, 216, *89*
Huber, R., 256, *263*
Hudlicky, T., 58, *8*, 71, *25*
Huebner, C.F., 40, *1*
Huflejt, M.E., 140, *78*, 144, *111*, 148, *78*, 169, *78*
Hug, I., 211, *12*, 214, *62*
Hughes, C., 130, *36*
Hughes, J., 130, *23*
Hughes, R.C., 143, *102*, 152, *172*
Hull, E., 44, *31*
Hull, R.D., 178, *268*
Hunter, T.J., 83, *54*, 85, *67*
Huxley, V.H., 150, *163*, 150, *164*, 162, *164*, 169, *163*, 169, *164*, 169, *225*

Huxtable, C.R., 91, *88*

I

Iacobelli, S., 144, *110*, 144, *111*, 151, *170*, 162, *195*, 164, *195*
Ichikawa, Y., 161, *194*
Iijima, K., 184, *287*
Ikami, T., 183, *284*
Ikawa, D.G., 40, *2*
Ikeda, Y., 145, *132*
Ilarregui, J.M., 141, *91*, 142, *91*, 144, *91*, 147, *91*
Ilg, K., 213, *51*
Ilk, N., 216, *93*, 216, *96*, 218, *122*, 233, *177*, 242, *177*, 243, *177*, 254, *122*, 255, *93*, 256, *122*
Im, J.H., 178, *270*
Imaeda, Y., 173, *248*
Imai, Y., 137, *58*
Imanaka, H., 91, *86*
Imaoka, S., 141, *81*, 145, *81*
Imperiali, B., 211, *22*, 213, *22*
Inagaki, H., 183, *284*
Inagaki, Y., 163, *202*
Ingragsia, L., 164, *203*
Ingrassia, L., 162, *196*
Inohara, H., 130, *33*, 143, *104*, 144, *116*, 148, *146*, 149, *155*
Inoue, M., 140, *70*, 142, *95*
Ipe, U., 182, *281*, 185, *281*
Iredale, J.P., 130, *23*
Irimura, T., 141, *81*, 141, *84*, 145, *81*
Isenmann, S., 185, *292*
Ishai-Michaeli, R., 177, *266*, 180, *266*, 184, *289*, 185, *289*
Ishida, A., 142, *95*
Ishida, H., 137, *58*, 183, *284*
Ishikawa, O., 141, *81*, 145, *81*
Ishikura, H., 146, *135*
Ishizuka, I., 175, *260*
Ito, E., 231, *171*, 232, *171*, 233, *171*, 253, *251*
Ito, H., 151, *65*
Ito, K., 147, *141*, 163, *200*, 163, *201*, 164, *141*
Itoh, S., 182, *283*
Itzkowitz, S.H., 137, *61*, 137, *62*, 140, *61*, 140, *70*
Iurisci, I., 151, *170*, 162, *195*, 164, *195*

Ivetic, A., 141, *86*
Iwanaga, T., 141, *81*, 145, *81*
Iyer, A., 102, *136*
Izumi, Y., 141, *84*

J

Jabeen, T., 156, *182*
Jablons, D.M., 147, *140*
Jackers, P., 142, *98*
Jackson, C.L., 149, *158*
Jacob, G.S., 182, *278*
Jacobsen, E.N., 63, *15*, 67, *19*
Jacoby, D., 145, *122*
Jain, R.K., 173, *252*, 175, *258*
Jakava-Viljanen, M., 217, *105*, 246, *105*
James, L.F., 91, *89*
James, P., 165, *208*, 182, *208*
Janesch, B., 219, *130*, 231, *167*, 239, *167*, 240, *167*, 241, *167*, 249, *130*, 250, *240*, 251, *130*, 252, *246*, 252, *247*, 253, *137*, 255, *247*, 256, *247*
Janssen, T., 141, *79*
Jansson, P.-E., 212, *40*, 217, *40*, 244, *217*
Jarosch, M., 215, *88*, 242, *210*
Jarrell, F., 217, *112*, 219, *112*, 230, *112*
Jarrell, H.C., 213, *45*
Jarrell, K.F., 230, *155*, 230, *156*, 230, *157*, 230, *158*, 230, *159*
Jarvis, G.A., 148, *145*
Jayaram, S., 130, *22*, 142, *22*, 151, *22*
Jayson, G.C., 145, *123*, 180, *123*
Jefferis, R., 256, *263*, 256, *264*
Jennings, M.P., 213, *54*
Jeschke, U., 148, *144*, 165, *144*, 167, *144*, 172, *241*, 185, *241*
Jiang, C.G., 145, *130*, 173, *130*, 172, *243*
Jiang, X., 164, *204*
Jiang, Y., 143, *108*
Jiménez-Barbero, J., 101, *130*, 182, *280*, 183, *285*
Jin, Y.-T., 147, *139*
Jing, Y.J., 172, *242*, 178, *242*
Joachimiak, A., 240, *203*
J.Ø.Duus, 235, *182*, 243, *182*, 244, *116*, 244, *182*
John, C.M., 148, *145*

Johnson, C.R., 71, *29*
Johnson, K.D., 150, *163*, 169, *163*
Johnson-Pais, T.L., 142, *95*
Joiner, C.S., 165, *206*
Joly, G.D., 67, *19*
Jomori, T., 183, *284*
Jones, C., 182, *281*, 185, *281*
Jones, E.Y., 127, *11*, 129, *19*
Jones, G.M., 217, *112*, 219, *112*, 230, *112*, 230, *157*, 230, *159*
Jones, M.G., 91, *95*
Jones, V.L., 151, *65*
Jose Hernandez, J., 182, *280*
Jouault, T., 159, *190*
Jourdian, G.W., 41, *6*, 41, *7*, 42, *7*, 43, *7*
Jubault, P., 183, *286*
Julien, S., 140, *75*, 141, *86*
Jung, T., 144, *112*
Junquera, F., 76, *39*

K

Kabuto, T., 141, *81*, 145, *81*
Kadoya, T., 163, *202*
Kagawa, S., 148, *146*
Kählig, H., 219, *131*, 219, *132*, 220, *131*, 220, *132*, 220, *139*, 223, *131*, 223, *132*, 223, *139*, 224, *132*, 224, *139*, 229, *131*, 229, *132*, 233, *180*, 234, *180*, 242, *180*, 243, *131*, 243, *139*, 244, *131*, 244, *132*, 244, *139*, 244, *180*, 245, *131*, 245, *132*, 245, *139*, 249, *131*
Kahl-Knutsson, B., 133, *45*, 148, *145*, 152, *45*, 152, *171*, 160, *192*
Kahsai, A.W., 170, *230*
Kaina, B., 172, *244*, 173, *249*
Kainz, B., 252, *247*, 255, *247*, 255, *257*, 256, *247*, 256, *257*
Kajiyama, K., 145, *132*
Kakigami, T., 183, *284*
Kalayci, O., 147, *137*
Kallio, R.E., 71, *26*
Kaltner, H., 127, *10*, 130, *23*, 133, *44*, 134, *44*, 144, *117*, 152, *44*, 165, *209*, 165, *210*, 165, *213*, 165, *214*, 165, *216*, 166, *209*, 166, *210*, 166, *216*, 166, *218*, 167, *218*, 167, *221*, 169, *221*
Kamerling, J.P., 167, *223*

Kameyama, M., 141, *81*, 145, *81*
Kaminski, L., 230, *150*, 230, *151*
Kamps, J., 185, *291*
Kanayama, H., 150, *162*
Kandiba, L., 217, *112*, 219, *112*, 230, *112*, 230, *153*
Kandler, O., 211, *25*, 214, *75*
Kang, S.-W., 115, *172*
Kanigsberg, A., 152, *173*, 156, *173*, 165, *173*
Kaniskan, H.U., 170, *230*
Kanitakis, J., 141, *79*
Kann, N., 130, *35*, 155, *35*
Kannagi, R., 141, *87*, 173, *87*
Kannenberg, E. (L.), 111, *168*, 236, *185*, 236, *186*, 236, *189*, 236, *190*, 236, *191*, 236, *190*, 237, *190*, 237, *191*, 238, *189*, 238, *190*, 244, *189*, 244, *190*, 253, *185*, 253, *186*, 253, *189*, 253, *252*
Kansas, G.S., 176, *261*
Kanwar, Y.S., 143, *103*
Kapadia, G., 130, *34*
Kaplan, J.B., 247, *229*
Kappelmayer, J., 145, *126*
Karamata, D., 232, *173*
Karlsson, A., 158, *185*
Karlyshev, A.V., 213, *48*
Karmakar, S., 148, *148*
Karsten, U., 148, *144*, 165, *144*, 167, *144*
Kasai, K., 130, *36*, 134, *46*
Kasai, K.I., 133, *37*, 143, *37*
Kasai, T., 150, *162*
Kato, H., 146, *135*
Kato, M., 176, *262*
Katoh, S., 144, *113*
Katsuyama, T., 145, *131*
Kaufman, B., 41, *8*, 41, *10*, 41, *11*, 41, *12*, 41, *14*, 41, *15*, 42, *8*, 42, *10*, 42, *11*, 42, *12*, 42, *14*, 42, *15*, 42, *23*, 43, *8*, 43, *10*, 43, *11*, 43, *12*, 44, *12*, 47, *53*
Kawabe, T., 173, *246*
Kawakami, T., 147, *137*
Kawamura, T., 212, *28*
Kawamura, Y.I., 175, *259*, 176, *259*
Kawase, J., 173, *248*
Kawashima, R., 175, *259*, 176, *259*
Kawata, T., 239, *195*

Kay, W., 244, *216*
Kayser, K., 127, *10*
Kazazian, H.H., 145, *126*
Kean, E.L., 42, *21*
Keenan, T.W., 44, *29*, 45, *29*, 46, *29*
Keller, W., 214, *86*, 215, *90*, 215, *91*
Kelly, J., 213, *45*, 230, *155*, 230, *159*
Kelly, J.F., 212, *33*, 213, *43*, 217, *33*, 230, *156*, 230, *158*
Kelly, R.C., 85, *66*, 101, *66*
Kelm, S., 127, *11*, 128, *14*, 128, *16*, 129, *16*, 129, *19*
Kennedy, L.J., 88, *81*
Kennedy, T., 145, *122*
Keranen, M.D., 83, *49*, 83, *50*, 84, *60*, 84, *61*
Kerbel, R.S., 140, *67*
Kereszt, A., 212, *41*, 217, *41*
Kern, J. (W.), 236, *183*, 236, *184*, 240, *203*, 253, *184*
Kern, V.J., 236, *183*
Keshavarz, T., 91, *87*
Keyhani, N.O., 49, *59*, 49, *60*, 49, *61*, 49, *62*
Khabut, A., 165, *212*
Khan, F., 45, *37*
Kharel, M.K., 107, *149*
Khatib, A.M., 172, *245*
Khorana, H., 42, *22*
Kidd, J., 165, *208*, 182, *208*
Kido, N., 249, *238*
Kieber-Emmons, T., 167, *220*
Kiermaier, E., 213, *50*
Kihara, S., 158, *187*
Kikuchi, H., 80, *47*
Kilburn, D.G., 256, *265*
Kim, H., 88, *79*, 88, *80*
Kim, H.C., 218, *125*, 257, *125*
Kim, H.J., 101, *123*
Kim, H.R.C., 143, *104*, 144, *118*, 148, *146*, 150, *162*
Kim, H.S., 133, *39*, 156, *39*
Kim, Y.J., 140, *68*, 144, *121*, 145, *121*
Kim, Y.S., 126, *2*, 137, *61*, 137, *62*, 140, *61*, 144, *2*, 145, *128*
Kimber, M.S., 229, *147*, 252, *245*
Kimura, C., 146, *135*
King, J.F., 101, *134*
Kipari, T., 130, *23*

Kiriakova, G., 150, *164*, 162, *164*, 162, *197*, 167, *197*, 169, *164*
Kirmaier, C., 182, *278*
Kirmse, R., 144, *112*
Kishida, K., 158, *187*
Kishimoto, T., 146, *135*
Kiso, M., 137, *58*, 165, *208*, 182, *208*, 183, *284*
Kiss, E., 212, *41*, 217, *41*
Kiss, R., 130, *23*, 130, *25*, 141, *79*, 147, *25*, 162, *196*, 164, *203*
Kitagawa, H., 140, *70*
Kjeldsen, T., 137, *62*
Klassen, J.S., 213, *55*, 214, *62*
Kleinert, H., 172, *244*
Klepp, O., 101, *124*
Kletzin, A., 214, *77*
Kleuser, B., 185, *292*
Klingbeil, P., 144, *112*
Klingl, A., 214, *77*
Kloog, Y., 144, *117*
Kneidinger, B., 247, *230*, 247, *233*, 247, *234*, 248, *230*, 255, *230*
Knirel, Y.A., 212, *39*, 212, *40*, 217, *40*, 231, *161*, 243, *215*
Ko, H.L., 146, *136*, 147, *136*, 151, *136*, 151, *169*
Ko, S., 59, *13*
Ko, S.Y., 59, *12*
Kobata, A., 211, *1*
Kobayashi, H., 249, *238*
Kobayashi, K., 173, *246*, 173, *248*
Koch, A., 240, *206*
Koch, A.E., 135, *50*
Koch, J.R., 71, *26*
Kocher, H.P., 185, *292*
Kochetkov, N.K., 243, *215*
Koenig, A., 177, *263*, 178, *263*
Koerdt, A., 216, *100*, 217, *100*, 219, *129*, 228, *100*, 231, *129*
Koganty, R.R., 140, *70*
Köhler, G., 215, *89*, 216, *89*
Kohno, S., 141, *89*
Kohsaka, M., 91, *86*
Koizumi, M., 141, *89*
Kojima, N., 253, *251*
Kokal, W.A., 137, *61*, 140, *61*

Kolanus, W., 141, *87*, 171, *238*, 173, *87*
Kolarich, D., 219, *132*, 220, *132*, 223, *132*, 224, *132*, 229, *132*, 244, *132*, 245, *132*
Kolb, H.C., 182, *282*
Kolbe, J., 224, *145*, 226, *145*, 243, *145*, 245, *218*
Kolset, S.O., 180, *271*
Komba, S., 137, *58*
Komiyama, K., 184, *287*
Kondorosi, Á., 212, *41*, 217, *41*
Kong, C., 149, *150*
König, H., 211, *25*, 214, *72*, 214, *75*, 214, *76*, 217, *72*, 230, *148*
Koning, G.A., 185, *291*
Konrad, Z., 230, *149*, 230, *151*
Konstantopoulos, K., 145, *125*, 180, *125*
Kontani, K., 144, *113*
Koong, A.C., 149, *150*
Kopitz, J., 165, *214*, 166, *218*, 167, *218*
Korhonen, T.K., 241, *208*
Kornfeld, R., 211, *3*
Kornfeld, S., 211, *3*
Kosh, K., 140, *77*
Kosma, P., 218, *119*, 219, *132*, 219, *133*, 220, *132*, 220, *133*, 223, *132*, 224, *132*, 224, *133*, 224, *144*, 225, *133*, 225, *144*, 229, *132*, 233, *177*, 235, *182*, 242, *177*, 243, *133*, 243, *177*, 243, *182*, 243, *213*, 244, *132*, 244, *182*, 245, *132*, 245, *222*, 246, *225*, 247, *225*, 247, *231*, 247, *232*, 247, *233*, 247, *234*, 248, *225*, 248, *231*, 248, *232*
Koths, K., 144, *116*
Kováč, P., 111, *165*, 111, *166*, 111, *167*
Kowalczyk, D., 182, *281*, 185, *281*
Kowalska, M.A., 178, *270*
Kowarik, M., 213, *49*, 213, *50*
Kozlowski, E.O., 181, *276*
Krapp, S., 256, *263*
Kresge, N., 48, *63*, 49, *63*
Kretzschmar, G., 182, *279*
Krzewinski, F., 237, *192*, 238, *192*
Kuchenbecker, K.M., 147, *140*
Kuchroo, V.K., 144, *113*
Kudo, S., 182, *283*
Kuen, B., 217, *101*, 217, *104*, 218, *118*, 219, *118*, 240, *106*, 246, *101*, 246, *104*

Kuhn, C., 148, *144*, 165, *144*, 167, *144*
Kumar, A., 143, *103*
Kumar, M., 214, *57*, 255, *57*
Kumar, M.V., 149, *158*
Kumar, N., 170, *236*, 171, *236*
Kundig, F.D., 46, *47*
Kundig, W., 46, *46*, 46, *47*, 46, *48*, 46, *49*, 47, *49*
Kunz, H., 137, *63*, 140, *63*, 182, *279*, 182, *281*, 185, *281*, 185, *293*
Küpcü, S., 255, *256*, 256, *256*
Kupper, T.S., 173, *251*
Kuriyama, H., 158, *187*
Kusumoto, S., 180, *272*
Kuwabara, I., 133, *44*, 134, *44*, 144, *111*, 152, *44*, 167, *221*, 169, *221*
Kuwabara, L., 165, *209*, 166, *209*
Kvarnström, I., 244, *217*
Kwok, S., 149, *150*

L

La Clair, J., 90, *84*
Laaberki, M.-H., 254, *253*
Laferte, S., 140, *67*, 140, *76*
Lahm, H., 130, *23*
Lal, A., 91, *91*
Lamed, R., 240, *198*
Landon, L.A., 167, *222*
Langenhan, J.M., 101, *128*, 102, *136*
Langlet, C., 174, *255*, 175, *255*
Langsford, M.L., 256, *265*
Lankester, A., 126, *1*
Lannigan, D.A., 94, *116*
Lanthier, P.H., 213, *45*
Largeau, C., 183, *286*
Larkin, M., 137, *58*
Larsen, R.D., 175, *256*
Larson, G., 130, *30*, 133, *30*
Larue, K., 229, *147*
Larumbe, A., 165, *212*
Lasky, L.A., 137, *58*, 145, *127*
Lau, K.S., 133, *43*
Lauber, K., 214, *77*
Laue, T.M., 184, *288*
Lavaud, C., 108, *154*
Lawson, A.M., 137, *58*
Le, D., 91, *111*, 93, *113*

Le, J., 91, *111*, 93, *113*
Le Mercier, M., 130, *25*, 147, *25*, 164, *203*
Le, Q.-T., 149, *150*
Leach, R.J., 142, *95*
Lechner, J., 211, *15*
Leclerc, E., 183, *286*
Lee, A.W.M., 59, *12*, 59, *13*
Lee, C., 148, *148*, 88, *79*, 88, *80*
Lee, R.T., 161, *194*
Lee, S.-W., 218, *125*, 257, *125*
Lee, Y.C., 49, *60*, 161, *194*
Leffler, H., 130, *20*, 130, *26*, 130, *28*, 130, *30*, 130, *34*, 130, *35*, 130, *36*, 133, *30*, 133, *40*, 133, *41*, 133, *42*, 133, *45*, 141, *41*, 147, *141*, 152, *45*, 152, 148, *145*, 149, 152, *171*, 152, *173*, 152, *174*, 152, *175*, 152, *176, 152*, 153, *175*, 154, *174*, 154, *180*, 155, *35*, 156, *40*, 156, *173*, 156, *176*, 156, *182*, 156, *184*, 157, *40*, 158, *20*, 158, *40*, 158, *152*, 158, *185*, 158, *188*, 159, *189*, 160, *191*, 160, *192*, 163, *176*,163, *199*, 164, *141*, 164, *176*, 164, *203*, 164, *204*, 164, *205*, 165, *173*, 165, *212*
Lefranc, F., 130, *25*, 147, *25*, 162, *196*
Lehr, J., 149, *157*, 151, *157*
Leitner, K., 242, *210*
Leloir, L., 45, *42*, 45, *43*
LeMarer, N., 143, *102*
Lemieux, R.U., 127, *8*
Lennarz, W.J., 213, *46*
Lensch, M., 165, *210*, 166, *210*, 169, *228*, 170, *228*
Lenter, M., 185, *292*
Leoff, C., 236, *185*, 236, *186*, 236, *190*, 236, *191*, 237, *190*, 237, *191*, 238, *190*, 244, *190*, 253, *185*, 253, *186*, 253, *252*
Leonidas, D.D., 130, *34*, 159, *190*
Leonori, D., 79, *40*, 79, *41*
Lerouge, P., 150, *161*
Lespieau, R., 58, *3*
Letts, P.J., 45, *45*
Levery, S.B., 45, *41*
Levinovitz, A., 185, *292*
Lewis, L.A., 143, *106*
Lewis, W.H., 247, *235*

AUTHOR INDEX

Li, C., 255, *259*
Li, C.C., 172, *244*
Li, D.H., 142, *96*
Li, F., 145, *130*, 173, *130*
Li, G.L., 149, *159*, 150, *159*
Li, H., 236, *188*
Li, J.B., 145, *130*, 172, *243*, 173, *130*
Li, M., 83, *56*, 83, *57*, 83, *58*, 109, *157*, 109, *158*
Li, N., 172, *239*, 176, *239*, 177, *239*, 179, *239*, 180, *239*, 180, *273*
Li, X., 236, *188*
Li, X.-B., 49, *59*
Li, Y.S., 145, *130*, 173, *130*
Li, Z., 45, *37*
Liao, D.I., 130, *34*
Liao, D.L., 130, *34*
Liao, Y.F., 91, *91*
Lichenstein, H.S., 135, *56*, 137, *56*
Lider, O., 133, *37*, 143, *37*
Lievin, J., 158, *186*
Lilly, C.M., 147, *137*
Lin, C.-l., 164, *204*
Lin, H.M., 144, *118*
Lin, Y., 172, *244*
Lin, Y.C., 147, *140*
Lindberg, B., 222, *142*, 244, *217*
Lindenthal, C., 212, *29*
Lindh, F., 244, *217*
Lindquist, K.C., 182, *278*
Lindsay, K.B., 91, *101*, 91, *103*
Linhardt, R., 177, *263*, 178, *263*
Link, K.P., 40, *2*
Link, P.K., 40, *1*
Linke, B., 218, *121*
Linton, D., 213, *47*, 213, *48*, 255, *47*
Liotta, D.C., 70, *23*
Lipkind, G.M., 243, *215*
Liskamp, R.M.J., 165, *209*, 165, *210*, 166, *209*, 166, *210*
Liu, B.C., 165, *213*
Liu, D., 250, *241*
Liu, F.R., 145, *130*, 172, *243*, 173, *130*
Liu, F.T., 130, *23*, 130, *24*, 130, *36*, 133, *44*, 134, *44*, 141, *24*, 142, *24*, 142, *98*, 144, *24*, 143, *104*, 144, *111*, 147, *137*, 152, *44*, 165, *209*, 166, *209*

Liu, F.-T., 130, *23*
Liu, H.-W., 101, *123*
Liu, J.C., 172, *244*
Liu, R., 174, *254*, 175, *254*
Liu, S.-Y., 240, *199*
Liu, T.-H., 166, *217*
Livingston, P.O., 137, *59*, 137, *60*, 140, *69*, 140, *70*, 140, *71*, 141, *71*, 140, *72*, 146, *134*
Lizak, C., 213, *52*, 255, *259*, 255, *262*
Lloyd, K.O., 140, *70*, 140, *71*, 140, *72*, 141, *71*
Lobsanov, Y.D., 130, *34*, 165, *212*
Locher, K.P., 213, *52*
Logan, S.M., 212, *30*, 212, *33*, 212, *34*, 212, *35*, 212, *36*, 212, *37*, 213, *43*, 217, *33*, 217, *36*, 230, *155*, 230, *156*, 230, *158*, 230, *159*
Loganathan, D., 169, *227*
Lombardo, D., 175, *257*
Longenecker, B.M., 140, *70*
Lönngren, J., 244, *217*
López-Lázaro, M., 101, *127*
Loren, M., 170, *231*
Lotan, R., 130, *32*, 130, *36*, 142, *32*, 142, *96*, 148, *143*, 150, *162*
Lott, J.R., 151, *167*
Lottspeich, F., 214, *77*
Low, L.Y., 236, *188*
Low, P., 147, *141*, 164, *141*
Lowary, T.L., 101, *135*
Lowe, J.B., 143, *105*, 175, *256*
Lowitz, K., 187, *294*
Lubitz, W., 217, *104*, 246, *104*, 240, *206*
Lucas, C.M., 135, *53*
Luche, J.L., 101, *133*
Ludowig, J., 40, *3*, 41, *3*
Ludwig, R.J., 180, *274*
Ludwig, T., 144, *112*
Lund, J., 256, *264*
Lunderberg, J.M., 254, *254*
Luo, J.Y., 176, *262*
Luo, P., 167, *220*
Luo, S., 180, *272*
Lupas, A., 239, *197*
Lynch, W.H., 231, *168*
Lynn, D.G., 69, *22*

Lyon, M., 145, *123*, 180, *123*
Lyons, D.E., 135, *56*, 137, *56*

M

M. Gluck, 141, *90*
Ma, J.C., 153, *177*
Ma, R., 43, *27*, 44, *27*, 45, *27*, 46, *27*
Ma, Y., 247, *235*
Mabry, T., 71, *27*
Macaluso, F., 165, *213*
MacDonald, I.C., 142, *97*
MacDonald, J.R., 170, *231*
Mackinnon, A.C., 130, *23*
MacLean, L.L., 247, *229*
MacLeod, A.M., 212, *27*
MacMillan, D.W.C., 68, *20*, 70, *24*
Macnab, R.M., 213, *44*
Mader, C., 235, *182*, 243, *182*, 244, *182*
Madigan, M.T., 214, *63*
Madsen, P., 130, *34*
Maenaka, K., 127, *11*
Maenaka, T., 127, *11*
Maes, E., 237, *192*, 238, *192*
Maffioli, C., 213, *51*
Magidovich, H., 230, *149*
Magnani, J.L., 141, *90*, 182, *279*
Magnusson, B.G., 152, *171*
Mahal, L.K., 217, *107*
Mahoney, J.A., 128, *14*
Maier, M.E., 69, *21*
Makristathis, A., 218, *118*, 219, *118*
Maljaars, C.E.P., 167, *223*
Mamourian, A.C., 178, *268*
Mancilla, E., 145, *122*
Mandal, P.K., 233, *176*, 240, *176*, 242, *176*, 244, *176*, 253, *176*
Mandel, U., 140, *75*
Mandl, F., 217, *101*, 246, *101*
Mangion, I.K., 70, *24*
Manna, S.K., 101, *125*
Mannori, G., 173, *250*
Manousos, G.A., 174, *254*, 175, *254*
Mansson, O., 141, *90*
Manuguerra, J.C., 128, *16*, 129, *16*
Marathe, D., 182, *278*
Marchetti, A., 144, *110*
Marcus, M.E., 130, *26*

Maresh, J., 69, *22*
Marhaba, R., 144, *112*
Mariano, P.S., 91, *100*
Marolda, C.L., 213, *49*
Marre, R., 212, *39*
Marrugat, R., 145, *129*, 172, *129*
Marston, C.K., 236, *190*, 237, *190*, 238, *190*, 244, *190*
Mårtensson, S., 45, *41*
Martin, R., 91, *99*
Martinek, T.A., 169, *229*, 170, *229*
Martinez, B., 241, *207*
Martinko, J.M., 214, *63*
Martin-Satue, M., 145, *129*, 172, *129*
Mas, E., 175, *257*
Masamune, S., 59, *12*, 59, *13*
Massiot, G., 108, *154*
Masuda, K., 239, *195*
Masuyer, G., 156, *182*
Matecko, I., 215, *87*, 234, *87*, 235, *87*, 242, *87*
Mathieu, S., 174, *255*, 175, *255*
Matsuda, M., 158, *187*
Matsukawa, Y., 158, *187*
Matsumoto, S., 173, *248*, 173, *246*
Matsumoto, Y., 183, *284*
Matsuo, O., 187, *294*
Matsushita, Y., 141, *81*, 145, *81*
Matsuzawa, Y., 158, *187*
Matta, K.L., 148, *148*, 148, *149*, 161, *194*, 173, *252*, 182, *278*
Mattanovich, D., 215, *88*
Matthews, L., 167, *222*
Matuschek, M., 240, *199*
Matuskova, J., 145, *126*
Mawhinney, T.P., 150, *163*, 150, *164*, 162, *164*, 169, *163*, 169, *164*, 169, *226*
May, A., 240, *202*
May, A.P., 129, *19*
Mayben, J.P., 174, *254*, 175, *254*
Mayer, H.F., 214, *84*, 217, *84*, 217, *101*, 217, *101*, 233, *177*, 235, *182*, 242, *177*, 243, *177*, 243, *182*, 244, *182*, 257, *84*
Mayer-Posner, F.J., 245, *219*
Mayo, K.H., 170, *231*, 170, *232*, 170, *233*, 170, *234*, 170, *235*, 170, *236*, 171, *236*
Mazurek, N., 143, *100*
McCarty, O.J.T., 145, *125*, 180, *125*

McDermott, K.M., 142, *94*
McDermott, T.R., 218, *119*
McEver, R.P., 135, *51*, 135, *54*, 135, *56*, 137, *56*, 145, *126*, 148, *148*, 184, *288*
McGuire, E.J., 47, *56*
McKenna, S.B., 103, *138*
McKenney, P.T., 211, *13*
McKillop, A., 79, *44*
McLellan, W.L., 126, *5*
McNally, D.J., 212, *36*, 217, *36*
McNeil, M.R., 247, *235*
Mechref, Y., 245, *223*
Medici, A., 75, *38*
Medow, N.D., 47, *52*
Megson, Z.A., 216, *100*, 217, *100*, 219, *129*, 228, *100*, 231, *129*,
Mehta, A.S., 111, *168*, 236, *186*, 253, *186*
Mehul, B., 152, *172*
Meier-Stauffer, K., 218, *118*, 219, *118*
Meldal, M., 154, *178*
Men, H., 88, *79*
Mentele, R., 214, *77*
Merchan, F.L., 76, *39*
Mercurio, A.M., 130, *36*
Merino, P., 76, *39*
Meromsky, L., 148, *143*
Merrick, J.M., 42, *19*
Merritt, J.H., 213, *53*, 255, *53*
Mescher, M.F., 211, *14*, 211, *23*, 217, *23*
Mesnage, S., 231, *164*, 233, *164*, 239, *164*, 240, *164*, 240, *164*, 240, *200*, 240, *201*, 241, *164*, 242, *164*, 252, *164*, 253, *164*
Messner, P., 247, *231*, 248, *231*, 211, *5*, 211, 7, 211, *11*, 211, *18*, 212, *18*, 214, *58*, 214, 69, 214, *80*, 214, *84*, 215, *87*, 216, *93*, 216, *98*, 216, *100*, 217, *18*, 217, *84*, 217, *100*, 217, *105*, 217, *106*, 217, *110*, 217, *111*, 218, *5*, 218, *11*, 218, *18*, 218, *58*, 218, *106*, 218, *111*, 218, *116*, 218, *117*, 218, *118*, 218, *119*, 218, *120*, 218, *122*, 218, *127*, 219, *11*, 219, *111*, 219, *117*, 219, *118*, 219, *128*, 219, *129*, 219, *130*, 219, *131*, 219, *132*, 219, *133*, 219, *134*, 219, *135*, 219, *136*, 220, *131*, 220, *132*, 220, *133*, 220, *134*, 220, *135*, 220, *136*, 220, *137*, 220, *139*, 220, *140*, 221, *120*, 221, *135*, 221, *137*, 221, *141*, 222, 7, 222, *106*, 222, *116*, 222, *120*, 222, *127*, 223, *131*, 223, *132*, 223, *136*, 223, *139*, 224, *132*, 224, *133*, 224, *134*, 224, *137*, 224, *139*, 225, *137*, 224, *144*, 224, *145*, 225, *133*, 225, *134*, 225, *144*, 226, *145*, 226, *146*, 227, *106*, 227, *120*, 227, *140*, 227, *141*, 228, *100*, 228, *106*, 228, *127*, 229, *111*, 229, *131*, 229, *132*, 229, *136*, 229, *137*, 231, *129*, 231, *162*, 234, *87*, 235, *87*, 239, *69*, 241, *117*, 242, 7, 242, *87*, 243, *131*, 243, *133*, 243, *136*, 243, *137*, 243, *139*, 243, *145*, 245, *136*, 244, *120*, 244, *131*, 244, *132*, 244, *137*, 244, *139*, 245, *106*, 245, *131*, 245, *132*, 245, *135*, 245, *139*, 245, *140*, 246, *105*, 246, *106*, 247, *106*, 247, *116*, 249, *130*, 249, *131*, 249, *134*, 251, *130*, 251, *135*, 252, *135*, 254, *11*, 254, *80*, 254, *111*, 254, *116*, 254, *122*, 255, *11*, 255, *93*, 255, *127*, 256, *106*, 256, *122*, 256, *127*, 257, *84*, 257, *106*, 257, *127*
Metcalf, J.B., 162, *197*, 167, *197*
Meterissian, S., 144, *119*, 172, *119*
Meyer, B. (H.), 217, *113*, 230, *113*, 231, *113*, 231, *160*, 231, *162*, 231, *163*, 232, *163*
Mezher, H.A., 91, *96*
Mi, D.H., 180, *275*
Miceli, M.C., 143, *106*
Michiels, K., 212, *32*
Mignot, T., 231, *164*, 233, *164*, 239, *164*, 240, *164*, 240, *164*, 241, *164*, 242, *164*, 252, *164*, 253, *164*
Mikami, I., 147, *140*
Miller, M.C., 170, *236*, 171, *236*
Miller, R.C. Jr., 256, *265*
Mills, D.C., 213, *55*
Mimata, H., 150, *165*
Mimura, Y., 256, *263*, 256, *264*
Minnaard, A.J., 88, *78*
Mischel, P.S., 144, *115*
Missert, J.R., 166, *217*
Missiakas, D.M., 254, *254*
Missiakis, D., 236, *183*
Mitchell, M., 91, *90*
Mitra, M.K., 216, *95*
Mitra, N., 142, *95*
Mitsiades, N., 149, *159*, 150, *159*
Miyahara, Y., 147, *138*

Miyake, M., 141, *83*, 144, *113*, 145, *83*
Miyamoto, D., 183, *284*
Mizuno, S., 230, *158*
Mizuno, T., 58, *4*
Mo, K.-F., 236, *188*
Mock, M., 111, *162*, 111, *163*, 231, *164*, 233, *164*, 239, *164*, 240, *164*, 240, *164*, 240, *200*, 240, *201*, 241, *164*, 242, *164*, 252, *164*, 253, *164*
Modin, G., 137, *61*, 140, *61*
Moens, S., 211, *6*, 212, *32*
Moffat, J.G., 42, *22*
Mohnen, D., 149, *158*
Moll, D., 215, *88*, 215, *89*, 216, *89*
Moll, H., 231, *161*
Molyneux, R.J., 91, *89*
Monia, B.P., 172, *245*
Monostori, E., 169, *229*, 170, *229*
Monsigny, M., 130, *36*
Montgomery, C.K., 137, *62*
Montreuil, J., 211, *2*
Monzavi-Karbassi, B., 167, *220*
Moore, F.D. Jr., 164, *204*
Moore, K.L., 135, *54*, 135, *56*, 137, *56*, 184, *288*
Mootoo, D.R., 91, *105*, 182, *280*, 183, *285*
Moran, M., 143, *106*
Mordoh, J., 141, *91*, 142, *91*, 144, *91*, 147, *91*
Moremen, K.W., 91, *91*
Morgan, J., 166, *217*
Mori, K., 80, *47*, 180, *272*
Mori, M., 145, *132*
Morishima, T., 173, *246*
Moritz, R.L., 44, *31*
Mormann, M., 245, *220*
Morre, J.D., 44, *29*, 45, *29*, 46, *29*
Morris, D.L., 173, *247*
Morris, H.R., 213, *47*, 213, *48*, 213, *49*, 213, *160*, 213, *162*, 255, *47*
Morris, V.J., 149, *160*
Morselt, H.W.M., 185, *291*
Möschl, A., 218, *120*, 221, *120*, 222, *120*, 227, *120*, 244, *120*
Moser, B., 256, *265*
Moses, F.E., 40, *3*, 41, *3*
Moskal, J.R., 43, *27*, 44, *27*, 45, *27*, 46, *27*

Mossine, V.V., 150, *163*, 150, *164*, 162, *164*, 162, *197*, 167, *197*, 167, *224*, 169, *163*, 169, *164*, 169, *224*, 169, *226*, 169, *225*
Mourelatos, Z., 185, *292*
Mousa, S.A., 145, *125*, 177, *264*, 180, *125*
Moxon, E.R., 213, *54*
Mrozowski, R.M., 94, *116*
Mucilli, F., 144, *110*
Mukhopadhyay, G., 129, *19*
Müller, J., 214, *81*
Müller, N., 215, *87*, 220, *137*, 221, *137*, 224, *137*, 225, *137*, 229, *137*, 233, *176*, 234, *87*, 234, *181*, 235, *87*, 240, *176*, 242, *87*, 242, *176*, 243, *137*, 244, *137*, 244, *176*, 244, *181*, 245, *181*, 253, *176*
Muller, W.E.G., 134, *46*
Munger, M.E., 144, *113*
Munn, D.H., 144, *113*
Murakami, Y., 217, *108*
Murata, T., 142, *95*
Murphy, G.F., 148, *149*
Murphy, W.J., 144, *113*
Murray, R.G.E., 214, *64*
Murruzzu, C., 91, *99*
Muschel, R.J., 178, *270*
Musselli, C., 137, *59*
Muthukumar, M., 216, *95*
Mutter, M., 101, *129*

N

Nadeau, D.R., 144, *120*, 145, *120*, 177, *120*
Nagahara, K., 144, *113*
Nagahata, S.-I., 144, *113*
Nagano, K., 217, *108*
Naggi, A., 177, *266*, 180, *266*
Naik, H., 149, *157*, 151, *157*
Nair, D.B., 217, *112*, 219, *112*, 230, *112*, 230, *157*
Naismith, J.H., 251, *242*, 255, *255*
Naito, A., 141, *89*
Nakabeppu, Y., 170, *232*
Nakada, H., 140, *70*, 142, *95*
Nakagawa, K., 175, *260*
Nakahara, S., 130, *33*, 143, *99*, 144, *99*
Nakajima, K., 141, *89*
Nakajima, T., 217, *109*
Nakamori, S., 141, *81*, 141, *84*, 145, *81*

Nakamura, H., 217, *108*
Nakamura, N., 145, *131*
Nakamura, T., 134, *46*, 158, *187*
Nakano, A., 141, *89*
Nakashima, T., 141, *89*
Nakayama, J., 145, *131*
Nakayama, O., 91, *86*
Nangia-Makker, P., 130, *27*, 130, *33*, 143, *100*, 143, *101*, 143, *104*, 144, *27*, 147, *27*, 148, *147*, 149, *153*, 150, *147*, 150, *162*, 151, *153*, 151, *166*
Naparstek, S., 217, *114*, 230, *114*, 230, *150*
Narita, T., 141, *88*
Nash, R.J., 91, *95*
Nashed, M., 165, *208*, 182, *208*
Natoli, C., 151, *170*
Naumova, I.B., 231, *172*
Navarre, W.W., 239, *193*
Nayaka, M.A.H., 130, *22*, 142, *22*, 151, *22*
Needham, L.K., 171, *238*
Neelamegham, S., 182, *278*
Nelson, A., 165, *206*
Nelson, D., 236, *187*
Nelson, R.M., 173, *250*
Neri, D., 255, *259*
Neri, P., 149, *159*, 150, *159*
Nesmelova, I., 170, *232*, 170, *233*, 170, *236*, 171, *236*
Neumann, L., 218, *127*, 221, *141*, 222, *127*, 227, *141*, 228, *127*, 255, *127*, 256, *127*, 257, *127*
Neuninger, C., 219, *135*, 220, *135*, 221, *135*, 224, *144*, 225, *144*, 245, *135*, 251, *135*, 252, *135*
Neuzil, J., 163, *201*
Newman, R.A., 101, *125*
Nguyen, H., 83, *49*, 84, *61*
Nguyen, J.T., 143, *103*
Nguyen, M., 135, *49*
Nguyen-Mau, S.-M., 254, *254*
Nicklin, P.L., 175, *258*
Nicotra, F., 101, *130*
Nifantiev, N. (E.), 151, *170*, 162, *195*, 164, *195*
Niisson, U.J., 158, *185*
Niki, T., 144, *113*
Nilsson, J., 130, *30*, 133, *30*

Nilsson, U.J., 130, *20*, 130, *30*, 130, *35*, 133, *30*, 133, *40*, 133, *45*, 147, *141*, 149, *152*, 152, *45*, 152, *171*, 152, *174*, 152, *175*, 152, *176*, 153, *175*, 154, *174*, 154, *180*, 155, *35*, 156, *40*, 156, *176*, 156, *182*, 156, *184*, 157, *40*, 158, *20*, 158, *40*, 158, *152*, 158, *188*, 159, *189*, 160, *191*, 160, *192*, 163, *176*, 163, *199*, 164, *141*, 164, *176*, 164, *203*, 164, *204*, 164, *205*, 165, *212*
Nimmich, W., 244, *217*
Nishi, N., 134, *46*
Nishikawa, A., 175, *260*
Nishikawa, Y., 44, *31*
Nishimura, S.I., 165, *216*, 166, *216*
Nishizawa, H., 158, *187*
Nita-Lazar, M., 213, *47*, 255, *47*
Niu, Y.C., 172, *244*
Niv, H., 144, *117*
Nobumoto, A., 144, *113*
Noel, J.P., 90, *84*
Noirot, P., 252, *249*
Nomura, T., 150, *165*
Nomura, Y., 150, *165*
Norberg, T., 133, *44*, 134, *44*, 152, *44*
Nordberg, R.E., 73, *30*
Norgard-Sumnicht, K. (E.), 137, *58*, 177, *263*, 178, *263*
North, S.J., 213, *47*, 255, *47*
Northrup, A.B., 68, *20*, 70, *24*
Norton, C.R., 144, *119*, 172, *119*
Nose, V., 164, *204*
Nothaft, H., 214, *60*, 255, *60*, 255, *262*
Noti, C., 74, *33*
Novotny, M.V., 245, *223*
Novotny, R., 245, *221*, 245, *222*, 246, *227*, 250, *221*, 255, *256*, 256, *256*
Nozawa, R., 163, *202*
Nshimyumukiza, P., 162, *196*
Nubel, T., 172, *244*, 173, *249*
Numao, S., 213, *50*, 213, *52*, 255, *259*
Numata, Y., 140, *70*
Nyberg, N.T., 244, *216*
Nygren, P.A., 255, *261*
Nyholm, P.G., 152, *175*, 153, *175*

O

Oberg, C.T., 130, *20*, 133, *40*, 152, *176*, 156, *40*, 156, *176*, 156, *182*, 157, *40*, 158, *20*, 158, *40*, 158, *185*, 160, *191*, 163, *176*, 164, *176*
Ochieng, J., 143, *101*
O'Doherty, G.A., 58, *9*, 83, *49*, 83, *50*, 83, *51*, 83, *54*, 83, *55*, 83, *56*, 83, *57*, 83, *58*, 83, *59*, 84, *60*, 84, *61*, 84, *63*, 84, *64*, 84, *65*, 85, *67*, 86, *68*, 86, *69*, 86, *70*, 86, *71*, 86, *72*, 86, *73*, 86, *74*, 87, *75*, 88, *76*, 88, *77*, 89, *82*, 90, *83*, 90, *84*, 91, *85*, 91, *109*, 91, *110*, 91, *111*, 93, *112*, 93, *113*, 94, *114*, 94, *115*, 94, *116*, 96, *117*, 96, *118*, 97, *119*, 98, *121*, 99, *122*, 101, *123*, 101, *131*, 101, *132*, 102, *136*, 102, *137*, 103, *138*, 103, *139*, 103, *140*, 103, *141*, 103, *142*, 103, *143*, 104, *144*, 104, *145*, 104, *146*, 106, *147*, 106, *148*, 107, *150*, 107, *151*, 108, *152*, 108, *153*, 109, *157*, 109, *158*, 111, *159*, 111, *160*, 111, *161*, 113, *171*, 115, *172*
Oeberg, C.T., 130, *30*, 133, *30*
Oette, K., 151, *169*
Ogata, S., 140, *70*
Oguchi, H., 151, *65*
Oh, Y.J., 214, *84*, 217, *84*, 257, *84*
Ohmori, K., 141, *87*, 173, *87*
Ohshiba, S., 175, *260*
Ohta, H., 141, *89*
Ohta, M., 142, *95*
Ohtsubo, K., 141, *93*
Ohyama, C., 145, *133*, 187, *294*
Oka, N., 143, *99*, 144, *99*, 150, *162*
Oka, T., 134, *46*
Okamoto, J., 147, *140*
Okamoto, T., 173, *246*
Okawa, Y., 249, *238*
Okcchukwu, P., 164, *203*
Olden, K., 91, *93*
Oldenburg, K.R., 169, *227*
Oling, F., 241, *207*
Omura, S., 184, *287*
O'Neil-Johnson, M., 108, *156*
Ong, E., 212, *27*
Oomizu, S., 144, *113*
Opperman, M.J., 148, *149*

Ornstein, D.L., 178, *268*
Ortiz Mellet, C., 166, *219*
Osbourn, J.M., 91, *111*, 93, *113*
Oscarson, S., 133, *44*, 134, *44*, 152, *44*, 165, *215*
Oseroff, A.R., 166, *217*
Oshima, M., 175, *260*
Ota, K., 143, *103*
Otsuji, E., 151, *65*
Overman, R.S., 40, *2*
Ozawa, S., 173, *248*

P

Paavonen, T., 145, *131*
Pabst, M., 217, *106*, 218, *106*, 218, *127*, 219, *130*, 221, *141*, 222, *106*, 222, *127*, 227, *106*, 227, *141*, 228, *106*, 228, *127*, 231, *162*, 235, *182*, 243, *182*, 244, *182*, 245, *106*, 246, *106*, 247, *106*, 255, *127*, 256, *106*, 256, *127*, 257, *106*, 257, *127*, 249, *130*, 251, *130*
Pace, K.E., 130, *31*, 143, *31*, 148, *148*
Palumbo, A., 255, *259*
Palva, A., 214, *71*, 217, *105*, 246, *105*
Pandey, R.K., 166, *217*
Pandey, S.K., 166, *217*
Panico, M., 213, *47*, 231, *160*, 231, *162*, 255, *47*
Panicot, L., 175, *257*
Papapostolou, D., 215, *92*
Papp, S.L., 141, *86*
Paraskeva, C., 145, *123*, 180, *123*
Pardo, D., 91, *108*
Pardon, E., 215, *92*
Parodi, A., 45, *43*
Parrish, A., 90, *84*
Partridge, E.A., 133, *43*
Pasqualini, R., 187, *294*
Passaniti, A., 140, *74*
Pasteels, J.L., 141, *79*
Pastor, N., 101, *127*
Patel, K., 245, *221*, 250, *221*
Patel, V.P., 143, *106*
Patnam, R., 154, *179*, 155, *179*, 160, *193*, 161, *193*, 164, *179*
Patrick, M., 91, *87*
Patterson, R.J., 130, *33*

Paul, G., 212, *31*, 230, *31*
Paulson, J.C., 127, *12*, 128, *12*, 129, *12*
Pavao, M.S.G., 181, *276*
Pavkov, T., 215, *90*
Pavkov-Keller, T., 214, *86*, 215, *91*
Paz, A., 144, *117*
Pearson, W.H., 63, *17*, 91, *102*
Pedersen, C., 243, *214*
Pedersen, H., 243, *214*
Péhau-Arnaudet, G., 215, *92*
Pelz, A., 128, *14*
Peng, G., 147, *138*
Peng, W., 147, *138*
Peretz, T., 185, *290*
Perez-Castells, J., 183, *285*
Peri, F., 101, *129*, 101, *130*
Pericas, M.A., 91, *99*
Perillo, N.L., 130, *26*, 130, *31*, 143, *31*, 143, *103*
Perlin, A.S., 243, *212*
Perrone, D., 75, *37*
Perry, M.B., 247, *229*
Person, M.D., 142, *96*
Petein, M., 141, *79*
Peter-Katalinic, J., 245, *220*, 249, *236*
Peters, J., 214, *70*, 239, *197*
Peters, N.R., 101, *128*
Peters, T., 182, *279*
Petersen, B.O., 235, *182*, 243, *182*, 244, *182*, 244, *216*
Petersen, L.J., 177, *264*
Petersen, P.J., 97, *120*
Peyfoon, E., 231, *160*, 231, *162*
Pfeffer, J., 254, *253*
Pföstl, A., 246, *226*, 246, *227*, 247, *226*, 247, *231*, 247, *232*, 248, *226*, 248, *231*, 248, *232*
Pfrengle, F., 74, *36*
Phansopa, C., 217, *109*
Phillips, G., 65, *18*
Phillips, J.A., 175, *258*
Phipps, B.M., 214, *78*
Piantadosi, C.A., 145, *122*
Picco, G., 140, *75*, 141, *86*
Piccolo, E., 151, *170*, 162, *195*, 164, *195*
Pienta, K.J., 130, *27*, 130, *33*, 144, *27*, 147, *27*, 149, *157*, 150, *163*, 151, *157*, 169, *163*, 169, *225*, 169, *226*

Pierry, C., 183, *286*
Pieters, C., 142, *98*
Pieters, R.J., 130, *29*, 146, *29*, 165, *209*, 165, *210*, 165, *211*, 166, *209*, 166, *210*, 169, *228*, 167, *221*, 169, *221*, 170, *228*
Pillai, S., 130, *36*
Pilobello, K.T., 217, *107*
Pinder, S., 140, *75*
Pinder, S.E., 141, *86*
Pineo, G.F., 178, *268*
Pinteric, L., 45, *45*
Pitzer, K.K., 58, *8*, 71, *25*
Platt, D., 149, *156*, 151, *156*
Plavner, N., 230, *149*
Podar, K., 149, *159*, 150, *159*
Podhajcer, O.L., 141, *91*, 142, *91*, 144, *91*, 147, *91*
Pohlentz, G., 249, *236*
Poirer, F., 130, *36*
Poirier, F., 130, *28*, 170, *232*
Poisson, J.F., 91, *106*
Polgar, J., 145, *126*
Polverini, P.J., 135, *50*
Pooley, H.M., 232, *173*
Postel, R., 170, *232*
Poulain, D., 159, *190*
Poulain, F., 183, *286*
Pouwels, P.H., 241, *207*
Pouyani, T., 135, *55*, 137, *55*
Powell, L.D., 128, *15*
Powers, D.A., 47, *51*
Powers, J., 252, *244*
Powers, J.D., 91, *102*
Pragani, R., 79, *45*
Presper, K.A., 45, *38*
Price, B.D., 164, *204*
Price, J.E., 150, *164*, 162, *164*, 162, *197*, 167, *197*, 167, *224*, 169, *164*, 169, *224*
Priem, B., 213, *51*
Priest, R., 182, *279*
Prieto, P.A., 175, *256*
Prodger, J.C., 182, *279*
Prorok, M., 174, *255*, 175, *255*
Prydz, K., 180, *271*
Puchberger, M., 233, *177*, 235, *182*, 242, *177*, 243, *177*, 243, *182*, 243, *213*, 244, *182*, 247, *233*, 247, *234*

AUTHOR INDEX

Pühler, A., 218, *121*
Pulverer, G., 146, *136*, 147, *136*, 151, *136*, 151, *169*
Pum, D., 214, *69*, 214, *83*, 214, *84*, 214, *85*, 216, *93*, 216, *94*, 216, *95*, 216, *96*, 216, *98*, 217, *84*, 218, *122*, 239, *69*, 214, *80*, 254, *80*, 254, *122*, 255, *93*, 256, *122*, 257, *84*
Pusztahelyi, T., 240, *202*
Putnoky, P., 212, *41*, 217, *41*
Pyne, S.G., 91, *101*, 91, *103*, 91, *107*

Q

Qasba, P.K., 174, *253*
Qian, Y., 130, *28*
Qian, Y.N., 152, *175*, 153, *175*
Qiao, L., 58, *7*
Qize, D., 141, *86*
Quinn, C.P., 111, *168*, 236, *185*, 236, *186*, 236, *188*, 236, *189*, 236, *190*, 236, *191*, 237, *190*, 237, *191*, 238, *189*, 238, *190*, 244, *189*, 244, *190*, 253, *185*, 253, *186*, 253, *189*, 253, *252*
Quinn, L., 84, *65*
Quinn, T.P., 140, *78*, 148, *78*, 169, *78*, 169, *225*
Quirion, J.C., 183, *286*

R

Rabinovich, G. (A.), 130, *23*, 130, *24*, 130, *26*, 133, *37*, 141, *24*, 141, *91*, 142, *24*, 142, *91*, 143, *37*, 143, *106*, 144, *24*, 144, *91*, 147, *91*, 147, *138*, 151, *170*, 162, *195*, 164, *195*
Rachel, R., 214, *77*
Raetz, C.R.H., 211, *10*, 230, *150*
Raetz, R.H., 218, *123*, 222, *123*, 227, *123*, 241, *123*, 250, *123*, 251, *123*
Ragupathi, G., 137, *59*, 140, *69*, 146, *134*
Rainey, F.A., 218, *118*, 219, *118*, 220, *139*, 223, *139*, 224, *139*, 243, *139*, 244, *139*, 245, *139*
Ralet, M.-C., 150, *161*
Ralph, S.J., 147, *141*, 163, 200, 163, *201*, 164, *141*
Ramakrishnan, B., 174, *253*

Rambaruth, N.D.S., 141, *92*
Rao, G., 145, *122*
Rao, N., 165, *208*, 182, *208*
Rapp, U., 245, *219*
Rauvolfova, J., 236, *186*, 253, *186*
Raymond, E., 170, *236*, 171, *236*
Raz, A., 126, *4*, 126, *5*, 127, *4*, 130, *27*, 130, *32*, 130, *33*, 130, *36*, 135, *4*, 142, *4*, 142, *32*, 142, *96*, 143, *99*, 143, *100*, 143, *101*, 143, *104*, 144, *27*, 144, *99*, 144, *116*, 144, *117*, 144, *118*, 147, *4*, 147, *27*, 148, *143*, 148, *146*, 148, *147*, 149, *153*, 149, *155*, 149, *157*, 150, *147*, 150, *162*, 150, *163*, 151, 149, 151, 149, *157*, 151, *153*, 151, *156*, 151, *157*, 151, *166*, 169, *163*, 175, *4*
Reading, C.A., 213, *53*, 255, *53*
Reddish, M., 140, *70*
Reed, III. L.A., 59, *12*
Reed, Y., 111, *168*
Reeves, P.R., 247, *228*, 250, *241*
Rehm, S., 141, *80*
Reid, C.L., 91, *94*
Reinhold, V.N., 133, *43*
Reissig, H.-U., 74, *35*, 74, *36*, 80, *48*
Reiz, B., 214, *62*
Remaut, H., 215, *92*
Ren, Y., 91, *102*
Renaudet, O., 166, *217*
Renkonen, J., 145, *131*
Renkonen, R., 145, *131*
Renz, M., 137, *58*
Replogle, T.S., 149, *157*, 151, *157*
Reuhs, B.L., 212, *41*, 217, *41*
Reuter, V.E., 140, *71*, 140, *72*, 141, *71*
Richardson, A.C., 91, *96*
Richardson, P., 149, *159*, 150, *159*
Richter, G.S., 254, *254*
Riedmann, E., 217, *104*, 246, *104*
Riera, A., 91, *99*
Riera, C.M., 143, *106*
Ries, W., 240, *204*
Rietschel, E.T., 212, *39*
Rigby, P.W.J., 130, *36*
Riley, L.W., 211, *8*, 214, *8*, 255, *8*
Rini, J.M., 130, *34*, 130, *36*, 152, *173*, 156, *173*, 160, *192*, 165, *173*, 165, *212*

Ristl, R., 217, *111*, 218, *111*, 219, *111*, 229, *111*, 252, *246*, 252, *247*, 254, *111*, 255, *247*, 256, *247*
Rittenhouse-Olson, K., 140, *78*, 148, *78*, 169, *78*
Rivera, H.N., 175, *256*
Rivier, A.S., 177, *265*
Robert, J.D., 91, *93*
Robinson, M.K., 141, *90*
Robinson, P., 140, *73*
Robinson, R.C., 129, *19*
Robotham, A., 230, *159*
Rogalsky, D.K., 102, *136*
Rohr, J., 107, *149*, 107, *151*
Rojanasakul, Y., 102, *136*, 103, *139*, 104, *145*, 104, *146*
Romana, L.K., 247, *228*
Rombouts, Y., 237, *192*, 238, *192*
Rooman, M., 158, *186*
Rosch, M., 185, *293*
Roseman, S., 40, *2*, 40, *3*, 40, *4*, 40, *5*, 41, *3*, 41, *6*, 41, *7*, 41, *10*, 41, *11*, 41, *12*, 41, *13*, 41, *14*, 41, *15*, 41, *17*, 42, *7*, 42, *10*, 42, *11*, 42, *12*, 42, *13*, 42, *14*, 42, *15*, 42, *17*, 42, *18*, 42, *19*, 42, *20*, 42, *21*, 42, *22*, 42, *23*, 42, *24*, 43, *7*, 43, *10*, 43, *11*, 43, *12*, 43, *13*, 43, *24*, 44, *12*, 44, *24*, 45, *24*, 45, *44*, 46, *17*, 47, *44*, 47, *49*, 47, *50*, 47, *51*, 47, *52*, 47, *53*, 47, *54*, 47, *55*, 47, *56*, 49, *57*, 49, *58*, 49, *59*, 49, *60*, 49, *61*, 49, *62*
Rosen, S.D., 137, *58*
Rostand, K.S., 173, *252*
Rostovtsev, V.V., 154, *178*
Roszkowski, K., 151, *169*
Rotblat, B., 144, *117*
Roth, S., 47, *56*
Rowlatt, C., 140, *73*
Roy, R., 154, *179*, 155, *179*, 156, *183*, 160, *193*, 161, *183*, 161, *193*, 162, *183*, 164, *179*, 165, *213*, 170, *183*
Roy, S., 217, *109*
Ruan, D.T., 164, *204*
Rubant, S., 180, *274*
Rubinstein, N., 141, *91*, 142, *91*, 144, *91*, 147, *91*, 147, *138*
Rueckert, A., 91, *104*
Rünzler, D., 215, *89*, 216, *89*

Ruoslahti, E., 187, *294*
Russwurm, R., 167, *221*, 169, *221*, 169, *228*, 170, *228*
Ryan, C., 236, *184*, 253, *184*
Rycroft, D.S., 108, *154*
Ryd, W., 151, *168*
Rydberg, H., 165, *212*

S

Sabesan, S., 165, *213*, 165, *215*
Sabet, M., 218, *125*, 257, *125*
Sacco, R., 144, *110*
Sachs, L., 47, *54*
Sackstein, R., 173, *251*
Sada, K., 127, *9*
Sadoulet, M.O., 175, *257*
Sadozai, K.K., 151, *65*
Saez, V., 182, *279*
Safaiyan, F., 180, *271*
Sah, N.K., 101, *125*
Saier, M.H., 47, *50*
Saile, E., 111, *168*, 236, *185*, 236, *186*, 236, *190*, 236, *191*, 237, *190*, 237, *191*, 238, *190*, 244, *190*, 253, *185*, 253, *186*, 253, *252*
Saitoh, O., 175, *260*
Sakahara, H., 141, *89*
Sakakibara, J., 217, *108*
Sakamoto, J., 141, *88*
Sako, D., 135, *55*, 137, *55*
Saksena, R., 111, *165*, 111, *166*, 111, *167*
Sakya, S.M., 97, *120*
Salameh, B., 158, *188*
Salameh, B.A., 152, *174*, 154, *174*, 154, *180*
Salk, P.L., 140, *73*
Salmivirta, M., 180, *271*
Salomonsson, E., 130, *30*, 133, *30*, 155, *181*, 163, *198*, 165, *212*
Salyan, M.E., 45, *41*
Salzer, J.L., 129, *19*
Samuelson, P., 255, *261*
Sandall, J.K., 140, *73*
Sandercock, L.E., 212, *27*
Sandhoff, K., 44, *30*
Sansone, F., 166, *217*, 166, *218*, 167, *218*
Santarius, U., 239, *197*
Santoro, D., 173, *250*

Sára, M., 214, *69*, 214, *80*, 214, *85*, 215, *88*,
    215, *89*, 216, *89*, 216, *98*, 217, *101*, 231,
    165, 232, *165*, 233, *177*, 235, *182*, 239,
    69, 239, *165*, 240, *204*, 240, *205*, 240,
    206, 241, *165*, 242, *165*, 242, *177*, 242,
    210, 243, *177*, 243, *182*, 244, *182*, 246,
    101, 253, *165*, 254, *80*
Sarkar, A.K., 172, *240*, 173, *240*, 173, *252*
Sarkar, M., 44, *31*
Sarvis, R., 130, *27*, 144, *27*, 147, *27*
Sasaki, O., 145, *132*
Sasaki, Y., 141, *81*, 145, *81*
Sathisha, U.V., 130, *22*, 142, *22*, 151, *22*
Satijn, S., 170, *232*
Sato, S., 154, *179*, 155, *179*, 156, *183*, 160,
    193, 161, *183*, 161, *193*, 162, *183*, 164,
    179, 170, *183*
Satoh, Y., 141, *88*
Saussez, S., 130, *23*
Savage, M.P., 159, *190*
Sawada, R., 145, *132*
Sawada, T., 141, *85*
Saxon, E., 255, *258*
Schachter, H., 42, *24*, 43, *24*, 44, *24*, 44, *31*,
    44, *32*, 45, *24*, 45, *32*, 45, *45*
Schäfer, G., 231, *161*
Schäffer, C., 211, *7*, 222, *7*, 242, *7*, 211, *18*,
    212, *18*, 217, *18*, 218, *18*, 214, *74*, 216,
    74, 218, *74*, 222, *74*, 240, *74*, 241, *74*,
    254, *74*, 214, *84*, 217, *84*, 257, *84*, 215,
    87, 234, *87*, 235, *87*, 242, *87*, 216, *93*,
    255, *93*, 216, *98*, 216, *100*, 217, *100*,
    228, *100*, 217, *106*, 218, *106*, 222, *106*,
    227, *106*, 228, *106*, 245, *106*, 246, *106*,
    247, *106*, 256, *106*, 257, *106*, 217, *110*,
    217, *111*, 218, *111*, 219, *111*, 229, *111*,
    254, *111*, 218, *115*, 220, *115*, 247, *115*,
    254, *115*, 218, *116*, 222, *116*, 247, *116*,
    254, *116*, 218, *119*, 218, *120*, 221, *120*,
    222, *120*, 227, *120*, 244, *120*, 218, *122*,
    254, *122*, 256, *122*, 218, *127*, 222, *127*,
    228, *127*, 255, *127*, 256, *127*, 257, *127*,
    219, *128*, 219, *129*, 231, *129*, 219, *129*,
    231, *129*, 219, *130*, 249, *130*, 251, *130*,
    219, *131*, 220, *131*, 223, *131*, 229, *131*,
    243, *131*, 244, *131*, 245, *131*, 249, *131*,
    219, *131*, 220, *131*, 223, *131*, 229, *131*,
    243, *131*, 244, *131*, 245, *131*, 249, *131*,
    219, *132*, 220, *132*, 223, *132*, 224, *132*,
    229, *132*, 244, *132*, 245, *132*, 220, *137*,
    221, *137*, 224, *137*, 225, *137*, 229, *137*,
    243, *137*, 244, *137*, 220, *139*, 223, *139*,
    224, *139*, 243, *139*, 244, *139*, 245, *139*,
    221, *141*, 227, *141*, 226, *146*, 231, *162*,
    231, *166*, 240, *166*, 242, *166*, 253, *166*,
    231, *167*, 239, *167*, 240, *167*, 241, *167*,
    253, *167*, 232, *175*, 242, *175*, 233, *176*,
    240, *176*, 242, *176*, 244, *176*, 253, *176*,
    233, *180*, 234, *180*, 242, *180*, 244, *180*
Schaffer, L., 141, *86*, 142, *95*
Schapira, M., 177, *265*
Schatton, T., 148, *149*
Schaub, R.G., 145, *126*
Schauer, R., 128, *14*, 128, *16*, 129, *16*, 129, *19*
Schaus, S.E., 63, *15*
Scheberl, A., 218, *118*, 218, *119*, 219, *118*,
    219, *131*, 219, *132*, 220, *131*, 220, *132*,
    220, *137*, 220, *139*, 221, *137*, 223, *131*,
    223, *132*, 223, *139*, 224, *132*, 224, *137*,
    224, *139*, 225, *137*, 229, *131*, 229, *132*,
    229, *137*, 243, *131*, 243, *137*, 243, *139*,
    244, *131*, 244, *132*, 224, *137*, 245, *131*,
    245, *132*, 247, *230*, 248, *230*, 249, *131*,
    255, *230*, 255, *256*, 256, *256*
Scheffers, D.J., 252, *249*
Scheffler, K., 182, *279*
Scherman, D., 183, *286*
Scherman, M.S., 247, *235*
Scherphof, G.L., 185, *291*
Schinazi, R.F., 70, *23*
Schindler, J., 149, *159*, 150, *159*
Schirm, M., 212, *34*, 212, *35*
Schirrmacher, V., 140, *73*, 146, *136*, 147, *136*,
    151, *136*
Schlott, B., 214, *72*, 217, *72*
Schlüter, A., 218, *121*
Schmid, E.R., 220, *140*, 227, *140*, 245, *140*
Schmidt, M.A., 211, *8*, 214, *8*, 255, *8*
Schmidt, R.R., 79, *43*
Schnaar, R.L., 128, *14*, 171, *238*
Schneewind, O., 236, *183*, 236, *184*, 239, *193*,
    240, *203*, 253, *184*, 254, *254*
Schneller, M., 127, *10*
Schocher, I., 240, *204*
Schoenhofen, C., 212, *36*, 212, *37*, 217, *36*
Schoenhofen, I.C., 212, *35*

Schoeppner, H.L., 142, *96*
Scholz, H.C., 217, *104*, 246, *104*
Schroederd, B.R., 94, *116*
Schuch, R., 236, *187*
Schulman, C., 141, *79*
Schultz, A., 47, *55*
Schultz, P.G., 169, *227*
Schulz, B.L., 213, *50*, 255, *259*
Schulz, G., 218, *120*, 219, *135*, 220, *135*, 220, *140*, 221, *120*, 221, *135*, 222, *120*, 224, *144*, 225, *144*, 227, *120*, 227, *140*, 233, *179*, 233, *180*, 234, *180*, 234, *179*, 242, *180*, 244, *120*, 244, *180*, 245, *135*, 245, *140*, 245, *218*, 251, *135*, 252, *135*
Schulze, S., 148, *144*, 165, *144*, 167, *144*
Schumacher, G., 180, *274*
Schuster, B., 214, *85*, 216, *93*, 216, *96*, 216, *98*, 218, *122*, 254, *122*, 255, *93*, 256, *122*
Schuster, K.C., 217, *101*, 246, *101*
Schuster-Kolbe, J., 219, *136*, 220, *136*, 220, *140*, 223, *136*, 227, *140*, 229, *136*, 243, *136*, 245, *136*, 245, *140*
Schützenmeister, N., 74, *33*
Schwab, H., 218, *121*
Schwarz, F., 255, *259*
Schwarz, H., 232, *174*, 239, *174*, 240, *174*, 241, *174*, 242, *174*, 252, *174*, 253, *174*
Schwarz, P., 91, *90*
Scialdone, M.A., 71, *29*
Scott, J.G., 83, *58*
Scott, S.A., 147, *141*, 163, *201*, 163, *202*, 164, *141*
Scudder, P., 182, *278*
Seberger P.J., 140, *77*
Seeberger, P.H., 245, *222*, 74, *32*, 74, *33*, 74, *34*, 79, *40*, 79, *41*, 79, *45*, 79, *46*, 111, *164*, 111, *170*
Seed, B., 135, *55*, 137, *55*, 141, *87*, 171, *238*, 173, *87*
Seetharaman, J., 152, *173*, 156, *173*, 165, *173*
Segmuller, B.E., 63, *17*
Seidel, V., 108, *155*
Seilhamer, J.J., 130, *31*, 143, *31*
Sekot, G., 214, *84*, 216, *100*, 217, *84*, 217, *100*, 219, *129*, 221, *141*, 227, *141*, 228, *100*, 231, *129*, 257, *84*
Sela, B.A., 126, *5*
Sela, B.-A., 47, *54*

Serova, M., 170, *236*, 171, *236*
Serre, A.L., 183, *286*
Sethi, T., 130, *23*
Settem, R.P., 217, *109*
Shah, M.R., 135, *50*
Shalom-Feuerstein, R., 144, *117*
Shan, M., 91, *110*, 94, *115*, 96, *117*, 96, *118*, 97, *119*, 103, *140*, 108, *153*
Shankar, V., 214, *57*, 255, *57*
Sharif, E.U., 103, *138*, 107, *151*, 108, *152*, 108, *153*
Sharma, A., 217, *109*, 161, *194*
Sharon, N., 126, *3*, 222, *143*
Sharpless, K.B., 154, *178*, 59, *12*, 73, *31*
Shashkov, A.S., 212, *40*, 217, *40*, 231, *172*, 243, *215*
Shaw, G.D., 135, *55*, 137, *55*, 182, *57*
Sheldon, H.K., 147, *137*
Shen, B., 97, *120*
Shen, J.J., 142, *96*
Shen, P.F., 144, *111*
Shen, Y.J., 129, *19*
Sherman, A.A., 151, *170*
Shi, P., 109, *157*
Shi, Y.H., 147, *140*
Shibata, M., 166, *217*
Shibata, T., 91, *86*
Shilatifard, A., 175, *256*
Shimazu, H., 182, *283*
Shimizu, T., 175, *259*, 176, *259*
Shimodaira, K., 145, *131*
Shimozato, K., 217, *108*
Shirley, R.B., 149, *158*
Shockman, G.D., 212, *28*
Shodai, T., 182, *283*
Shoham, Y., 240, *198*
Shono, N., 150, *162*
Siebert, H.C., 130, *23*, 169, *228*, 170, *228*
Sillanpää, L.J., 241, *208*
Silva, M., 109, *157*
Silvescu, C.I., 133, *43*
Simoni, R., 48, *63*, 49, *63*
Simoni, R.D., 47, *50*
Simpson, R.J., 44, *31*
Singer, M.S., 137, *58*
Singh, B., 256, *265*
Singh, P.K., 142, *94*
Singhal, A.K., 140, *71*, 141, *71*

Sinha, A., 174, *253*
Sirois, S., 156, *183*, 161, *183*, 162, *183*, 170, *183*
Skinner, M.P., 135, *53*
Skogman, F., 159, *189*
Skomedal, H., 151, *168*
Slaaby, R., 152, *173*, 156, *173*, 165, *173*
Sleytr, U.B., 211, *5*, 218, *5*, 211, *24*, 214, *74*, 214, *83*, 214, *85*, 214, *88*, 214, *90*, 214, *91*, 216, *93*, 255, *93*, 216, *94*, 216, *95*, 216, *96*, 216, *98*, 217, *24*, 218, *24*, 214, *67*, 218, *74*, 222, *74*, 239, *67*, 240, *74*, 241, *74*, 254, *67*, 254, *74*, 214, *69*, 239, *69*, 214, *80*, 254, *80*, 214, *73*, 214, *79*, 217, *101*, 246, *101*, 217, *105*, 246, *105*, 218, *117*, 219, *117*, 241, *117*, 218, *118*, 219, *118*, 218, *120*, 221, *120*, 222, *120*, 227, *120*, 244, *120*, 218, *122*, 254, *122*, 256, *122*, 219, *134*, 220, *134*, 224, *134*, 225, *134*, 249, *134*, 219, *136*, 220, *136*, 223, *136*, 229, *136*, 243, *136*, 245, *136*, 220, *140*, 227, *140*, 245, *140*, 224, *145*, 226, *145*, 243, *145*, 232, *175*, 242, *175*, 233, *179*, 234, *179*, 235, *182*, 243, *182*, 244, *182*, 239, *194*, 240, *204*, 240, *205*, 242, *210*, 245, *218*, 246, *224*, 255, *256*, 256, *256*
Smit, E., 241, *207*
Smith, C., 91, *95*
Smith, C.W., 141, *84*
Smith, D., 158, *185*
Smith, D.F., 175, *256*
Smith, P.W., 91, *97*
Smrecki, V., 234, *181*, 244, *181*, 245, *181*
Sobel, E., 142, *98*
Sobel, M., 180, *272*
Sobel, M.E., 142, *98*
Sohma, Y., 163, *202*
Solf, R., 91, *92*
Somers, W.S., 135, *57*, 182, *57*
Sondengam, B.L., 108, *154*
Sondermann, P., 256, *263*
Song, B.H., 172, *242*, 178, *242*
Song, L., 91, *100*
Soo, E.C., 212, *34*
Sordat, B., 130, *23*
Sørensen, O.W., 244, *216*

Sorme, P., 133, *45*, 152, *45*, 152, *171*, 152, *175*, 153, *175*, 160, *192*
Sowa, M., 141, *85*
Sparrow, C.P., 133, *41*, 141, *41*
Spero, J., 40, *2*
Spertini, O., 177, *265*
Spigset, O., 101, *124*
Spiro, R., 211, *20*
Spivak, C.T., 42, *20*
Spohr, U., 127, *8*
Sprengard, U., 182, *279*
Springer, G.F., 137, *62*, 141, *79*
Sproviero, D., 141, *86*
Srikhanta, Y.N., 213, *54*
Stabile, M.R., 71, *25*
Stafford, G.P., 217, *109*
Stahl, B., 219, *136*, 220, *136*, 223, *136*, 229, *136*, 243, *136*, 245, *136*
Stahl, D., 214, *63*
Ståhl, S., 255, *261*
Stahl, W., 182, *279*
Stahmann, M.A., 40, *1*, 40, *2*
Stahn, R., 171, *237*, 172, *241*, 185, *237*, 185, *237*, 185, *241*
Stallforth, P., 74, *34*, 79, *42*, 79, *46*
Stanek, Jr. J., 58, *5*
Stanley, P., 211, *4*, 219, *4*, 222, *4*
Stannard, K.A., 147, *141*, 164, *141*
Starosotnikov, A.M., 83, *56*
Steeber, D.A., 135, *47*
Steeghs, L., 213, *54*
Steegmaier, M., 185, *292*
Steele, J.G., 140, *73*
Steensma, D.H., 71, *29*
Steigerwald, J.C., 42, *23*
Steindl, C., 215, *87*, 234, *87*, 234, *181*, 235, *87*, 242, *87*, 244, *181*, 245, *181*
Steiner, K., 217, *111*, 218, *111*, 218, *116*, 219, *111*, 222, *116*, 229, *111*, 245, *220*, 245, *221*, 245, *222*, 246, *225*, 247, *116*, 247, *225*, 248, *225*, 249, *236*, 249, *237*, 250, *221*, 251, *237*, 251, *242*, 254, *111*, 254, *116*, 254, *257*, 256, *257*
Steininger, C.N., 182, *278*
Stephens, S., 212, *41*, 217, *41*
Steplewska-Mazur, K., 141, *90*
Stern, R.J., 247, *235*

Sternberg, L.R., 143, *100*
Stevenson, J.L., 135, *52*, 177, *52*, 177, *267*, 178, *52*, 179, *52*, 181, *52*
Stewart, P.L., 148, *148*
Steyaert, J., 215, *92*
Stillman, B.N., 144, *115*
St.Michael, F., 213, *45*
Stoddart, J.F., 165, *206*, 166, *217*
Stoffel, M.T., 149, *158*
Stoffyn, A., 44, *28*, 45, *28*
Stoffyn, P., 44, *28*, 45, *28*
Stoll, M.S., 137, *58*
Stowell, S.R., 148, *148*
Strauss, J., 246, *225*
Strecker, G., 130, *34*
Strominger, J.L., 211, *23*, 217, *23*
Stroud, M.R., 45, *41*
Strubel, N.A., 135, *49*
Stuart, A.C., 137, *58*
Stueckle, T.A., 103, *139*
Sud, S., 169, *226*
Suda, Y., 180, *272*
Sue, D., 253, *252*
Sugimachi, K., 145, *132*
Sugiyama, T., 249, *238*
Sujatha, M.S., 133, *38*
Sullivan, E.M., 147, *141*, 164, *141*
Sullivan, F.X., 137, *58*
Sullivan, W.R., 40, *2*
Sumper, M., 211, *16*, 211, *26*, 212, *31*, 217, *16*, 230, *16*, 230, *31*
Sun, Y.X., 172, *244*
Sundin, A., 130, *35*, 154, *180*, 155, *35*, 155, *181*, 158, *185*, 163, *198*, 163, *199*, 165, *212*
Supko, J.G., 148, *149*
Surolia, A., 165, *215*
Sutherland, I.W., 241, *209*
Suzuki, J., 182, *283*
Suzuki, T., 141, *89*
Suzuki, Y., 183, *284*
Svensson, I., 148, *145*
Svergun, D.I., 215, *90*
Swaminathan, G.J., 159, *190*
Swanson, B.J., 142, *94*
Sylvestre, P., 111, *163*
Szczepanowski, R., 218, *121*
Szolnoki, E., 169, *229*, 170, *229*
Szymanski, C.M., 213, *45*, 214, *59*, 214, *60*, 255, *60*
Szymanski, M., 211, *9*, 212, *42*, 213, *9*, 213, *42*, 255, *9*

T

Tabak, L.A., 149, *151*
Tafel, J., 57, *1*, 58, *1*
Tai, G., 149, *154*, 150, *154*
Tai, G.H., 172, *239*, 176, *239*, 177, *239*, 179, *239*, 180, *239*
Tailor, H., 143, *106*
Tait, L., 149, *153*, 149, *157*, 151, *153*, 151, *157*
Takada, A., 141, *87*, 173, *87*
Takagi, H., 141, *88*
Takahashi, H.K., 137, *62*
Takahashi, M., 158, *187*
Takahashi, N., 141, *87*, 173, *87*
Takahashi, T., 146, *135*
Takamatsu, S., 141, *93*
Takamiya, R., 141, *93*
Takatsuka, S., 141, *85*
Takenaka, Y., 148, *146*, 150, *162*
Taki, T., 141, *83*, 145, *83*
Tanaka, H., 183, *284*
Tanaka, N., 140, *70*
Tané, P., 108, *154*
Tang, J., 214, *83*, 135, *57*, 182, *57*
Tang, S., 128, *14*, 129, *19*
Taniguchi, N., 141, *93*
Taniuchi, Y., 141, *84*
Tao, T.W., 126, *5*
Tateno, H., 144, *113*
Taylor, M.E., 126, *3*
Taylor, W.H., 172, *240*, 173, *240*
Taylor-Papadimitriou, J., 140, *75*, 141, *86*
Tedder, T.F., 135, *47*
Tegtmeyer, H., 141, *79*
Tejero, T., 76, *39*
Tejler, J., 149, *152*, 158, *152*, 158, *188*, 159, *189*, 164, *205*, 165, *212*
Ten Hagen, K.G., 149, *151*
Terada, M., 182, *283*
Teranishi, T., 128, *17*, 129, *17*

Terano, H., 91, *86*
Tesarz, M., 215, *90*, 215, *91*
Tessier, L., 213, *45*
Theobald, L.K., 149, *158*
Theriot, J.A., 236, *183*
Thibault, J.-F., 150, *161*
Thibault, P., 212, *33*, 212, *34*, 212, *43*, 212, *35*, 217, *33*
Thijssen, V.L.J.L., 170, *232*
Thomas, C.J., 165, *215*
Thomas, S., 162, *196*
Thompson, E., 143, *101*
Thorne, K.J.I., 211, *24*, 217, *24*, 218, *24*
Thornley, M.J., 214, *66*
Thorpe, A.J., 58, *8*, 71, *25*
Thorson, J.S., 101, *128*
Tian, E., 149, *151*
Tian, M.H., 180, *273*
Tibrewal, N., 107, *151*
Tieslau, C., 126, *1*
Timmer, M.M., 79, *42*
Timmer, M.S.M., 74, *32*, 74, *33*
Timoshenko, A.V., 127, *10*
Tinari, N., 144, *110*, 144, *111*, 151, *170*, 162, *195*, 164, *195*
Tisnes, P., 162, *196*
Toca-Herrera, J.-L., 214, *83*
Tochino, Y., 158, *187*
Toda, M., 142, *95*
Tokuhara, M., 175, *259*, 176, *259*
Toma, N., 180, *272*
Tomcik, D.J., 83, *54*, 85, *67*
Tomiya, N., 183, *284*
Tony, K.A., 182, *280*, 183, *285*
Tor, Y., 174, *253*
Torizuka, Serum K., 141, *89*
Tornoe, C.W., 154, *178*
Torri, G., 177, *266*, 180, *266*, 184, *289*, 185, *289*
Toscano, M A., 141, *91*, 142, *91*, 144, *91*, 147, *91*
Toscano, M.A., 147, *138*
Tosi-Couture, E., 240, *200*, 240, *201*
Totani, M., 175, *260*
Toyama-Sorimachi, N., 175, *259*, 176, *259*
Toyokuni, T., 151, *65*
Tran, N.M., 149, *158*
Traxler, P., 84, *62*
Tretli, S., 101, *124*
Truant, R., 252, *248*
Trust, T.J., 212, *33*, 212, *42*, 213, *42*, 217, *33*
Tscheliessnig, R., 216, *94*, 216, *96*
Tsuboi, S., 145, *132*, 145, *133*
Tsuji, T., 141, *84*, 182, *283*
Tsurumi, Y., 91, *86*
Tsuruta, N., 183, *284*
Tsuyuoka, K., 141, *87*, 173, *87*
Tsvetkov, Y.E., 151, *170*, 212, *40*, 217, *40*
Tubak, V., 169, *229*, 170, *229*
Tullberg, E., 164, *205*, 165, *212*
Turk, J.R., 150, *163*, 150, *164*, 162, *164*, 169, *163*, 169, *164*, 169, *225*
Turnbull, J.E., 145, *123*, 180, *123*
Tutt, A., 141, *86*
Tvaroska, I., 101, *130*
Tyrrell, D., 165, *208*, 182, *208*
Tzeng, Y.L., 69, *22*

U

Uch, R., 174, *255*, 175, *255*
Uchida, K., 230, *159*
Ueda, M., 173, *248*
Uhl, J., 165, *214*
Uhlenbruck, G., 146, *136*, 147, *136*, 151, *136*, 151, *169*
Uittenbogaart, C.H., 143, *103*
Ullrich, A., 144, *110*
Um, H.-S., 218, *125*, 257, *125*
Umemoto, S., 173, *248*
Ungaro, R., 166, *217*, 166, *218*, 167, *218*
Unger, B., 220, *139*, 223, *139*, 224, *139*, 243, *139*, 244, *139*, 245, *139*
Unger, F.M., 233, *179*, 234, *179*
Upreti, R.K., 214, *57*, 255, *57*
Urashima, T., 134, *46*
Ushiyama, S., 184, *288*
Usui, T., 142, *95*, 183, *284*

V

Vaczi, B., 169, *229*, 170, *229*
Valderrama-Rincon, J.D., 213, *53*, 255, *53*
Valentine, K.A., 178, *268*

Valentini, P., 165, *210*, 166, *210*, 169, *228*, 170, *228*
Valvano, M.A., 213, *48*, 213, *49*, 245, *221*, 250, *221*
van Aalten, D.M.F., 128, *18*
van den Brule, F. (A.), 142, *98*, 144, *114*
van den Nieuwenhof, I.M., 127, *12*, 128, *12*, 129, *12*
van der Ley, P., 213, *54*
van der Merwe, P.A., 129, *19*
van der Schaft, D.W.J., 170, *233*
van Eijk, L.I., 170, *233*, 170, *235*
Van Gerven, N., 215, *92*
Van Laar, E.S., 170, *231*
Van Lanen, S.G., 107, *151*
van Soest, R., 162, *196*
Vanderleyden, J., 211, *6*, 212, *32*
Vanderslice, P., 182, *278*
VanDyke, D.J., 230, *158*
Vanegas, J.P., 141, *79*
VanLeer, P., 141, *79*
VanRheenen, V., 85, *66*, 101, *66*
VanVelthoven, R., 141, *79*
Varki, A., 127, *13*, 211, *4*, 211, *19*, 219, *4*, 222, *4*, 128, *15*, 135, *48*, 135, *52*, 137, *58*, 140, *68*, 142, *95*, 144, *120*, 144, *121*, 145, *120*, 145, *121*, 145, *124*, 176, *124*, 177, *52*, 177, *120*, 177, *263*, 177, *267*, 178, *52*, 178, *263*, 179, *52*, 180, *124*,, 181, *52*
Varki, N.M., 137, *58*, 142, *95*, 144, *121*, 144, *120*, 145, *120*, 145, *121*, 145, *124*, 176, *124*, 177, *120*, 180, *124*
Varseev, G.N., 69, *21*
Vasta, G.R., 130, *34*
Vatzaki, E.H., 130, *34*
Veenstra, R.G., 144, *113*
Veith, A., 214, *77*
Vemula, R., 94, *116*
Ventura, V.V., 230, *150*
Verhofstad, N., 170, *232*
Vestweber, D., 182, *281*, 185, *281*, 185, *292*, 185, *293*
Vidal, S., 165, *206*
Villa, O.F., 218, *126*, 228, *126*, 255, *126*, 257, *126*
Vinogradov, E., 212, *36*, 212, *37*, 217, *36*, 245, *221*, 250, *221*

Vinogradov, Y., 244, *216*
Vinogradova, O., 111, *169*
Vinson, M., 129, *19*
Virji, M., 213, *54*
Vismara, E., 184, *289*, 185, *289*
Vissa, V.D., 247, *235*
Vlodavsky, I., 177, *266*, 180, *266*, 184, *289*, 185, *289*, 185, *290*
Voisin, S., 230, *155*
Volker, S., 239, *197*
von der Lieth, C.W., 130, *23*
von Reitzenstein, C., 165, *214*
Vorum, H., 130, *34*
Vosbeck, K., 91, *92*
Vrasidas, I., 165, *209*, 165, *210*, 166, *209*, 166, *210*

W

Wacker, M., 213, *47*, 213, *49*, 255, *47*
Wada, J., 143, *103*
Waghorne, C., 140, *67*
Wagner, D.D., 145, *126*
Wakarchuk, W.W., 212, *36*, 217, *36*
Walberg, L.A., 140, *70*
Walcheck, B., 173, *251*
Waldron, K.C., *35*, 212
Walek, D., 170, *233*
Walker, F.J., 59, *12*
Wallner, E.I., 143, *103*
Walz, G., 141, *87*, 171, *238*, 173, *87*
Wan, Q., 140, *69*
Wancewicz, E.V., 172, *245*
Wang, H., 178, *270*
Wang, H.M., 176, *262*
Wang, H.Y., 147, *138*
Wang, H.-Y.L., 103, *139*, 103, *138*, 103, *141*, 103, *142*, 104, *145*, 104, *146*, 109, *157*, 111, *161*, 113, *171*
Wang, J.L., 130, *33*, 130, *36*
Wang, L., 130, *33*, 102, *136*, 247, *228*, 250, *241*
Wang, L.C., 173, *252*
Wang, L.-X., 49, *60*, 255, *259*
Wang, R.F., 181, *277*
Wang, R.-F., 147, *138*
Wang, X., 149, *154*, 150, *154*
Wang, X.Y., 172, *241*, 185, *241*

Wang, Y., 148, *147*, 150, *147*
Wang, Y.-T., 254, *254*
Warnock, B., 91, *90*
Warrenm, R.A.J., 212, *27*, 256, *265*
Watanabe, K., 144, *113*
Watanabe, M., 142, *95*
Watanabe, T., 141, *88*
Waterman, P.G., 108, *155*
Waters, S.J., 170, *231*
Watson, A.A., 91, *95*
Watson, D., 41, *6*, 230, *155*
Watson, D.C., 213, *45*
Webber, J., 170, *231*
Weber, E., 169, *229*, 170, *229*
Weerapana, E., 211, *22*, 213, *22*
Wei, M., 172, *239*, 176, *239*, 177, *239*, 178, *269*, 179, *239*, 180, *239*, 180, *273*, 181, *277*
Weigel, B.J., 144, *113*
Weigel, N., 46, *49*, 47, *49*, 47, *51*
Weisemann, R., 182, *279*
Weiss, A.H., 58, *4*
Wellmar, U., 133, *45*, 152, *45*, 152, *171*
Welply, J.K., 182, *278*
Wen, X.C., 172, *244*
Wenzel, K., 171, *237*, 185, *237*, 185, *237*
Werz, D.B., 245, *222*, 111, *164*, 111, *170*
Wessel, H.P., 177, *265*
Weston, B.W., 174, *254*, 175, *254*
Whang, E.E., 164, *204*
Whisenant, T., 173, *252*
White, M.C., 58, *11*
White, S.L., 91, *93*
Whited, G.M., 71, *25*
Whitfield, C., 218, *123*, 218, *124*, 222, *123*, 227, *123*, 229, *147*, 241, *123*, 245, *221*, 250, *123*, 250, *221*, 250, *239*, 251, *123*, 252, *243*, 252, *244*, 252, *245*
Whitfield, D. (M.), 212, *36*, 212, *37*, 217, *36*
Whittal, R.M., 214, *62*
Whorton, R., 145, *122*
Wibberg, D., 218, *121*
Widom, A., 145, *126*
Wiegel, J., 240, *199*
Wieland, F. (T.), 211, *15*, 211, *16*, 212, *31*, 217, *16*, 230, *16*, 230, *31*
Wiest, I., 148, *144*, 165, *144*, 167, *144*
Wild, M. (K.), 182, *281*, 185, *281*, 185, *293*
Wildhaber, I., 214, *78*
Wilkins, P. (P.), 135, *54*, 236, *185*, 253, *185*, 253, *252*
Williamson, R.T., 97, *120*
Wilmanowski, R., 172, *241*, 185, *241*
Wilson, L.J., 70, *23*
Wilson, M.E., 247, *229*
Wilton, R., 240, *203*
Winans, K.A., 149, *151*
Wintjens, R., 158, *186*
Witte, A., 217, *104*, 246, *104*
Wlodarczak, L., 162, *196*
Wolf, E., 130, *23*
Wolitzky, B.A., 144, *119*, 172, *119*
Wong, C., 73, *31*
Wong, C.-H., 58, *7*
Wong, R., 144, *120*, 145, *120*, 177, *120*
Wong, W.T., 182, *279*
Wood, R., 41, *6*
Woods, N., 108, *154*
Wren, B.W., 211, *9*, 213, *9*, 213, *47*, 213, *48*, 255, *9*, 255, *47*
Wu, B., 94, *116*, 101, *123*, 104, *145*, 109, *157*, 109, *158*
Wu, J., 230, *158*, 230, *159*
Wu, M.-H., 147, *139*
Wu, T., 172, *243*
Wugeditsch, T., 215, *87*, 219, *131*, 219, *133*, 220, *133*, 220, *131*, 220, *137*, 221, *137*, 223, *131*, 224, *133*, 225, *133*, 224, *137*, 225, *137*, 229, *131*, 229, *137*, 234, *87*, 235, *87*, 242, *87*, 243, *131*, 243, *133*, 243, *137*, 243, *213*, 244, *131*, 244, *137*, 245, *131*, 249, *131*
Würtz, P., 244, *216*
Wylezol, M., 141, *90*

X

Xia, J., 182, *278*
Xia, L., 143, *108*
Xia, L.J., 145, *126*

AUTHOR INDEX

Xie, X., 177, *265*
Xin, W., 103, *139*
Xing, Y., 90, *84*, 103, *140*
Xu, H.M., 172, *243*
Xu, X.C., 142, *96*
Xu, Y., 143, *108*
Xu, Z.D., 147, *140*
Xue, H., 149, *154*, 150, *154*

Y

Yagi, F., 134, *46*
Yago, A., 141, *87*, 173, *87*
Yagui-Beltran, A., 147, *140*
Yamada, N., 141, *85*
Yamaguchi, K., 175, *260*
Yamaji, T., 128, *17*, 129, *17*
Yamamura, H., 127, *9*
Yamasaki, M., 150, *165*
Yamashina, I., 140, *70*
Yamashita, K., 142, *95*
Yamauchi, A., 144, *113*
Yamazaki, K., 149, *157*, 151, *157*
Yan, P.S., 130, *32*, 142, *32*
Yan, Q., 213, *46*
Yan, W., 247, *235*
Yang, F., 174, *253*
Yang, H.Y., 97, *120*
Yang, J., 218, *125*, 257, *125*
Yang, R.Y., 143, *104*
Yao, R., 212, *42*, 213, *42*
Yapo, B.M., 150, *161*
Yavuz, E., 213, *51*
Yeola, S., 70, *23*
Yogeeswaran, G., 140, *73*
Yokochi, T., 249, *238*
Yoshii, T., 148, *146*
Yoshiki, T., 146, *135*
Yoshimura, F., 217, *108*
Yoshioka, H., 71, *27*
You, L., 147, *140*
Young, N.M., 127, *7*, 212, *36*, 213, *45*
   217, *36*
Young, V.G., 83, *49*, 84, *61*
Yu, C., 149, *151*
Yu, F., 150, *162*
Yu, L., 147, *137*
Yu, M., 172, *243*

Yu, X., 58, *9*, 106, *147*
Yuan, M., 137, *62*
Yue, F., 143, *108*
Yue, L.L., 172, *244*
Yuen, C.T., 137, *58*
Yunker, C.K., 143, *100*
Yurist-Doutsch, S., 230, *149*, 230, *152*

Z

Zaccai, N.R., 127, *11*
Zachara, N.E., 243, *213*
Zacharski, L.R., 178, *268*
Zähringer, U., 212, *39*, 212, *40*, 217, *40*, 231,
   161
Zajecki, W., 141, *90*
Zakrzewicz, A., 177, *265*
Zalipsky, S., 185, *291*
Zamojski, A., 58, *6*
Zarschler, K., 217, *111*, 218, *111*, 218, *116*,
   219, *111*, 219, *130*, 222, *116*, 229, *111*,
   245, *222*, 247, *116*, 249, *130*, 250, *240*,
   251, *130*, 252, *247*, 254, *111*, 254, *116*,
   255, *247*, 256, *247*
Zayni, S., 214, *84*, 217, *84*, 217, *111*, 218,
   111, 219, *111*, 219, *131*, 219, *132*,
   219, *133*, 220, *131*, 220, *133*, 220, *132*,
   223, *131*, 223, *132*, 224, *132*, 224, *133*,
   225, *133*, 229, *111*, 229, *131*, 229,
   132, 233, *176*, 233, *180*, 234, *180*, 240,
   176, 242, *176*, 242, *180*, 242, *210*, 243,
   131, 243, *133*, 244, *131*, 244, *132*, 244,
   176, 244, *180*, 245, *131*, 245, *132*, 246,
   225, *247*, 225, *247*, 232, *247*, 234, *248*,
   225, *248*, 232, *249*, *131*, 250, *240*,
   253, *176*, 254, *111*, 255, *256*, 256, *256*,
   257, *84*
Zcharia, E., 185, *290*
Zedde, C., 162, *196*
Zeisig, R., 171, *237*, 185, *237*, 185, *237*
Zeng, X.L., 172, *239*, 176, *239*, 177, *239*, 177,
   265, 178, *269*, 179, *239*, 180, *239*, 180,
   273, 180, *275*, 181, *277*
Zenita, K., 141, *87*, 173, *87*
Zhan, Q., 148, *149*
Zhang, B., 45, *37*
Zhang, F., 247, *235*
Zhang, H., 146, *134*

Zhang, H.S., 140, *69*, 140, *71*, 140, *72*, 141, *71*
Zhang, J., 69, *22*, 149, *150*
Zhang, Q., 94, *116*, 104, *145*
Zhang, R., 240, *203*
Zhang, S.L., 140, *69*, 140, *70*, 140, *71*, 140, *72*, 141, *71*, 146, *134*
Zhang, T., 149, *154*, 150, *154*
Zhang, Y., 170, *231*, 170, *235*
Zhang Y., 86, *69*
Zhao, H., 91, *105*
Zhao, K.W., 143, *108*
Zhao, X.Y., 143, *108*
Zheng, B., 214, *62*
Zheng, S., 178, *269*, 180, *275*
Zheng, X., 166, *217*
Zhi, Y., 149, *154*, 150, *154*
Zhong, W., 111, *168*, 236, *186*, 253, *186*
Zhong, Y., 106, *148*

Zhou, M., 91, *85*, 101, *131*, 101, *132*, 102, *136*, 102, *137*, 103, *139*, 104, *144*, 107, *150*, 115, *172*
Zhou, Q., 144, *113*
Zhou, Y., 149, *154*, 150, *154*
Zhou, Y.F., 172, *239*, 176, *239*, 177, *239*, 179, *239*, 180, *239*
Zhou, Z., 130, *34*
Zhu, J., 142, *96*
Zhu, J.L., 140, *69*
Zhu, W., 218, *125*, 257, *125*
Zimmermann, P., 79, *43*
Zoeller, M., 144, *112*
Zolghadr, B., 214, *77*, 231, *162*
Zou, J., 167, *222*
Zuberi, R.I., 147, *137*
Zwirner, N.W., 141, *91*, 142, *91*, 144, *91*, 147, *91*
Zylstra, G.J., 71, *28*

SUBJECT INDEX

Note: Page numbers followed by "*f*" indicate figures, "*t*" indicate tables and "*s*" indicate schemes.

A
Aberrant glycosylations
  and cancer metastasis
    endothelial selectins, 141
    ST3Gal-I sialyltransferase, 140
    sTn expression, 140
    TF antigen, 140–141
  in tumors
    blood group-related antigens, 137, 139*f*
    gangliosides, 137, 138*f*
    mucins, 140
    β-1,6-*N*-acetylglucosaminyltransferase (TF), 137–140
    sialyltransferases, 137–140
N-Acetylgalactosaminyltransferases (GalNAcTs), 42
N-Acetylglucosaminyltransferases (GlcNAcTs), 42
Achmatowicz approach
  alcohols and amines, 83
  Noyori hydrogen-transfer reaction, 83
  papulacandin ring system, 82*s*, 84
  talo and *gulo* diastereomers, 83–84
Allosteric inhibitors, galectins
  anginex and 6DBF7, 170
  calixarene 0118, 170–171, 171*f*
  DX-52-1 and HUK-921, 170
Angyal, Stephen John Charles
  acetonation of *myo*-inositol, 5–6
  chromium trioxide in acetic acid, application of, 8
  cis-inositol, 5–6
  Conformational Analysis, 5, 6–7
  epi-inositol, 5–6
  Fischer glycosidation, 8
  interaction between the inositol hydroxyl groups and sodium borate in aqueous solution, 6
  levo-inositol, 5–6
  methyl aldofuranosides, paper on, 8
  myo-inositol, 5–6
  neo-inositol, 5–6
  nine inositols, in the Mills projection, 7*s*
  polysubstituted six-membered ring systems, 8–9
  review on the cyclitols, 6
  Varian A60 nuclear magnetic resonance (NMR), 7–8
  work on the inositols, 5–6
Archaea
  Gram-positive and negative, 214
  methanogenics, 230
  N-glycosylation, 231
Archaeal S-layer glycoproteins
  cytochrome $b_{558/566}$, *Sl. acidocaldarius*, 230–231, 231*f*
  functional characterization, carbohydrate-active proteins, 229–230
  gene cluster, 231
  glycosylated flagellins, 230
  haloarchaea and methanogens, 229–230
  Har. marismortui, 229–230
  Hfx. volcanii, 229–230
  β-ManNAcA, 230
  Mco.maripaludis, 230
  protein *N*-glycosylation in *Haloarcula marismortui*, 229–230

SUBJECT INDEX

Archaeal S-layer glycoproteins (Cont.)
  SCWPs, 231
  "waiting" lipid moiety, 229–230
Asymmetric oxidative biocatalytic aldol
    approach
  chlorobenzene, 71
  2,4-dideoxyhexoses, 71, 72s
  enantiomerically pure monoacetate, 71, 72s
  enzymatic desymmetrization, 71–73
  mannose derivatives, 71
  ozonolysis, 73
  protected D-mannose, 71
  sugar derivatives, 71, 73s

B

Bacteria
  Austrian beet-sugar factories, 218
  eukaryotes, 211
  Gram-positive and negative, 214
  humanized glycoproteins, 213
  Neisseria meningitidis and *N. gonorrhoeae*, 213–214
  prokaryotic domains, 214
  S-layer glycans, 222–228
Bacterial cell-envelope glycoconjugates
  degradation reactions, 242–243
  description, 210
  glycan biosynthesis, 246–254
  glycan engineering, 254–257
  nonclassical secondary cell-envelope
      polysaccharides, 231–241
  polysaccharides and glycopeptides
      isolation, 241–242
  protein glycosylation, 211–217
  surface-layer glycoproteins, 217–231
Bacterial protein glycosylation
  biological functions, 216–217
  cell envelope architectures
    Campylobacter jejuni, 213
    cellular glycan structures, 214
    3-D crystallization experiments, 214–215, 216
    equilibrium unfolding profiles, 216
    gram–positive bacteria and archaea, 214
    location and ultrastructure, S-layers, 214
    Monte Carlo method, 216
    N-linked glycans, 213
    two-dimensional (2-D), 214–215

eukaryotic glycoproteins, 211
glycan–protein linkages, 211
glycoconjugate, 211
gram-positive bacteria and archaea, 214
in vitro biosynthesis machinery, 213
N-linked glycans, 211
non-S-layer glycoproteins, 212
O-linked glycans, 211
PglL, 213–214
prokaryotic glycoproteins, 211–212
sialic acid-like sugars, 212
type IV pili, 213–214
Bacterial S-layer glycoproteins
  A. thermoaerophilus DSM 10155, 218, 219f
  Bacillaceae, 218
  core structure, 218–222
  Gram-negative bacterial cells, 218
  mesophilic Gram-positive bacteria, 218
  S-layer glycans (*see* S-layer glycans)
  terminal groups and noncarbohydrate, 229, 229f
  T. forsythia S-layer oligosaccharide, 218
  two-dimensional crystalline arrays, 218
Basu–Roseman pathway, 44–45
Buchanan, John Grant
  chemistry of "active sulfate", 23
  chiral synthons ("chirons"), 26
  stereoselectivity of epoxide cleavages, 25
  structure of Vitamin B12, work on, 23–24
  structures of pneumococcal antigens, 25
  synthesis of *C*-nucleoside antibiotics, 26
  "teichoic acids", 23
  undergraduate studies, 23

C

Cell-envelope glycan biosynthesis
  genetics, S-layer glycoprotein, 246–247
  glycosyltransferases, 249–250
  nucleotide-sugar, 247–248
  SCWP glycosylation gene clusters, 252–254
Cell-envelope polysaccharides and
    glycopeptides
  chaotropic agents, 241–242
  characterization, 241–242
  HPAEC PAD/chromatography-mass-
      spectrometric analysis, 241–242

O-glycans, 241–242
O-linked, 241–242
SCWPs, 242
Cell-wall-targeting mechanisms
  csaAB operon, 239–240
  G. stearothermophilus PV72/p2 and
    P. alvei CCM 2051^T, 240
  net-neutral charge, 240–241
  nonconserved character, S-layer-binding
    mechanisms, 240–241
  pyruvate transferase CsaB, 240
  SCWP structures, *Bacillaceae*, 241
  S-layer proteins, 239–240
  SLH domains, 239–240
  Thermus thermophilus, 240
Cytidine 5′-(glycerol diphosphate), 24–25
Cytidine 5′-(ribitol diphosphate), 24–25

D

Danishefsky Hetereo-Diels–Alder approach
  chromium(III) complexes, 65–67
  (+/−)-3-deoxy-manno-oct-
    2-ulopyranosonate (Kdo) derivative,
    63, 65*s*
  1,3-dialkoxydienes, 60–63
  dihydroxylation, double bond, 63
  Jacobsen chiral catalyst, 65–67
  Lewis acids, 63
  papulacandin D, 63, 65, 66*s*
  stoichiometric chiral auxiliary, 60–63, 64*s*
Degradation reactions
  HF treatment, SCWP, 242–243
  mass spectrometry, 245
  NMR spectroscopy, 243–245
  Smith degradation, 243
De novo asymmetric synthesis, pyranoses
  asymmetric oxidative biocatalytic aldol
    approach, 71–73
  carbohydrate synthesis, 57
  Danishefsky Hetereo-Diels–Alder
    approach, 60–67
  MacMillan Iterative Aldol approach, 67–71
  monosaccharide sugars, 57
  Msamune–Sharpless approach, 58–60
  non-*de novo* asymmetric approaches,
    74–80
  O'Doherty *de novo* approach, 74–80
  oligosaccharides, 57

Sharpless dihydroxylation/enzymatic aldol
    approach 2-ketoses, 73–74
  unnatural sugars, 57–58
Dondoni thiazole approach
  bis-acetonation, 74–75
  D-glyceraldehyde to higher sugars, 75
  Fischer approach, 74
  homologation strategy, 76
  N-acetyl-D-mannosamine, 74–75

G

Galactosyltransferases (GalTs), 42
Galectins
  galectins-1 and-3 bind type, 133, 134*f*
  inflammatory function, 130
  inhibition
    allosteric inhibitors, 170–171
    anti-galectin mAb, 147–148
    B16 melanoma knockdown transfectant
      cells, 147
    curcumin, 147, 148*f*
    glycopeptides and peptidomimetics,
      167–170
    macrophages, 146–147
    MCP (*see* Modified citrus pectin
      (MCP))
    multivalent inhibitors, 162–167
    siRNA silencing, 147
    small carbohydrate-based inhibitors,
      151–162
    swallow-root pectic polysaccharide (PP),
      151
    T-cell immune function, 147
    therapeutic agents, 146–147
    tumor-associated carbohydrate antigens,
      148–149
  LacNAc-II-containing saccharides,
    133–134
  O-and N-linked glycans, 130
  prototype, tandem repeat type and chimera
    type, 130
  tumor development
    colon cancer cell line (LS174T),
      142–143
    galectin-3 expression, MDA-MB-231
      and HeLa cells, 142–143
    human breast carcinoma lines, 142–143
    hypoxia, 143–144

Galectins (Cont.)
  LGalS3BP, 144
  stages, 144
  thymocyte population, 143
  X-ray crystal structures, 130–133, 131f
Glycan engineering
  glycosylation
    B. fragilis O-glycan, 256–257
    cross-glycosylation of nonnative proteins, 256–257
    heterologous proteins, 255–256
    "immobilization matrix", 255–256
    mass spectrometric analysis, native B. fragilis, 256–257
    nanobiotechnology applications, 255–256
    S-layer glycoproteins, 256
    stabilizing functions, 256–257
    T. forsythia O-glycan, 256–257
  S-layer glycobiology toolbox
    carbohydrate-active enzymes, 254–255
    L-rhamnose, 254–255
    nanobiotechnology, 255
    nanometer-scale cell-surface display, 254
    prokaryotic cell surfaces, 254
    thermostable Rml enzymes, 254–255
Glycopeptides and peptidomimetics
  antiapoptotic effects, 169
  glycosylamines, 167, 169f
  H-bonding and hydrophobic interactions, 167
  Lac-L-Leu, 167–169
  pulmonary metastases, 169
  solid-phase synthesis, 167
  Tyr-Xxx-Tyr motif, 169–170
Glycoprotein
  FlaA, 212–213
  glycans, 221–222
  S-layer (see S-layer glycoproteins)
  TibA, 212
Glycosyltransferases
  in vitro analysis, S-layer glycan biosynthesis, 249
  methylation, 249–250
  P. alvei CCM 2051^T, 249
  polysaccharides, E. coli O8 and O9a, 249–250
  WsaE, 249

H

Heparins
  adhesion molecules, 178
  adverse effects, 177, 180–181
  anticoagulant activity, 176
  cancer metastasis, 177
  COLO320 cells, 176–177
  doses, 179
  E-selectins, 176–177
  Fondaparinux, 179, 179f
  inhibition, P-selectin, 179–180
  low-molecular-weight, 177–178
  N,6-O-sulfated and 6-O-sulfated, 180
  2-O-sulfated iduronic acid residues, 180
  "RO-heparin", 177–178
  tetrasaccharide fragment, 176
  treatment, cancer cells, 178
^{1}H NMR spectroscopy, 7–8
Human umbilical-vein endothelial cells (HUVECs)
  capillary-tube formation, 151
  cDNA treatment, 175–176
  colon carcinoma cells, 172, 173
  cytotoxic agents, 172–173
  Macrosphelide B, 184
  tumor-associated blood vessels, 149
HUVECs. See Human umbilical-vein endothelial cells (HUVECs)

I

Iterative dihydroxylation, dienoates
  D-and L-deoxymannojirimycin, 87–88
  galactono-1,4-lactones and deoxy sugars, 82s, 84–85
  galactono-1,5-lactones, 85–86
  galacto-papulacandin analogues, 86–87
  L-talono-1,4-lactone derivative, 86, 87s
  4-substituted aldono-1,5-lactone, 86

L

Lectin–carbohydrate interactions
  abnormal glycosylation (see Aberrant glycosylations)
  anticancer approaches
    antibodies, 146
    binding affinities, 146
    galectin inhibitors, 146–171

pharmacokinetics, 146
selectin inhibitors, 171–187
variation, glycosylation, 145–146
biological processes, 126
cytotoxicity, 126
discovered, plants, 126
galectins, 130–134, 142–144
hydrogen bonding, 126–127
phosphorylation triggers, 127
"rolling" process, 126–127
selectins, 134–137, 144–145
siglecs, 127–129, 141–142

M

MacMillan iterative aldol approach
aldol reactions to synthesize sugars, 67–68
altro-papulacandin, 70–71
catalysts, reaction, 67–68
Chandrasekhar's synthesis, 68–69
glucose, mannose and allose derivatives, 68, 69s
hydroxyacetophenone derivative, 69
α-hydroxy aldehydes, 68
L-proline, 67–68
Mukaiyama-type Lewis acid-catalyzed reaction, 67–68
stereoisomers, papulacandin sugar framework, 69–70
Masamune–Sharpless approach, hexoses
allylic alcohols, 58
C-2 to C-5 stereocenters, 59
diastereomeric sulfides, 60
eight L-hexoses, 60, 62s
Fischer approach, 58
four hexose precursors, 60, 61s
SAE, 59
tetrose fragment, 59
Mass spectrometry
ESI Q-TOF, 245
LC-ESI-MS/MS, 245
MALDI, 245
MALDI-TOF, 245
T. forsythia S-layer glycan, 245
MCP. See Modified citrus pectin (MCP)
Modified citrus pectin (MCP)
characteristics, 149
cytotoxic agents, 150

galacturonic acids, 149
human breast carcinoma cells (MDA-MB-435), 151
HUVECs, 149
melanoma B16-F1 cells, 151
β-(1→4)-galactan, rhamnogalacturonan I, 149–150, 150f
Multivalent inhibitors, galectins
arrangements, CRDs, 165
calixarenes and oligopeptides, 166–167, 168f
mono-, di-and trivalent lactoside derivatives, 164
neutral saccharide ligands, 167
N-terminal domain, galectin-3, 165
TDGs (see Thiodigalactosides (TDGs))
tetravalent lactoside, 165–166, 166f
TF-polyacrylamide, 165

N

Noncarbohydrate and terminal groups, 229, 229f
Nonclassical SCWPs
cell-wall-targeting mechanism, 239–241
common and variable features, *Bacillaceae*
A. thermoaerophilus DSM 10155, 234, 235f
B. anthracis, 235
B. cereus ATCC 10987, 236, 237f
B. cereus ATCC 14579, 237, 238f
B. cereus strains G92141 and 03BB87, 236, 237f
classification, 232–233
β-D-galactosyl moiety, 236
G. stearothermophilus NRS 2004/3a, 233–234, 234f
G. stearothermophilus PV72/p2, 234–235, 235f
HF-treated *B. anthracis*, 235, 236f
ManNAc-GlcNAc backbone disaccharide motif, 233
P. alvei CCM 2051^T, 233, 233f
pentasaccharide repeating unit, 235
pyruvate substitution, 238
Thm. thermosaccharolyticum strains E207-71, 233, 234f
PG, 232

Nonclassical SCWPs (Cont.)
and S-layer interactions, *Bacillaceae*, 239
structure, 238–239
Nonclassical secondary cell-envelope
polysaccharides
classification, 231–232
PG meshwork, 232
SCWPs (*see* Secondary cell-wall polymers
(SCWPs))
Non-*de novo* asymmetric approaches
Dondoni thiazole approach, 74–76
Reissig approaches, 80
Seeberger approaches, 76–80
Nuclear magnetic resonance (NMR)
spectroscopy
anomeric proton, 244
diphosphate unit, 244
glycopeptide, 245
glycose constituents, 243–244
methylation, 244
pyruvate groups and peptidoglycan, 244
S-layer glycans and SCWPS, 243–244
structural elucidation of end groups, 244
tyrosine-connected sugars, 245
WATERGATE pulse-sequence, 245
Nucleotide-sugar biosynthesis, 247–248, 248f

O

O'Doherty *de novo* approach
Achmatowicz approach, 83–84
asymmetric-reduction approaches to chiral furan alcohols, 80–83
D-enantiomers, *manno*-, *gulo*-and *talo*-pyranoses, 80–83
dihydroxylation, 81s, 83
enantioselective synthesis
alactono-1,4-lactones and deoxy sugars, 80–83
papulacandin ring system, 80–83
iterative dihydroxylation, dienoates, 84–88
oligosaccharides, 80–83
oxidation approaches to chiral furan alcohols, 80–83
palladium-catalyzed glycosylation, 88–89
synthesis and medicinal chemistry
8a-*epi*-swainsonine, 93

anthrax tetrasaccharide, 113, 114–115
anthrose glycosyl donor, 111–113
antibacterial activity, cleistriosides and cleistetrosides, 110, 110t
anticancer activity, cleistrioside and cleistetroside, 110, 111t
C-5'-alkyl side-chains and oligodigitoxin analogues, 103–104, 105s
carba sugar analogue, inhibitor, 96, 97s
cleistetroside-2, 108–109
cleistrioside and cleistetroside families, 108–109, 110s
C-6-substituted analogues, inhibitor, 96–97, 98s
cyclitols (5a-carba pyranoses), 94–95, 96s
cytotoxicity evaluation, digitoxin analogues, 103–104, 104t, 105t
daumone and analogues, 89–90
dideoxy-D-swainsonine and D-swainsonine, 91–92
digitoxin and digitoxigenin mono-and di-digitoxosides, 101, 102s
digitoxin monosaccharide analogues, 102–103, 104s
glycosylated methymycin analogues, 99–101
glycosylated tyrosine portion, 97–99
homo-adenosine, 90–91
jadomycin B, 107–108
landomycin E, trisaccharide portion, 104–107
Noyori reduction steps, 89–90
PI-080, trisaccharide portion, 104–107
postglycosylation and glycosylation sequence, 113–114
rhamnose, oligosaccharides, 111–112
(1r4),(1r6)-heptasaccharides, 114s, 115
RSK inhibitor, 94, 95s
trehalose analogues, 93–94
trisaccharides and digitoxin methoxyamino neoglycosides, 101–102, 103s

P

Palladium-catalyzed glycosylation, 88–89

SUBJECT INDEX

R

Reissig approaches, 80
Roseman, Saul
  bacterial phosphotransferase system (PTS)
    discovery of PTS, 46
    enolpyruvate phosphate (PEP), 46s
    Enzyme I and Enzyme II, 46–47
    histidine-containing protein, 46–47
    sugar transport by PTS, 46s
  Basu–Roseman pathway, 39
  biosynthesis of cell-surface
      macromolecules, 39
  biosynthesis of hexosamines and
      nucleotide sugars, 42
  biosynthesis of mucin glycoproteins and
      brain gangliosides
    glycoproteins and gangliosides,
        biosynthesis in vitro, 42
    glycosyltransferases, 42
  biosynthesis of oligoglycosyl-
      glycoproteins
    branching of oligoglycosyl moieties, 45–46
    studies of glycosyltransferases, 45–46
  biosynthesis of the GD1a ganglioside, 39
  glycobiologist, 39, 49
  glycosyltransferases (GLTs), 41
  intercellular adhesion and fish theory
    cell-surface hydrogen bonding and
        enzyme–substrate interaction, 48s
    "chicken factor", 47–48
    "intercellular adhesion" group, 47
  "Marine Chitin Degradation" project, 48, 49s
  "Monday Night Research Group (MNRG)
      Meetings", 50
  N-acetylmannosamine from degradation of
      sialic acid (Neu5Ac), 40
  publications, 41–42
  sialic acid structure, 41
  sialyltransferase, discovery
    Basu–Roseman pathway, 44–45
    biosynthesis of GD1a ganglioside from
        ceramide, 44s
    biosynthesis of GD3 ganglioside, 44–45
    biosynthetic pathway for GD1a
        ganglioside, 43
    biosynthetic steps for autoimmune
        antigen, 44–45
    GD1a ganglioside from
        lactosylceramide, 43
    sialyl-lactose (SL), 43
    theory of "One Enzyme for One
        Linkage", 43–44

S

SAE. *See* Sharpless asymmetric epoxidation
    (SAE)
Secondary cell-wall polymers (SCWPs)
  glycosylation gene clusters
    acetylation, 253–254
    anchoring function, 252–253
    B. anthracis, 253
    csaAB, 253
    ManNAc-GlcNAc backbone
        disaccharide motif, 252
    Orf1 might act, 252
    putative acyl-transferase systems,
        253–254
    TagA and TagO, 252
    upstream and downstream, 252
  nonclassical, 232–241
Seeberger approaches
  de novo asymmetric approaches, 79
  2-deoxy-2-nitro-hexopyranoside, 76–79
  D-galacturonic acid derivative, 76–79
  Evans aldol reaction, 79
  L-colitose and 2-*epi*-colitose, 76–79
  L-glucuronic, L-iduronic, and L-altruronic
      acid building blocks, 76–79
  monosaccharide building block, 76–79
  pyranose ring, 80
  Upjohn-type dihydroxylation, 79
Selectins
  B16-F1 cells, 145
  E-, L-and P-selectins, 135
  emboli, 145
  glycosylated mucin-like structures,
      135–137
  inhibition
    andrographolide, 172, 172*f*
    antisense techniques, 175
    anti-sLe^a Abs, HUVECs, 173
    antiviral agent NMSO3, 182–183, 182*f*
    C-glycosyl compounds, 183–184, 183*f*
    chitosans, 181–182, 181*f*

Selectins (Cont.)
    cimetidine, 172–173, 172f
    cytotoxic agents, 172–173
    dermatan sulfates, 181, 181f
    fucose and sialic acid residues, 182
    fucosyltransferase I (FUT1) expression, 174–175
    β-GlcNAc-(1→3)-β-Gal-O-naphthalenemethanol, 173–174, 174f
    glycoproteins, 185
    heparins (see Heparins)
    liposomes, glycans, 185
    lovastatin, 172, 172f
    Macrosphelide B, 184
    peptidomimetics, 185–187
    Sd^a antigen, 175–176, 175f
    soluble E-selectin, 173
    substrates, 171–172
    sulfated trimannose C-C-linked dimers, 184–185, 184f
    treatment, tumor cells, 171–172
    trNOE NMR experiments, 182
    P-and L-selectins, carcinoma metastasis, 144–145
    primary tumors, 144
    sialyl Lewis X (sLe^x) and sialyl Lewis A (sLe^a), 134–135, 135f
    sLe^a and sLe^x, 145
    thrombin, 145
    X-ray crystal structure, PSGL-1, 135–137, 136f
Sharpless asymmetric epoxidation (SAE), 59
Sharpless dihydroxylation/enzymatic aldol approach 2-ketoses, 73–74
Sialyltransferases (SATs), 42
Siglecs
    description, 127
    expression, siglec-4a (MAG), 142
    macrophages, siglec-15, 141–142
    Neu5Ac and Neu5Gc, 127–128, 128f
    sTn binding, TGF-β secretion, 141–142
    X-ray structure, siglec-2, siglec-5 and Neu5Ac, 128–129, 129f
S-layer glycans
    assembly, 251–252
    A. thermoaerophilus

DSM 10155, 223–224, 225f
L420-91^T and GS4-97, 224, 225f
bacterium–host cross talk, 216–217
B. fragilis glycan, 228, 228f
core oligosaccharides, 219–220
    A. thermoaerophilus L420-91, 220, 221f
    P.alvei CCM 2051 S-layer glycan, 219, 221f
    Rhamnosyl, 219, 220f
    Thermoanaerobacterium thermosaccharolyticum D120-70, 220
dTDP-L-Rha, 254–255
glycosidic linkage types, 222
Gram-positive bacteria, 222
G. stearothermophilus
    NRS 2004/3a, 223, 223f
    NRS 2004/3a and G. tepidamans GS5-97^T, 247
G. tepidamans
    GS5-97, 229
    GS5-97^T, 223–224, 224f, 229, 229f
Haloferax volcanii, 229–230
in vitro analysis, biosynthesis, 249
Lb.buchneri, 227
lipopolysaccharides, 226–227
MALDI MS, 245
monosaccharide, 222
2-O-methyl groups, 229
P. alvei CCM 2051^T, 224, 225f
and SCWPs, 243–244
S-layer polypeptide backbone, 218–219
slg gene regions, 246–247
T. forsythia, 216–217, 227–228, 228f, 245, 256–257
Thb. thermohydrosufuricus L77-66 (DSM 569), 224, 226f
Thb. thermohydrosulfuricus
    L111-69 and L110-69, 223, 223f
    S102-70, 227, 227f
Thm. thermosaccharolyticum
    D120-70, 223, 224f
    E207-71, 226, 226f
transfer, protein, 252
tripartite structure, 222
without core oligosaccharides
    Bacteroidales, 221–222

D-glucohomooligomers, 221
Lb. buchneri, 221, 222f
O-glycosylation, 221
S-layer glycoproteins
  archaeal, 229–231
  bacterial, 218–229
  biosynthesis
    assembly, S-layer glycan, 251–252
    genetics, 246–247
    initiation, 250
    transfer, S-layer glycan, 252
  core units and linkage, 244
  3-D crystallization experiments, 214–215
  description, 217–218
  glycans (see S-layer glycans)
  membrane-spanning domain, 229–230
  N-linked pentasaccharide and employ dolichyl phosphate, 229–230
  posttranslational modification, 229–230
  structure-function relationships, 216–217
Small carbohydrate-based inhibitors, galectins
  allyl β-Lac and methyl β-LacNAc, 151–152
  anionic inhibitors
    Arg and His residues, galactose, 160
    2-O-benzylphosphate galactosides, 160, 160f
  anomeric oximes, 158–159
  C-2 and C-4 double epimerizations
    anomeric position, triazoles, 159
    galactoside analogues, 159
    triazolyl β-D-mannopyranosides, 159, 159f
  C-1-aromatization, 156
  C-2-epimerization
    Arg and His residues, 156–157
    2-O-functionalized methyl 3-O-(4-toluoyl)-β-D-talopyranosides, 157, 157f
  talopyranosides, 157–158

C-Galactosyl compounds, 160–161, 161f
glycosidic linkage, 152
ionic and polar amino acid residues, 152
2-O-aromatization, 155–156, 155f
3-O-aromatization
  amide groups, 152
  3′-benzamido-LacNAc derivatives, 152–154, 153f
  3-deoxy-3-(triazol-1-yl)-galactopyranoside, 154, 154f
  isoxazoles, 155
  3-O-alkynylbenzyl galactosides, 154–155
  3′-p-hydroxybenzamido-LacNAc derivative, 152–154, 153f
4-OH and 6-OH groups, galactopyranosides, 152
1-thiogalactosides, 161–162, 162f
Smith degradation, 243
Sugar phospho-transport system (PTS), 39

T

TDGs. See Thiodigalactosides (TDGs)
Thiodigalactosides (TDGs)
  antiapoptotic effect, galectin-1, 163–164
  Arg144 and Arg186, ester moieties, 164
  Balb/c nude mice, 163–164
  3,3′-Ditriazolyl-thiodigalactosides, 163–164
  lung metastasis, 163–164
  nanomolar binding affinities, 162–163, 163f
  oligovalent lactulose amines, 164

U

Ungar, Dr. Andrew, 9–10

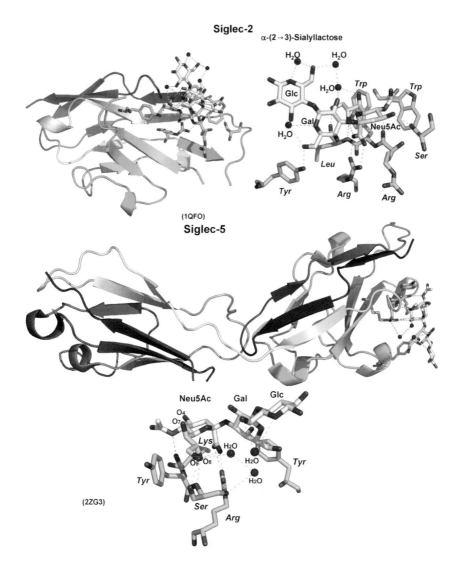

PLATE 1. The X-ray crystal structures of siglec-2 and siglec-5, which illustrate the key interactions between the binding site and Neu5Ac.

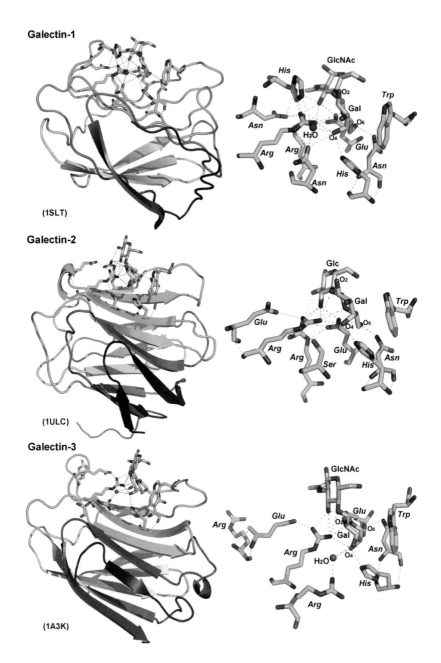

PLATE 2—*Continued*

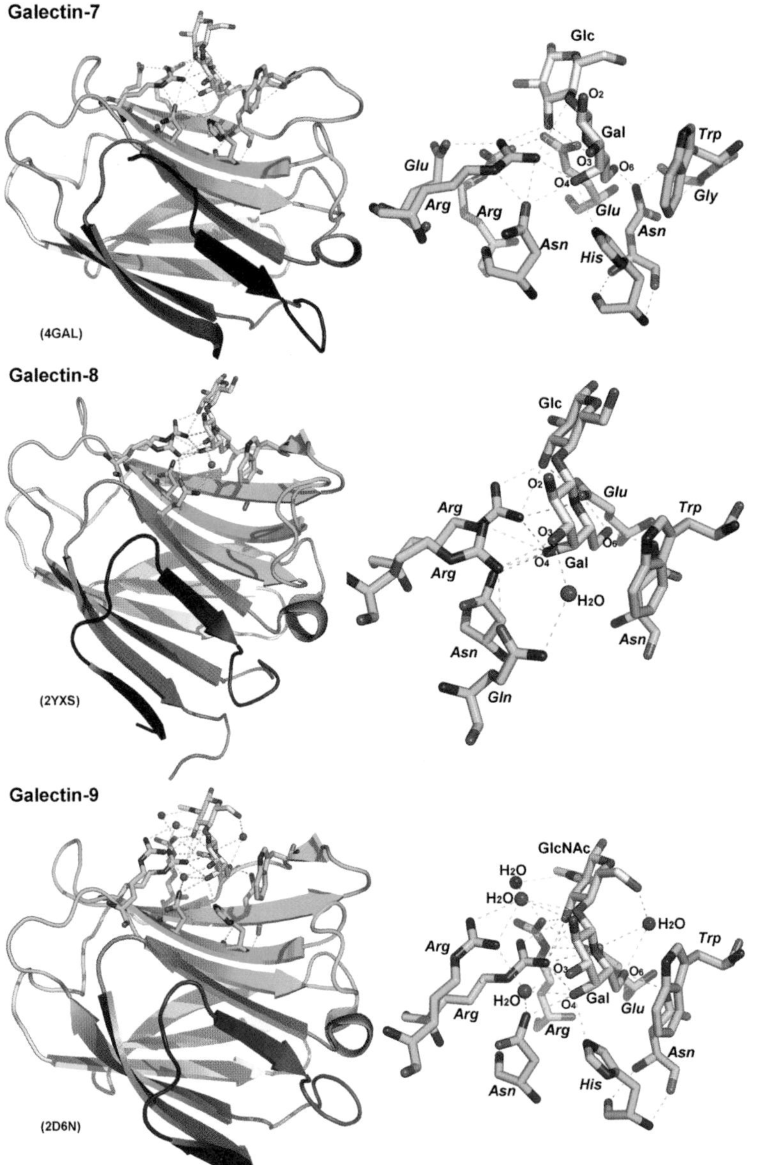

PLATE 2. The X-ray crystal structures of substrate-bound galectins-1, -2, -3, -7, -8, and -9.

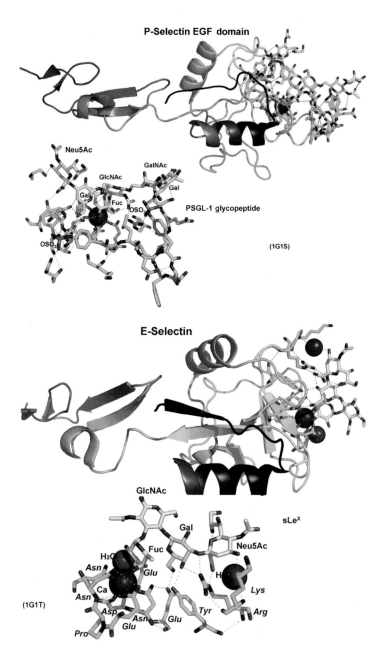

PLATE 3. The crystal structures of substrate-bound E- and P-selectins.

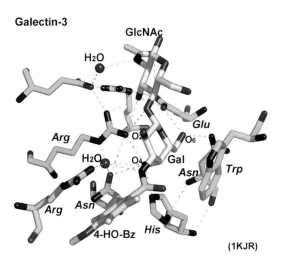

PLATE 4. X-ray structure of galectin-3 with a 3'-*p*-hydroxybenzamido-LacNAc derivative. The aromatic ring is closely facing the guanidinium group of an arginine residue to engage in a cation–π interaction.

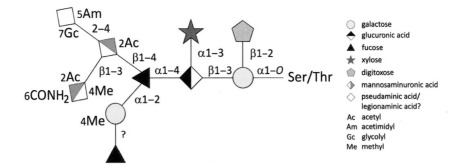

PLATE 5. Schematic structure of the *T. forsythia* S-layer glycan. Adapted from the open access journal *Biomolecules*; © 2012 by the authors, licensee MDPI, Basel, Switzerland; http:/creativecommons.org/licenses/by/3.0.